European Dietary Surveys: What's on the Menu?

European Dietary Surveys: What's on the Menu?

Editors

Murielle Bochud
Igor Pravst

MDPI

Basel • Beijing • Wuhan • Barcelona • Belgrade • Novi Sad • Cluj • Manchester

Editors
Murielle Bochud
University of Lausanne
Lausanne
Switzerland

Igor Pravst
Nutrition Institute
Ljubljana
Slovenia

Editorial Office
MDPI
St. Alban-Anlage 66
4052 Basel, Switzerland

This is a reprint of articles from the Special Issue published online in the open access journal *Nutrients* (ISSN 2072-6643) (available at: https://www.mdpi.com/journal/nutrients/special_issues/european_dietary).

For citation purposes, cite each article independently as indicated on the article page online and as indicated below:

Lastname, A.A.; Lastname, B.B. Article Title. *Journal Name* **Year**, *Volume Number*, Page Range.

ISBN 978-3-7258-0303-3 (Hbk)
ISBN 978-3-7258-0304-0 (PDF)
doi.org/10.3390/books978-3-7258-0304-0

Contents

nutrients

MDPI

Article

Evaluation of Dietary Intake and Anthropometric Status in 1–9-Year-Old Children Living in Serbia: National Food Consumption Survey according to the EU Menu Methodology

Milica Zeković [1,*], Jelena Milešević [1], Marija Takić [1], Marija Knez [1], Ivana Šarac [1], Agnes Kadvan [2], Mirjana Gurinović [1] and Marija Glibetić [1]

1 Centre of Research Excellence in Nutrition and Metabolism, Institute for Medical Research, National Institute of Republic of Serbia, University of Belgrade, 11000 Belgrade, Serbia; jelena.milesevic@gmail.com (J.M.); marijapo2001@gmail.com (M.T.); marijaknez186@gmail.com (M.K.); ivanasarac@yahoo.com (I.Š.); mirjana.gurinovic@gmail.com (M.G.); mglibetic@gmail.com (M.G.)

2 Capacity Development Network in Nutrition in Central and Eastern Europe (CAPNUTRA), 11000 Belgrade, Serbia; k_agi@yahoo.com

* Correspondence: zekovicmilica@gmail.com

Abstract: The Serbian Food Consumption Survey among 1–9-year-old-children was conceptualized and conducted in compliance with the principles, established protocols, and guidelines of the EU Menu project between 2017 and 2021. Valid data were collected for 576 individuals (290 1–3-year-old toddlers and 276 3–9-year-old children). Regardless of age and gender category, the majority (68.80%) of children had normal weights according to the Body Mass Index-for-age classification system. The median daily energy intake was 1406.71 kcal with no differences between the settlement types. The overall median contributions of carbohydrates, protein, and fat to the total energy intake were 47.54%, 14.06%, and 37.88%, respectively. The proportions of the macronutrient intake deviated from the dietary reference values with compliance to the recommendations being particularly poor for fat and fiber. The consumption of energy-dense food groups such as meat and meat products, fat and oil, sugar, and confections was more pronounced among older children. The survey results provide a valuable insight into the nutritional status and dietary habits of toddlers and children 1–9 years old living in Serbia. They may serve as an evidence platform for public health programs, a valuable asset for decision-makers, and a reliable reference to guide nutritional policies, diet monitoring, and interventions targeting this population group in the future.

Keywords: dietary intake assessment; anthropometric status; food consumption; children nutrition

Citation: Zeković, M.; Milešević, J.; Takić, M.; Knez, M.; Šarac, I.; Kadvan, A.; Gurinović, M.; Glibetić, M. Evaluation of Dietary Intake and Anthropometric Status in 1–9-Year-Old Children Living in Serbia: National Food Consumption Survey according to the EU Menu Methodology. *Nutrients* **2022**, *14*, 3091. https://doi.org/10.3390/nu14153091

Academic Editor: Bernadette P. Marriott

Received: 12 June 2022
Accepted: 26 July 2022
Published: 27 July 2022

1. Introduction

A nutritionally appropriate, diversified, balanced, and safe diet is of paramount importance for health promotion and preservation over the course of one's life [1]. A plethora of factors including socioeconomic status, cultural and environmental circumstances, food availability and affordability, knowledge, attitudes, and personal preferences interact in a dynamic and complex manner forming an individual's time-specific dietary pattern. In childhood, optimal nutrition promotes intensive growth and fosters proper physical and psychological development, cognitive operability, and academic performance. Accordingly, food choices, nutrient intake, and dietary behavior in this delicate period may have a profound impact on shaping nutrition-related health trajectories [2].

The detrimental effects of early-life malnutrition are acknowledged from both short-term and long-term perspectives. The consequences of an inadequate diet during childhood include stunting, wasting, overweight, developmental delays, increased susceptibility to infections, and an intensified overall risk of adverse health outcomes [3]. Furthermore, these early-life disadvantages may interact with and aggravate lifestyle and nutrition factors later in life, leading to prolonged and persistent repercussions with particular emphasis on non-communicable diseases. Therefore, childhood is a period of critical vulnerability, but also an opportunity for establishing and consolidating healthy eating habits [4].

National food consumption surveys serve as an indispensable evidence platform for informing relevant policy decisions. They provide valuable population-level data for food intake, dietary patterns and nutrient adequacy evaluations, diet-related risk assessments, and the development of guidelines [5]. Furthermore, these surveys are significant instruments for malnutrition screening, the identification of disparities and critical areas of concern, and for the monitoring of applied interventions and their impact and effectiveness [6]. Within the European Food and Nutrition Action Plan, the WHO explicitly stimulates member-states to strengthen local surveillance programs and undertake nationwide nutrition surveys, thereby highlighting the need for valid, representative, and harmonized data [7]. Nevertheless, the provision of such surveys across Europe is rather inconsistent. A recent review reported that less than two-thirds of WHO European countries conducted national diet surveys since 1990, highlighting the Central and Eastern European region as the major data-gap area [6].

The Multiple Indicator Cluster Survey (MICS), carried out in Serbia in 2019 on a nationally representative sample and a sample of Roma households, provides high-quality, statistically robust, comparable, and internationally standardized data on the critical indicators of well-being among women, children, and adolescents [8]. The MICS program is an invaluable policy instrument and a source of credible information for the evaluation of the progress achieved towards nationally-defined objectives and global commitments, such as the Sustainable Development Goals (SDGs). Although addressing the nutritional status of children under five according to basic anthropometric indices, the duration of exclusive breastfeeding, and the prevalence of minimal dietary diversity, the scope and methodological frame of these surveys do not cover all-encompassing data regarding dietary patterns, food consumption, and nutrient intake among children. A Yugoslav study of atherosclerosis precursors in schoolchildren in Serbia, conducted from 1998 to 2003, was the first and until now remained the only country-level survey in Serbia based on comprehensive dietary assessment methods [9]. Therefore, there was an evident need for up-to-date reliable and robust dietary intake data and related information collected in accordance with internationally acknowledged, validated indicators and standardized methodologies [10].

Within the "What's on the Menu in Europe?" (EU Menu) framework, the European Food Safety Authority (EFSA) has been leading a collective effort to establish a harmonized pan-European Food Consumption database featuring aligned methodological approaches, consistent information coding, and solid data-quality criteria with the potential of intercountry comparisons and analyses. The National Food Consumption Survey in Serbia, supported by the EFSA, was launched in 2017, with an aim to provide insight into children's food consumption, nutrient adequacy, intake risks of potentially hazardous substances, and dietary trends [11]. The objective of the present study is the overview of the energy, macronutrient, and food group intake among one- to nine-year-old-children living in Serbia.

2. Materials and Methods

2.1. Survey Design and Study Population

The Serbian Food Consumption Survey methodology details have been previously published [11]. In brief, the study was conceptualized and conducted in compliance with principles, established protocols, and guidelines of the EU Menu project between 2017 and 2021 [12,13]. The nationally representative sample comprised children residing in private households in the territory of the Republic of Serbia and was classified into two subgroups: (1) one- to three-year-old toddlers, and (2) three to nine year-old-children. A national population register updated with relevant population projections was used as the sampling frame. The recruitment was performed at the household level with age, sex, and residential region applied as three stratification layers. Only one subject was selected from each participating household and institutionalized individuals were considered ineligible. Study subjects were evenly distributed over four seasons and weekdays in order to capture the inter-seasonal variability in dietary patterns and day-to-day consumption fluctuations. The survey schedule covered the whole year including the festive calendar, i.e., national holidays and religious celebrations.

Informed written consent was obtained from all recruited subjects. The study was conducted in accordance with the Institutional Ethical board standards (Approval number: EO 123/2017, 8 December 2017, Institute for Medical Research, National Institute of the Republic of Serbia, University of Belgrade, Serbia) and the principles of the Helsinki declaration.

2.2. Data Collection

Data collection was performed by trained interviewers with a project-specific national questionnaire compilation approved by EFSA. Acknowledging the relevance of family and social contextual factors in nutrition research, the general questionnaire collected data on sociodemographic variables for the parents/caretakers, as well as the general health and food-allergy status for children. Pre-defined answer options were arranged in compliance with the EU Menu methodology standards, classification systems, and nomenclature. Dietary intake assessment was based on twenty-four-hour food diaries reported for two non-consecutive days with at least one-week time distance to account for intraindividual food consumption variance. After each recording period, within debriefing face-to-face sessions, interviewers carefully reviewed the diaries with the parent/caretaker to clarify uncertainties, resolve possible omissions, probe for additional information or make necessary amendments, and thereby ensure adequate quality of the obtained data. The structural format of the diary was organized according to the stepwise procedure of the multiple-pass interview approach. All items were reported prospectively, i.e., at the point of consumption. A comprehensive set of collected information included meal type, place and time of each consumption occasion, food name and category (simple foods vs recipes), disaggregated ingredients in case of composite dishes, preparation procedures, estimated portion size, and if applicable, supplementary details regarding particular qualitative features such as fat and/or sugar content. Diverse approaches were applied to quantify food consumption depending on the available information and respondents' preferences. In addition to natural units, standardized portions, and calibrated household measures, participants were encouraged to use the previously validated photographic Food Atlas specifically developed for the Balkan region [14]. Furthermore, precise mass (g) or volume (mL) were determined in case the exact data were presented on the label, or when the parent/caretaker measured the food quantity during the preparation or serving (e.g., infant formulae).

2.3. Data Processing and Dietary Intake Assessment

Advanced software platform for nutrition research Diet Assess and Plan (DAP) was applied for data storage, processing, and dietary intake assessment. Encompassing electronic versions of standard food consumption and general questionnaires, portion size estimation picture books, Serbian food composition database (FCDB), and appropriate nutrient recommendation datasets, DAP was verified by the EFSA in the ring trial that took place within the initiative "Dietary monitoring tools for risk assessment" in 2014, and was used in numerous national, regional, and international projects [15–17]. The integrated FoodEx2 coding system enabled food matching between dietary questionnaires and the food composition database [18]. Nutrient estimates were based on the mean daily values calculated from the two nonconsecutive twenty-four-hour food diary replicates. Food group classification referred to categories proposed by EuroFIR and comprised milk/milk products, eggs/egg products, meat/meat products, seafood and related products, fat/oil, grains/grain products, nuts/seeds/kernel products, vegetable/vegetable products, fruit/fruit products, sugar/sugar products, beverages (non-milk), miscellaneous food products, and products for special nutritional use. Estimated energy intake was expressed in both kcal and MJ, protein intake in absolute values (g) and relative to body weight (g/kg body mass), and percentage contribution to total energy intake was calculated for all macronutrients and saturated, monounsaturated, and polyunsaturated fatty acids. Nutrient adequacy assessment was performed against both Serbian national [19] and EFSA [20] age-appropriate sets of recommendations as summarized in Table 1.

Table 1. Comparative overview of EFSA and Serbian daily requirements for carbohydrates, protein, and fat for 1–9-year-old children.

	EFSA			RS	
		Total carbohydrates (g/day)			
Age		RI		RDI	RI
1–17 years		45–60 TE%	1–3 years	150–180 g	50–60 TE%
			3–7 years	200–240 g	50–60 TE%
		Total fat (g/day)			
1 year		35–40 TE%	1–3 years	40–47 g	30–35 TE%
2–3 years		35–40 TE%	3–7 years	53–62 g	
4–17 years		20–35 TE%			
Age	AI				
		Dietary fiber (g/day)			
1–3 years	10				
4–6 years	14				
7–10 years	16				
		Protein (g/kg bw per day)			
Age	AR	PRI		RDI	
12–17 months	0.95	1.14	1–2 years	1.20	
18–23 months	0.85	1.03			
2 years	0.79	0.97	2–3 years	1.15	
3 years	0.73	0.90			
4 years	0.69	0.86	3–5 years	1.10	
5 years	0.69	0.85			
6 years	0.72	0.89	6–7 years	1.00	
7 years	0.74	0.91			
8 years	0.75	0.92			
9 years	0.75	0.92			

RS—Serbian national age-appropriate set of recommendations; EFSA—European Food Safety Authority dietary reference values; TE—total energy intake; AR average requirements; AI—adequate intake; RDI—recommended daily intake; PRI—population reference intake.

2.4. Anthropometric Assessment

Prior to fieldwork, all the participating interviewers completed the training course for performing anthropometric measurements of the population of children and used identical equipment. Weight and length/height evaluation was performed during the personal interview and followed the WHO child growth assessment protocols. Portable stadiometers (Seca 213, Secagmbh & Co., Hamburg, Germany) with 0.1 cm accuracy were applied for height measurement. Weight was measured in light clothing using a digital balance with taring capability, calibrated to 0.1 kg (Tanita BC-545N, Tanita, Tokyo, Japan). Body mass index (BMI) was calculated as the ratio of weight (in kg) and recumbent length or standing height (in m). Subjects were classified based on nutrition conditions using the BMI-for-age standardized charts and z-score- (i.e., standard deviations (SD)) cut-off points proposed by the WHO [21]. The following BMI-for-age categories were determined: severely underweight—$z_{sd} \leq -3$; underweight—$-3 < z_{sd} \leq -2$; normal weight—$-2 < z_{sd} \leq 1$; possible risk of overweight—$1 < z_{sd} \leq 2$ for children below 60 months of age; overweight—$2 < z_{sd} \leq 3$ and $1 < z_{sd} \leq 2$ for subjects bellow 60 months of age and older children, respectively; and finally, obese—$z_{sd} > 3$ and—$z_{sd} > 2$ for subjects bellow 60 months of age and 5 to 9-year-old children, respectively. For children under

60 months of age, weight-for-length/height indicator was applied as a complementary metric for overweight/obesity estimates in accordance with the WHO age and sex appropriate standards [21].

2.5. Statistical Analyses

Statistical data analysis was performed with IBM SPSS Statistics 22 (*SPSS Inc.* Chicago, IL, USA). In this study, measurable data were characterized by measures of variability (the mean and standard deviation), whereas attributive data were in absolute numbers and frequencies. The normality of the variable distribution was explored with the *Shapiro–Wilk* test. The differences in continuous data with normal distribution were assessed with the Student's *t*-test; otherwise, in case of skewed distribution, the *Mann–Whitney* test was applied. For energy and relevant macronutrients, results are presented in terms of the mean and standard deviation as well as the distribution of the estimated intake per day (percentiles 5, 25, 50, 75, and 95). The *Kruskal–WallisH* test was applied to explore the differences in macronutrient and energy intake between the geographical areas. The inferential *Chi-square* test was used to explore the relationships between categorical variables. Statistical hypotheses were analyzed at the 0.05 significance level.

3. Results

A total of 774 eligible subjects (403 girls and 371 boys) were approached during the recruitment process and parents or legal guardians of 580 of the prospective subjects voluntarily agreed to participate. Complying with the study protocol and EFSA guidelines, subjects with only one twenty-four-hour food diary available (n = 4) were disqualified and excluded from the analyses. Accordingly, valid data were collected for 576 individuals (290 1–3-year-old toddlers and 276 3–9-year-old children), yielding an overall response rate of 74.41%. No statistically significant difference in the response rate was observed regarding the age subgroup or gender classification. The regional distribution of subjects corresponded with the a priori defined strata and the national geographical population density statistics. Nevertheless, subjects residing in urban households were overrepresented compared to those living in rural areas (85.42% vs. 14.58%). The vast majority of children (98.26%) followed a conventional (omnivorous) diet. Based on the parental/caretaker reports, only one participant in each age sub-category practiced veganism/vegetarianism, while less than 2% had a diet adjusted to a specific health condition such as celiac disease or diabetes. The prevalence of confirmed food allergies ranged from 2.07% in the toddlers' group to 5.59% among older children with an even gender-based distribution. The overview of the study sample characteristics including parental education, occupation, and professional profile is presented in Table 2.

Based on anthropometric measurements, the median length/height and weight of the children were 101.0 cm, range: 53.0–155.0 cm (under 60 months of age: 92.0 cm, range: 53.0–128.0 cm and 5–9-year-old: 128.0 cm, range: 83.0–155.0 cm) and 16.0 kg, range: 8.0–51.0 kg (under 60 months of age: 14.0 kg; range: 8.0–30.0 kg, and 5–9-year-old: 25.0 kg, range: 14.0–51.0 kg), respectively. A sample nutritional status overview based on the BMI-for-age categories is displayed in Table 3. Regardless of age and gender category, the majority (68.80% on a total sample level) of children had normal weights according to the BMI-for-age classification system. A total of 63 subjects (10.94%) were characterized as overweight, 36 (6.25%) as obese, and for an additional 68 (11.80%) aged under 60 months the possible risk of being overweight was determined. Based on the weight-for-length/height indicator, 72 individuals under 60 months of age (19.1% of age subsample, 40 girls and 32 boys) were assigned in the overweight risk category. Furthermore, 29 individuals (16 girls and 13 boys) were overweight and 20 (6 girls and 15 boys) were obese, accounting for 7.9% and 5.3% of the age subsample, respectively. The proportion of subjects allocated in these categories corresponded well with the estimates based on the BMI-for-age classification system. Most misclassifications occurred near the percentile cut-off points, with no differences regarding sex and age subcategories. No association was observed between gender and overweight/obesity prevalence regardless of the metric applied. Concerning the sample's geographical distribution, the proportion of overweight/obese subjects in Western Serbia and the Šumadija region (23.8%) was significantly higher than in other analyzed areas ($p < 0.05$). Furthermore, overweight and obese individuals were more prevalent in urban (26.8%) in comparison with rural households (15.7%) ($p < 0.05$). Nevertheless, the proportion of underweight children was also significantly higher in urban compared to rural areas (15.7% vs. 1.2%, $p < 0.05$).

Table 2. Characteristics of the study population.

Variable	Toddlers $n = 290$	Children $n = 286$	Total Sample $n = 576$
Gender, n (%)			
Female	141 (48.62%)	139 (48.60%)	280 (48.61%)
Male	149 (51.38%)	147 (51.40%)	296 (51.39%)
Age in years, $\bar{x} \pm$ SD	1.98 ± 0.64	6.01 ± 1.87	3.98 ± 2.43
Response rate, n (%)	73.98%	74.87%	74.42%
Distribution per geographical region, n (%)			
Belgrade (capital city) region	72 (24.93%)	64 (22.38%)	136 (23.61%)
Vojvodina region	78 (26.90%)	78 (27.27%)	156 (27.08%)
Region of Šumadija and Western Serbia	84 (28.97%)	84 (29.37%)	168 (29.17%)
South-Eastern Serbia region	56 (19.31%)	60 (20.98%)	116 (20.14%)
Dietary pattern of a child, n (%)			
Conventional	287 (98.97%)	279 (97.55%)	566 (98.26%)
Vegan/vegetarian	1 (0.34%)	1 (0.35%)	2 (0.35%)
Diet related to health conditions (e.g., celiac disease, diabetes)	2 (0.69%)	6 (2.10%)	8 (1.39%)
Prevalence of confirmed food allergies, n (%)	6 (2.07%)	16 (5.59%)	21 (3.65%)
Household size and composition, $\bar{x} \pm$ SD			
People per household	3.79 ± 0.97	4.08 ± 1.06	3.94 ± 1.03
Household members ≥ 18 years old	2.19 ± 0.71	2.33 ± 0.84	2.26 ± 0.78
Household members 10–17 years old	0.16 ± 0.42	0.43 ± 0.62	0.30 ± 0.55
Household members < 10 years	1.51 ± 0.62	1.49 ± 0.61	1.50 ± 0.62
Child lives with, n (%)			
Both parents	272 (93.80%)	263 (91.96%)	535 (92.88%)
Only mother	17 (5.86%)	20 (6.99%)	37 (6.42%)
Only father	0 (0.00%)	1 (0.35%)	1 (0.17%)
Other	1 (0.34%)	2 (0.70%)	3 (0.52%)
Highest level of formal education- mother; father, n (%)			
ISCED 0: less than primary education attained	0 (0.00%); 5 (1.72%)	2 (0.70%); 4 (1.40%)	2 (0.34%); 9 (1.56%)
ISCED 1: Primary education	1 (0.34%); 3 (1.03%)	1 (0.35%); 7 (2.45%)	2 (0.34%); 10 (1.74%)
ISCED 2: Lower secondary education	11 (3.79%); 9 (3.10%)	14 (4.90%); 14 (4.90%)	25 (4.34%); 23 (3.99%)
ISCED 3: Upper secondary education	78 (26.9%); 100 (34.48%)	94 (32.87%); 112 (39.16%)	172 (29.86%); 212 (36.81%)
ISCED 4/5: Post-secondary/Short-cycle tertiary education	25 (8.62%); 14 (4.83%)	23 (8.04%); 30 (10.49%)	48 (8.33%); 44 (7.64%)
ISCED 6: Bachelor's or equivalent level	109 (37.59%); 107 (36.90%)	101 (35.31%); 77 (26.92%)	210 (36.46%); 184 (31.94%)
ISCED 7/8: Master's/Doctoral or equivalent level	66 (22.76%); 52 (17.93%)	51 (17.83%); 41 (14.33%)	117 (20.31%); 93 (16.15%)
Employment status—mother; father, n(%)			
Unemployed	42 (14.48%); 8 (2.76%)	39 (13.64%); 16 (5.59%)	81 (14.06%); 24 (4.17%)
Working for pay or profit	193 (66.55%); 268 (92.41%)	217 (75.87%); 252 (88.11%)	410 (71.18%); 520 (90.28%)
Pupil, student, further training, unpaid work experience	4 (1.38%); 1 (0.34%)	1 (0.35%), 0 (0.00%)	5 (0.87%); 1 (0.17%)
In retirement or early retirement or has given up business	0 (0.00%); 0 (0.00%)	0 (0.00%), 2 (0.70%)	0 (0.00%); 2 (0.35%)
Maternity, parental, or sick leave	38 (13.10%); 0 (0.00%)	17 (5.94%); 0 (0.00 %)	55 (9.55%); 0 (0.00%)
Permanently disabled	0 (0.00%); 0 (0.00%)	0 (0.00%), 1 (0.35%)	0 (0.00%); 1 (0.17%)
In compulsory military or community service	0 (0.00%); 1 (0.34%)	1 (0.35%), 1 (0.35%)	1 (0.17%); 2 (0.35%)
Fulfilling domestic tasks	10 (3.45%); 0 (0.00%)	7 (2.45%), 1 (0.35%)	17 (2.95%); 1 (0.17%)
Not applicable/Other	3 (1.03%); 12 (4.14%)	4 (1.40%); 13 (4.55%)	7 (1.22%); 25 (4.34%)

Table 2. *Cont.*

Variable	Toddlers n = 290	Children n = 286	Total Sample n = 576
Professional profile—mother; father, n (%)			
Managers	19 (6.55%); 39 (13.45%)	20 (6.99%); 41 (14.34%)	39 (6.77%); 80 (13.88%)
Professionals	92 (31.72%); 71 (24.48%)	72 (25.17%); 47 (16.43%)	164 (28.47%); 118 (20.49%)
Technicians and associate professionals	29 (10.00%); 39 (13.45%)	25 (8.74%); 43 (15.03%)	54 (9.38%); 82 (14.24%)
Clerical support workers	45 (15.51%); 11 (3.79%)	50 (17.48%); 14 (4.90%)	95 (16.49%); 25 (4.34%)
Service and sales workers	26 (8.97%); 31 (10.69%)	32 (11.19%); 38 (13.29%)	58 (10.07%); 69 (11.98%)
Skilled agricultural, forestry, and fishery workers	1 (0.34%); 5 (1.72%)	0 (0.00%); 1 (0.35%)	1 (0.17%); 6 (1.04%)
Craft and related trades workers	11 (3.79%); 8 (2.76%)	12 (4.20%); 21 (7.43%)	23 (3.99%); 29 (5.03%)
Plant and machine operators, and assemblers	2 (0.69%); 22 (7.59%)	0 (0.00%), 13 (4.55%)	2 (0.35%); 35 (6.08%)
Elementary occupations	8 (2.76%); 6 (2.07%)	14 (4.90%); 11 (3.85%)	22 (3.82%); 17 (2.95%)
Armed forces occupations	3 (1.03%); 6 (2.07%)	1 (0.35%); 3 (1.05%)	4 (0.69%); 9 (1.56%)
Other	54 (18.62%); 52 (17.93%)	60 (20.98%); 54 (18.88%)	114 (19.79%); 106 (18.40%)
Settlement type, n (%)			
Urban	249 (85.86%)	243 (84.97%)	492 (85.42%)
Rural	41 (14.14%)	43 (15.03%)	84 (14.58%)

Table 3. Nutritional status overview based on Body mass index (BMI)-for-age categories for a nationally representative sample of children aged 1–9 years living in Serbia (n = 576).

Sample Group (Age Categories)	BMI-for-Age Categories *					
	Severely Underweight	Underweight	Normal Weight	Possible Risk of Overweight	Overweight	Obese
Children <60 moths (n = 376)	8 (2.1%)	15 (4.5%)	230 (61.2%)	68 (18.1%)	32 (8.5%)	23 (6.1%)
Girls (n =176)	3 (1.7%)	8 (4.5%)	104 (59.1%)	38 (21.6%)	15 (8.5%)	8 (4.5%)
Boys (n =200)	5 (2.5%)	7 (3.5%)	126 (63.0%)	30 (15.0%)	17 (8.5%)	15 (7.5%)
Children 5–9 years (n = 200)	3 (1.5%)	4 (2.0%)	149 (74.5%)	NA	31 (15.5%)	13 (6.5%)
Girls (n =176)	0 (0.0%)	2 (1.9%)	83 (79.8%)	NA	16 (15.4%)	3 (2.9%)
Boys (n =200)	3 (2.1%)	3 (3.1%)	66 (66.8%)	NA	15 (15.6%)	10 (10.4%)

* severely underweight: $z_{sd} \leq -3$; underweight: $-3 < z_{sd} \leq -2$; normal weight: $-2 < z_{sd} \leq 1$; possible risk of overweight: $1 < z_{sd} \leq 2$ for children below 60 months of age; overweight: $2 < z_{sd} \leq 3$ and $1 < z_{sd} \leq 2$ for subjects bellow 60 months of age and 5–9 years old children, respectively; obese: $z_{sd} > 3$ and: $z_{sd} > 2$ for subjects bellow 60 months of age and 5–9 years old children, respectively; NA—not applicable.

The distribution of estimated energy and macronutrient intake (including mean values and percentiles) across age and gender groups is presented in Table 4. Median daily energy intake was 1406.71 kcal, range: 495.27–3156.36 kcal with no differences between settlement types. The participants residing in the South-Eastern Serbia region had a significantly higher energy intake compared to other geographical strata ($\chi(3) = 21.892$, $p < 0.001$) (Table 5). Although not reaching the statistical significance threshold neither at the total sample level nor within the age-subgroup analyses, boys had a higher energy intake than girls (1513.51 ± 474.08 kcal vs.1473.86 ± 462.27 kcal). No gender-wise differences were observed concerning the estimated intake of total carbohydrates, fat, and protein. Nevertheless, the intake of protein relative to body weight was significantly higher among female subjects aged 1–3 years compared to their male peers (3.67 ± 1.13 g/kg body weight vs. 3.38 ± 1.13 g/kg body weight, $p < 0.05$).Among toddlers, expectedly, the absolute intake of energy, carbohydrates, protein, fat, and fiber increased with age (r = 0.401, r = 0.334, r = 0.393, and r = 0.278, all $p < 0.001$). Analogously, in 3–9-year-old children, a positive correlation was recorded between the absolute intake of energy, carbohydrates, protein, fat, and fiber and age (r = 0.326, r = 0.243, r = 0.281, r = 0.342 and r = 0.128, all $p < 0.001$). Nevertheless, in this group, the protein intake relative to body weight decreased with age (r= −0.392, $p < 0.01$). The percentage contribution of macronutrients to total energy remained consistent across the subgroup age span except for fat (1–3 year-old toddlers = 0.153, $p < 0.01$; 3–9 year-old children: r = 0.186, $p < 0.01$). The overall median contributions of carbohydrates, protein, and fat to total energy intake were 47.54%, 14.06%, and 37.88%, respectively. A total of 356 (61.81%) subjects had 45–60% of their energy derived from their intake of carbohydrates, while for 30.00% of toddlers and 38.11% of 3–9-year-old children the respective contribution was below the lower reference intake limit of 45% with an even distribution among girls (35.35%) and boys (34.78%). The median contribution of fat to total energy intake among toddlers was 36.68%, with 24.13% and 32.07% of participants falling into the age-appropriate national and EFSA reference intake ranges, respectively. The fat intake was even higher in the older age sub-group with more than two-thirds (76.20%) of subjects exceeding the 35% of energy derived from fat. The contribution of carbohydrates and monounsaturated fatty acids was significantly higher in Vojvodina (northern autonomous region of Serbia) in comparison with other geographical areas ($\chi(3) = 12.246$, $p < 0.01$, $\chi(3) = 11.905$, $p < 0.01$, respectively) (Table 5). The protein intake relative to body weight was above the age-specific average requirements defined by the EFSA in all the participants, with estimated values almost triply exceeding the population reference intake thresholds. The median dietary fiber intake per day among toddlers was 10.68 g (range: 1.11–38.56 g), surpassing the age-specific adequate intake (AI) level of 10 g/day, as proposed by the EFSA. A total of 171 individuals (58.66%) in this age group reached that threshold with no differences between the female and male participants. The median estimated intake of fiber among 4–6-year-olds (13.00 g, range: 4.32–29.23 g) was below the AI level for that age subgroup (i.e., 14 g/day) with a slightly higher proportion of boys achieving the reference cut-off point (n = 32, 43.24%) compared to their female peers (n = 27, 40.90%). Similar observations were made in the oldest subgroup, i.e., among those7–9 years old: although showing an age-related increasing trend, the median fiber intake (13.83 g, range: 5.77–44.71 g) remained below the AI level, with only 36.73% (girls: 34.25%and boys: 39.73%) of participants achieving the 16 g/day target. Children living in rural households consumed more dietary fiber daily than those residing in urban areas (12.31 g, range: 1.11–44.71 g vs. 11.09, range: 2.20–28.58 g, respectively, $p < 0.05$). Moreover, the intake of dietary fiber was significantly higher in the South-Eastern Serbia region than in other analyzed areas ($\chi(3) = 29.372$, $p < 0.001$).

The contribution of the food groups to the total energy intake, energy intake from carbohydrates, protein, fat, dietary fiber, and saturated fatty acids across age and gender categories are summarized in Table 6. Based on the dietary records, grains and grain products were the most commonly consumed food group (eaten by almost 100% of the study participants), followed by milk and milk products (96.87%), vegetables and vegetable products (95.31%), and fruits and fruit products (93.58%).The share of the consumers of meat and meat products among the participants were fourfold higher compared to the consumers of seafood and related products (84.90% vs. 20.31%). Sugar and related products occurred in dietary records of approximately two thirds of respondents, while less than 10% consumed nuts, seeds, and kernel products. Significant gender-wise differences in the food group consumption patterns were not observed in the analyzed sample (data not shown). Overall, grains and grain products accounted for more than one-third of the total energy intake. Furthermore, these food items were the dominant source of energy intake for carbohydrates and dietary fiber. In addition to grains, fruit, vegetables, confections, and associated products were the dominant sources of carbohydrates. Unsurprisingly, energy and nutrient-dense animal source food groups, i.e., milk, meat, and their related products were significant contributors to the protein and fat intake. Moreover, the percentage of energy intake from saturated fatty acids was most apparent in the milk and milk product category. The food group consumption patterns fluctuated between the two age sub-groups. The consumption of milk and milk products, fruit and fruit products, and products for special nutritional use (formulas) was more pronounced among toddlers. In contrast, 3–9-year-oldchildren had a greater energy intake derived from meat and meat products, fat and oil, sugar and sugar products, and miscellaneous foods.

Table 4. Distribution of daily energy and macronutrients intake across age and gender categories in a nationally representative sample of children 1–9 years of age living in Serbia (n = 576).

Variable	Girls	Boys	Girls and Boys					
	$\bar{x} \pm$ SD	$\bar{x} \pm$ SD	$\bar{x} \pm$ SD	P5	P25	P50	P75	P95
Energy (kcal)								
toddlers	1267.44 ± 361.01	1299.53 ± 395.17	1283.93 ± 378.63	733.69	1038.14	1237.71	1456.06	2005.46
children	1683.24 ± 460.58	1730.41 ± 470.20	1707.48 ± 465.33	1016.57	1377.17	1652.01	1991.16	2600.88
total sample	1473.86 ± 462.27	1513.51 ± 474.08	1494.24 ± 473.61	865.72	1176.07	1406.71	1739.80	2341.23
Energy (MJ)								
toddlers	5.30 ± 1.51	5.43 ± 1.65	5.37 ± 1.58	3.07	4.34	5.18	6.09	8.39
children	7.04 ± 1.93	7.24 ± 1.97	7.14 ± 1.95	4.25	5.76	6.91	8.33	10.88
total sample	6.17 ± 1.93	6.33 ± 2.03	6.25 ± 1.98	3.62	4.92	5.89	7.28	9.80
Carbohydrates (g)								
toddlers	149.97 ± 45.91	158.15 ± 48.48	154.17 ± 47.34	84.76	124.65	146.61	179.14	233.30
children	193.42 ± 56.05	201.01 ± 56.53	197.32 ± 56.33	116.98	159.83	191.86	227.00	302.74
total sample	171.54 ± 55.54	179.44 ± 56.76	175.60 ± 56.26	102.08	137.41	167.87	204.66	286.31
Carbohydrates (%TE)								
toddlers	47.62 ± 7.79	49.01 ± 7.07	48.33 ± 7.44	36.64	43.85	48.37	52.83	60.45
children	46.13 ± 5.97	46.91 ± 6.91	46.53 ± 6.47	35.95	42.64	46.73	50.45	56.84
total sample	46.88 ± 6.97	47.97 ± 7.06	47.44 ± 7.03	36.14	43.27	47.54	51.57	59.11
Protein (g)								
toddlers	46.26 ± 13.51	46.20 ± 15.72	46.23 ± 14.67	25.68	36.72	44.13	53.91	73.96
children	60.51 ± 17.55	61.28 ± 19.31	60.90 ± 18.45	34.89	48.47	58.72	71.83	91.88
total sample	53.33 ± 17.17	53.69 ± 19.12	53.52 ± 18.18	29.61	40.58	51.08	63.99	86.46
Protein (g/kg body mass)								
toddlers	3.67 ± 1.13	3.38 ± 1.13	3.52 ± 1.14	1.91	2.76	3.40	4.13	5.52
children	2.66 ± 0.90	2.69 ± 0.93	2.68 ± 0.91	1.42	2.03	2.56	3.22	4.27
total sample	3.17 ± 1.14	3.04 ± 1.09	3.10 ± 1.11	1.54	2.32	2.97	3.69	5.08
Protein (%TE)								
toddlers	14.76 ± 2.49	14.22 ± 2.47	14.49 ± 2.49	10.91	12.83	14.06	15.99	18.96
children	14.50 ± 2.35	14.16 ± 2.32	14.33 ± 2.34	10.94	12.83	14.06	15.62	18.17
total sample	14.64 ± 2.42	14.19 ± 2.40	14.41 ± 2.42	10.87	12.82	14.06	15.84	18.63
Fat (g)								
toddlers	53.61 ± 21.40	53.57 ± 20.35	53.59 ± 20.84	25.60	40.35	50.00	62.70	91.00
children	74.17 ± 24.67	75.69 ± 25.53	74.95 ± 25.08	40.45	57.24	72.52	90.26	121.91
total sample	63.82 ± 25.24	64.56 ± 25.56	64.20 ± 25.38	30.68	45.35	60.05	76.43	112.34
Fat (%TE)								
toddlers	37.62 ± 7.44	36.76 ± 6.53	37.18 ± 6.99	27.05	32.81	36.68	41.07	48.52
children	39.36 ± 5.59	38.93 ± 5.90	39.14 ± 5.74	29.55	35.30	39.15	42.65	48.69
total sample	38.48 ± 6.63	37.84 ± 6.31	38.15 ± 6.47	27.73	34.11	37.88	41.79	48.65
Saturated fatty acids (%TE)								
toddlers	13.15 ± 3.71	13.57 ± 3.52	13.37 ± 3.62	7.69	11.26	13.10	15.59	19.19
children	14.04 ± 2.81	13.77 ± 3.09	13.90 ± 2.96	9.42	12.04	13.92	15.88	18.50
total sample	13.59 ± 3.32	13.67 ± 3.31	13.63 ± 3.31	8.36	11.59	13.60	15.73	18.96
Monounsaturated fatty acids (%TE)								
toddlers	11.47 ± 3.74	11.68 ± 2.80	11.58 ± 3.29	6.97	9.75	11.25	12.98	16.40
children	12.61 ± 2.58	12.64 ± 2.51	12.63 ± 2.53	8.82	11.01	12.44	14.11	17.02
total sample	12.04 ± 3.25	12.16 ± 2.70	12.10 ± 2.98	7.79	10.31	11.84	13.48	16.81
Polyunsaturated fatty acids (TE%)								
toddlers	7.56 ± 2.32	7.76 ± 2.67	7.66 ± 2.50	3.97	5.86	7.57	9.05	11.84
children	8.56 ± 2.60	8.37 ± 2.38	8.46 ± 2.49	4.55	6.81	8.46	9.94	12.53
total sample	8.06 ± 2.51	8.06 ± 2.55	8.06 ± 2.53	4.14	6.32	7.98	9.61	12.37
Dietary fiber (g)								
toddlers	11.47 ± 4.90	11.97 ± 4.87	11.73 ± 4.88	5.65	8.84	10.68	13.76	20.16
children	14.23 ± 5.43	14.22 ± 5.44	14.23 ± 5.42	7.23	10.57	13.38	17.11	23.91
total sample	12.84 ± 5.34	13.09 ± 5.27	12.97 ± 5.30	6.48	9.38	12.09	15.74	22.17

TE—total energy, P—percentile.

Table 5. Distribution of daily energy and macronutrient intake across geographical areas in a nationally representative sample of children 1–9 years of age living in Serbia (n = 576).

Variable Median (Range)	Belgrade (Capital City) Region	Vojvodina Region	Šumadija and Western Serbia Region	Southeastern Serbia Region
Energy (kcal)	1418.77 (594.89–2365.11)	1391.46 (648.82–2929.27)	1322.39 (595.27–2966.27)	1608.46 (738.51–3156.36) ***
Carbohydrates (%TE)	48.30 (32.08–67.91)	45.93 (14.84–75.60) **	47.78 (26.11–65.01)	48.07 (33.42–63.04)
Protein (%TE)	13.92 (8.26–21.51)	14.46 (7.09–23.24)	13.82 (8.35–22.86)	13.94 (9.18–19.13)
Fat (%TE)	37.20 (20.70–56.48)	38.46 (17.30–68.11)	37.79 (21.59–58.65)	37.41 (25.89–49.71)
Saturated fatty acids (%TE)	13.15 (3.46–20.95)	13.92 (3.49–24.88)	13.67 (5.20–28.27)	13.41 (8.69–24.56)
Monounsaturated fatty acids (%TE)	11.43 (4.27–22.24)	12.48 (5.48–29.71) **	11.73 (4.92–20.49)	11.73 (7.13–19.89)
Polyunsaturated fatty acids (TE%)	7.79 (1.83–15.49)	7.90 (2.59–16.19)	7.66 (3.18–21.52)	7.65 (2.71–15.87)
Dietary fiber (g)	11.38 (4.05–30.70)	11.53 (2.20–44.71)	10.78 (1.10–26.52)	13.72 (4.75–38.56) ***

** $p < 0.01$, *** $p < 0.001$, The Kruskal-Wallis H test; TE—total energy.

Table 6. Contribution of food groups to total energy intake and energy intake from carbohydrates, protein, fat, dietary fiber and saturated fatty acids across age and gender categories in a nationally representative sample of 1–3-year-old toddlers (n = 290) and 3–9-year-old children (n = 286) living in Serbia.

Food Groups ($\bar{x} \pm$ SD)	%TE	Carbohydrates (%TE)	Protein (%TE)	Fat (%TE)	Dietary Fibre (%TE)	Saturated FA (%TE)
Milk/milk products						
Toddlers	18.70 (13.77–24.56)	10.28 (6.15–15.13)	28.59 ± 11.84	26.63 (20.92–36.57)	0.04 ± 0.10	45.48 ± 15.64
Children	15.38 (10.62–19.66) ***	6.93 (4.48–11.09) ***	22.86 ± 9.38 ***	21.73 (14.96–29.48) ***	0.02 ± 0.06 *	38.46 ± 15.18 ***
Total sample	16.94 (12.21–21.86)	8.49 (4.96–12.97)	25.75 ± 11.06	24.33 (17.49–31.99)	0.03 ± 0.08	42.00 ± 15.79
Eggs/egg products						
Toddlers	3.07 (0.55–5.13)	0.14 (0.02–0.21)	7.60 (1.46–11.71)	5.14 (1.00–8.70)	0.01 ± 0.02	3.76 (0.73–6.93)
Children	2.86 (0.91–5.03)	0.13 (0.04–0.22)	6.84 (2.40–11.76)	4.47 (1.63–8.64)	0.01 ± 0.01	3.33 (1.19–6.42)
Total sample	3.00 (0.73–5.08)	0.13 (0.04–0.21)	7.22 (2.01–11.69)	4.86 (1.24–8.62)	0.01 ± 0.01	3.59 (0.87–6.64)
Meat/meat products						
Toddlers	9.46 (5.39–13.58)	0.05 (0.00–0.18)	24.13 ± 13.04	15.76 (8.70–24.29)	0.06 ± 0.10	14.87 (7.27–24.63)
Children	11.28 (7.14–15.81) ***	0.11 (0.01–0.24) ***	28.76 ± 12.58 ***	18.68 (10.22–26.75) **	0.08 ± 0.13 *	18.26 (10.09–28.25) **
Total sample	10.27 (6.12–14.93)	0.09 (0.00–0.22)	26.43 ± 13.01	17.08 (9.27–25.36)	0.07 ± 0.12	16.76 (8.49–26.45)
Seafood and related products						
Toddlers	1.30 ± 3.00	0.18 ± 0.54	3.11 ± 7.40	1.80 ± 4.62	0.01 ± 0.04	0.27 ± 0.69
Children	1.13 ± 2.80	0.23 ± 0.99 *	2.72 ± 6.55	1.40 ± 3.52	0.01 ± 0.03	0.23 ± 0.66
Total sample	1.23 ± 2.90	0.21 ± 0.80	2.92 ± 7.05	1.50 ± 4.32	0.01 ± 0.04	0.25 ± 0.67
Fat/oil						
Toddlers	9.22 ± 4.64	0.00 (0.00–0.01)	0.01 (0.00–0.04)	25.35 ± 11.55	–	12.32 (7.88–17.92)
Children	11.15 ± 4.89 ***	0.00 (0.00–0.02)	0.02 (0.00–0.08)	29.01 ± 11.64 ***	–	13.99 (8.75–21.30) *
Total sample	10.18 ± 4.86	0.00 (0.00–0.01)	0.01 (0.00–0.05)	27.17 ± 11.73	–	12.97 (8.51–19.93)
Grains/grain products						
Toddlers	30.27 ± 9.15	46.80 ± 12.23	23.08 ± 8.42	10.47 (5.59–16.39)	39.31 ± 14.50	6.54 (3.37–12.23)
Children	30.82 ± 8.84	49.81 ± 11.79 **	23.23 ± 7.86	8.48 (4.74–14.28) *	43.96 ± 15.45 ***	5.47 (2.27–9.71)*
Total sample	30.55 ± 8.99	48.29 ± 12.09	23.15 ± 8.14	9.60 (5.13–15.70)	41.62 ± 15.14	5.95 (2.82–10.58)
Nuts/seeds/kernel products						
Toddlers	0.04 (0.00–0.29)	0.05 (0.00–0.19)	0.03 (0.00–0.26)	0.01 (0.00–0.14)	0.21 (0.00–1.61)	0.02 (0.00–0.20)
Children	0.09 (0.02–0.77)	0.10 (0.03–0.35) ***	0.07 (0.02–0.55) ***	0.03 (0.01–0.53) *	0.62 (0.08–2.46) ***	0.03 (0.01–0.47) **
Total sample	0.05 (0.01–0.59)	0.07 (0.01–0.26)	0.05 (0.01–0.45)	0.02 (0.00–0.40)	0.42 (0.02–2.14)	0.02 (0.00–0.33)
Vegetables/vegetable products						
Toddlers	4.70 (3.06–6.94)	7.22 (4.26–10.86)	5.93 (3.58–9.17)	0.80 (0.50–1.25)	20.95 (13.63–31.24)	0.32 (0.17–0.55)
Children	5.11 (3.62–7.52)	8.31 (5.73–12.24) **	5.76 (3.33–8.95)	0.85 (0.52–1.31)	23.84 (16.00–33.79)	0.36 (0.20–0.58)
Total sample	4.91 (3.21–7.21)	7.96 (4.94–11.72)	5.80 (3.44–9.01)	0.81 (0.51–1.28)	22.31 (14.61–32.78)	0.34 (0.18–0.56)
Fruits/fruit products						
Toddlers	11.41 (6.94–14.80)	21.06 ± 11.29	2.93 (1.69–4.28)	1.08 (0.68–1.77)	31.65 ± 15.02	0.70 (0.36–1.13)
Children	7.90 (4.52–12.18) ***	16.07 ± 10.60 ***	2.07 (1.08–3.64) ***	0.67 (0.33–1.22) ***	23.52 ± 15.00 ***	0.41 (0.14–0.85) ***
Total sample	9.50 (5.35–13.86)	18.58 ± 11.22	2.59 (1.36–3.96)	0.88 (0.45–1.57)	27.61 ± 15.54	0.59 (0.26–1.02)
Sugar/sugar products						
Toddlers	2.69 (0.15–6.96)	4.80 (0.28–9.68)	0.12 (0.00–2.04)	0.47 (0.00–6.00)	0.00 (0.00–1.78)	0.74 (0.00–8.09)
Children	5.40 (2.22–9.81) ***	7.49 (3.45–13.17) ***	1.27 (0.05–3.20) ***	3.75 (0.00–9.47) ***	0.75 (0.00–2.59) ***	4.52 (0.00–13.42) ***
Total sample	4.31 (1.09–8.53)	6.09 (2.10–11.69)	0.67 (0.00–2.78)	2.10 (0.00–7.79)	0.23 (0.00–2.42)	2.89 (0.00–10.56)
Beverages (non–milk)						
Toddlers	1.50 ± 3.20	2.91 ± 5.62	0.08 ± 0.44	0.06 ± 0/36	0.01 ± 0.02	0.02 ± 0.20
Children	1.89 ± 2.92	3.88 ± 5.74	0.08 ± 0.31 *	0.05 ± 0.20	0.01 ± 0.01	0.01 ± 0.08
Total sample	1.75 ± 3.15	3.33 ± 5.74	0.08 ± 0.38	0.06 ± 0.29	0.01 ± 0.01	0.02 ± 0.16
Miscellaneous food products						
Toddlers	0.02 (0.00–0.13)	0.03 (0.00–0.19)	0.00 (0.00–0.33)	0.00 (0.00–0.01)	0.00 (0.00–0.38)	0.00 (0.00–0.00)
Children	0.16 (0.03–4.31) ***	0.26 (0.04–4.72) ***	0.40 (0.00–2.67) ***	0.03 (0.00–2.43) ***	0.28 (0.00–3.91) ***	0.00 (0.00–1.66) ***
Total sample	0.06 (0.00–1.25)	0.08 (0.00–1.71)	0.12 (0.00–1.23)	0.00 (0.00–0.67)	0.00 (0.00–0.95)	0.00 (0.00–0.46)
Products for special nutritional use						
Toddlers	1.90 ± 7.50	2.04 ± 7.74	1.73 ± 7.72	1.70 ± 7.47	0.09 ± 0.36	0.55 ± 2.47
Children	0.16 ± 1.76 ***	0.20 ± 2.34	0.04 ± 0.53 **	0.12 ± 1.57 **	0.01 ± 0.19	0.06 ± 0.92 ***
Total sample	1.03 ± 5.50	1.12 ± 5.84	0.89 ± 5.57	0.96 ± 5.41	0.04 ± 0.29	0.30 ± 1.88

TE—total energy intake; FA—fatty acids. Data are presented as average ± standard deviation for normally distributed data, and as median (range) for skewed data; differences between toddler and children groups were tested with Student's t-test for normally distributed data and Mann–Whitney test in case of skewed distribution; * $p < 0.05$, ** $p < 0.01$, *** $p < 0.001$.

4. Discussion

The Serbian Food Consumption Survey, conducted on a nationally representative sample (n = 576) in line with the rigorous standards and methodological framework of the EU Menu project established by the EFSA, provides much-needed insight into the nutritional status and dietary habits of toddlers and children 1–9 years old living in Serbia.

An individual's overall health is influenced by genetic and epigenetic legacies; intrauterine stressors; environmental exposures; behavioral factors; interpersonal relationships; sociopolitical, cultural, and economic structures; norms; and opportunities. Acknowledging the importance of proper childhood development in the pursuit of health across individuals' lifespan, the World Health Organization (WHO) advocates for the adoption and implementation of a life-course approach accentuating the need for "early, timely, appropriate and collective" actions [22]. Practical, sustainable, and evidence-based strategies and policies targeting the optimization of nutrition among children could make a major contribution to mitigating the humanistic, clinical, and financial impact of acute complications and late sequelae of young-age malnutrition. A multisectoral food system approach systematically and comprehensively engaging various stakeholders with coordinated, coherent, and complementary actions across governmental and social structure levels is the most effective manner for addressing the complexity of nutritional issues while concomitantly ensuring safe, affordable, and sustainable diets [23–25].

Approximately two-thirds of the study participants had a normal weight according to the WHO BMI-for-age classification system. Nevertheless, 10.94% of children were characterized as overweight, 6.25% as obese, and for an additional 11.80% of participants under 5 years of age the anthropometric measurements suggested a possible risk of becoming overweight. These results are in accordance with the data reported in the Multiple Indicator Cluster Survey findings conducted in Serbia in 2019 [26]. Furthermore, the findings of our study corresponded well with the UNICEF/WHO/World Bank Group Joint Child Malnutrition Estimates of the prevalence of overweight among children in Serbia. With the modeled estimate value of 10.8%, Serbia was allocated in the high overweight prevalence category based on the country-level threshold classification system proposed by the WHO-UNICEF Technical Advisory Group on Nutrition Monitoring [27]. Excess body weight is a complex disorder with a rising prevalence worldwide, implying complex and interconnected genetic, metabolic, psychological, socioeconomic, and environmental etiopathogenetic factors [28]. Paradoxically concomitant with undernutrition, the prevalence rates of overweight and obesity in childhood have reached alarming levels, especially in developed and developing countries [29]. Furthermore, with respect to obesity during childhood, and in particular during adolescence, there is a high risk for the persistence of the condition into adulthood [30]. Obesity at a young age has been associated with an increased morbidity burden and numerous medical and socio-emotional repercussions. These conditions include, but are not limited to: cardiovascular diseases, metabolic disturbances, diabetes mellitus, sleep apnea, hepatic steatosis, musculoskeletal issues, dermatologic problems, menstrual abnormalities, asthma, depression, poor socialization, and anxiety [31]. Given the multifaceted nature of both the causes and consequences of childhood obesity, the prevention and management of this problem require actions across a variety of contexts such as family, school, community, the healthcare system, and governmental public health services [32]. The identification of children with excess body weight and the application of age-appropriate, culturally sensitive evidence-based measures and interventions may lower the obesity rates as well as the public health and societal burden of obesity in adulthood [33–36].

Analyses of the urban-rural disparities in nutritional status revealed that the deviation from normal body weight towards both ends of the spectrum was more prevalent in urban settings compared to rural households. This may be attributed to a westernized dietary pattern, less physical activity, increased exposure to highly processed food with poor nutritional quality, distancing from the traditional food consumption model, and meal plan irregularities [37]. Worldwide data suggest that compared to rural circumstances, urban dietary trends incline towards energy-dense high-fat food; an increased consumption of refined, more polished, and milled grains; and a higher intake of animal-source products, sugar, salt, and (ultra)processed food [38]. Furthermore, there is an additional impact of intensified susceptibility and responsiveness to the mass media and marketing activities of the commercial food sector and an increased consumption of food prepared outside of the home. The "nutrition transition" phenomenon encompasses concurrent changes in dietary and energy expenditure patterns, coinciding with economic development, urbanization, demographic shifts, modernization, epidemiological factors, food system changes, and technological advancements [8]. Early-life exposure to such practices is of particular concern since it may aggravate the adverse health effects of an unfavorable energy balance and diet quality [39]. An improved comprehension of the drivers and outcomes of the discrepancies between the eating habits, lifestyle, nutritional status, and body composition among urban and rural residents may contribute to an enhanced assessment of their health demands and the development of interventions tailored according to their respective settings [40].

The mean estimated energy daily intake of 6.25 $\pm$ 1.98 MJ was comparable with reports from other European countries, and expectedly, higher values were recorded for boys and older children. The consumption of energy-dense food groups such as meat and meat products, fat and oil, sugar, and confections was more pronounced among older children. These trends may be related to physiological factors, increased requirements due to developmental and physical activity factors, diet diversification, the broadening of food repertoires, and the increased eating autonomy that develops with aging [41,42]. There were no intake discrepancies regarding particular food groups between girls and boys, suggesting that gender-associated food–preference differences may occur later in life [40]. The nutritional composition analysis revealed certain deviations regarding the proportions of the macronutrient intake and dietary reference values. Congruently with the comprehensive overview of the nationally representative dietary surveys encompassing 53 countries in the WHO's Europe remit, the compliance with the recommendations was particularly poor for fat and fiber intake [41]. With the overall median estimation of 37.88% for the energy derived from fat, more than two-thirds of the study participants exceeded the 35% threshold. Fat and energy intake are closely correlated, which makes it perplexing to isolate their individual effects on bodyweight measures. Although high-quality long-term trials and properly designed prospective studies are warranted to fully elucidate the health effects of a high-fat diet in the population of children, there is a substantial body of evidence associating such eating patterns with cardiometabolic risk factors and adiposity indices in children [43]. Along with other complimentary measures, conducted at the individual, community, and population levels, an adjustment of fat consumption may contribute to the maintenance of regulated energy balance for a healthy weight and the prevention of overweight and obesity in youth [3].

Although showing an age-related increasing trend in terms of absolute daily intake values, the proportion of subjects reaching the dietary fiber AI threshold decreased from toddlers towards older subgroups in the analyzed children cohort. Only in the toddlers' group did the median estimated intake per day surpass the age-specific AI level. In both succeeding subgroups, namely the 4–6- and 7–9-year-old participants, more than 60% remained below the AI value, with a slightly higher proportion of boys reaching the reference cut-off point than their female counterparts. The observed trends might be partially attributed to differences in the organizational framework and practical perspectives for collective child nutrition in Serbia. There are official, legally endorsed recommendations for coordinating, monitoring, and implementing the collective feeding programs for children in kindergartens and primary schools [19,44]. In these institutions, meal planning is performed by certified professionals in accordance with the age-appropriate nutritional normative standards. Although not including specific values for dietary fiber, the recommendations for the representation of food groups in daily/weekly menus as well as the distribution of energy and macronutrients across the meals adjusted for the duration of the child's stay are laid down in official guidelines for both institution levels. The kindergarten program encompasses three mandatory meals (i.e., breakfast, snack, and lunch), and one additional discretionary snack, presumably providing 75%of the total daily energy needs. However, in primary schools, there is wide variability regarding the meal-offer coverage, which depends on the operational, logistical, and organizational resources. Furthermore, while kindergarten feeding programs include all the enrolled children, organized nutrition in primary schools is optional, and the decentralized nature of school-meal programs in Serbia hinders the availability of precise data regarding the proportion of students who benefit from school feeding. Dietary fiber, as an essential nutrient, confers a plethora of both short-term and long-term functional and physiological health benefits. The most common effect associated with a sufficient fiber intake is the improvement of the digestive process and overall gastrointestinal health, including laxation. Nonetheless, fiber benefits extend beyond the gut function, and include cholesterol reduction, glycemic control, the prevention of cancer and cardiovascular diseases, and weight management, as well as supporting the immune system and proper cognition in children [45–47]. The majority of official guidelines propose recommendations only in terms of quantity, neglecting to address the relevant physico-chemical properties, fermentability, bulking, and dietary sources. However, the diversity of fiber types in food, the heterogeneity of these types' physiological effects, and the complexity of their functional properties provide the scientific rationale for a consideration of the qualitative features in addition to the simple total daily consumption amount [48]. The data retrieved from numerous national dietary surveys suggest that children in most industrialized countries in Europe, North America, and Oceania fail to achieve the recommended intake levels [49,50]. Given the potential adverse effects of a low-fiber intake on pediatric diets, parents/caretakers should be counseled to encourage children to consume a diverse diet including more whole-grain products, fruits, vegetables, legumes, nuts, seeds, and other fiber-rich food sources. Furthermore, efforts should be made to provide general public health guidance on dietary fiber sources, recommendations, and the importance of health maintenance and disease prevention.

The physical and social environment are significant determinants of the eating patterns in the pediatric population. Parents' nutritional behaviors and attitudes, as well as peers' dietary habits, shape children's food appeal and feeding style. The availability and accessibility of and repeated exposure to healthy foods are crucial to developing familiarity and acceptance [51]. It is challenging to assess the exact adequacy of the dietary intake of the food groups in the analyzed cohort as country-specific Food-Based Dietary Guidelines (FBDG) for the Serbian population do not exist at this stage [52]. The development of such a tool would help to better understand the quality of the diets of Serbian children and toddlers, identify potential nutritional challenges in this vulnerable population, and enable the timely addressalof specific inconsistencies. The WHO, FAO, and EFSA encourages member states do develop national FBDG using a multi-disciplinary stepwise approach with a particular focus on the diet–disease relationships relevant to the population of interest [53]. Well-designed, carefully structured dietary guidelines that take into account generally established nutritional principles, environmental impact, and current scientific evidence are a prerequisite for the development of policies aimed at shifting the consumption patterns of children towards healthier directions, thereby ensuring long-term beneficial effects and the prevention of diet-related diseases later in life [54–56].

Certain limitations of the present study should be acknowledged. The cross-sectional design prevents us from drawing any causal inferences, and further longitudinal research is warranted to acquire a better understanding of the potential nutritional challenges and health determinants among children living in Serbia. The intricacy of the study protocol, the high degree of parental/caretaker involvement required, the lack of financial participation incentives, and the obsolescence of the Census data were major concerns that hindered recruitment. Despite the effort invested in the nationally representative sampling, given the voluntary nature of participation in this study, selection bias to some extent cannot be completely precluded, thus limiting the generalizability of the presented findings. Among the key issues in nutritional epidemiology are measurement error and the inherent limitations of self and proxy-reported dietary consumption data. Nevertheless, the data collection was conducted by trained professionals with expertise in nutrition research in accordance with internationally recognized methodological guidance in a prospective manner, taking into account seasonal, weekly, and intraindividual dietary variability. Furthermore, comprehensive quality assurance procedures were employed across the study's critical points including sampling/recruitment, fieldwork management, and data cleaning.

5. Conclusions

Given the scarcity of previously available data, the presented Serbian Food Consumption Survey may serve as a fundamental evidence platform for public health programs, a valuable asset for decisionmakers, and a reliable reference to guide nutritional policies, diet monitoring, and interventions targeting the population of toddlers and children in the future. Furthermore, the application of a harmonized methodological approach established under the EU Menu project facilitates comparative analyses across the continent and contributes to a common pan-European food consumption database.

Author Contributions: Conceptualization, M.Z., J.M., M.G. (Mirjana Gurinović) and M.G. (Marija Glibetić); methodology, M.Z., J.M., M.T., M.K. and I.Š.; software and IT support, A.K.; formal analysis, M.Z., J.M., M.T. and I.Š.; writing—original draft preparation, M.Z.; writing—review and editing, J.M., M.T., M.K., I.Š. and A.K.; supervision, M.G. (Mirjana Gurinović) and M.G. (Marija Glibetić); project administration, M.Z., J.M., M.G. (Mirjana Gurinović) and M.G. (Marija Glibetić). All authors have read and agreed to the published version of the manuscript.

Funding: This research was funded by the Ministry of Education, Science and Technological Development of the Republic of Serbia, Agreement on Realization and Financing of Scientific Research work of NIO for 2022, No.451-03-68/2022-14/200015. In addition, research was funded by the European Food Safety Authority OC/EFSA/DATA/2016/02-CTO1.

Institutional Review Board Statement: The study was conducted in accordance with the guidelines laid down in the Declaration of Helsinki and all procedures involving human subjects were approved by the Institute for Medical Research Ethics Committee in Serbia on 8 December 2017 (EO 123/2017).

Informed Consent Statement: Informed consent was obtained from all subjects involved in the study.

Data Availability Statement: Results attained in this study are included in the manuscript. Individual data are not available due to official legal, organizational and data security policies and ethical restrictions.

Conflicts of Interest: The authors declare no conflict of interest.

References

1. Jacob, C.M.; Baird, J.; Barker, M.; Cooper, C.; Hanson, M. *The Importance of a Life Course Approach to Health: Chronic Disease Risk from Preconception through Adolescence and Adulthood: White Paper*; World Health Organization: Geneva, Switzerland, 2017; p. 41.
2. Taylor-Baer, M.; Herman, D. From epidemiology to epigenetics: Evidence for the importance of nutrition to optimal health development across the life course. In *Handbook of Life Course Health Development*; Halfon, N., Forrest, C.B., Lerner, R.M., Faustman, E., Eds.; Springer: Cham, Switzerland, 2017; pp. 431–462.
3. UNICEF; WHO; World Bank. *Levels and Trends in Child Malnutrition: UNICEF/WHO/World Bank Group-Joint Child Malnutrition Estimates*, 2021st ed.; UNICEF: New York, NY, USA; WHO: Geneva, Switzerland; World Bank: Washington, DC, USA, 2021; 32p.
4. Kupka, R.; Siekmans, K.; Beal, T. The diets of children: Overview of available data for children and adolescents. *Glob. Food Secur.* **2020**, *27*, 100442. [CrossRef]
5. Zhang, L.; Geelen, A.; Boshuizen, H.C.; Ferreira, J.; Ocké, M.C. Importance of details in food descriptions in estimating population nutrient intake distributions. *Nutr. J.* **2019**, *18*, 17. [CrossRef] [PubMed]
6. Rippin, H.L.; Hutchinson, J.; Evans, C.E.L.; Jewell, J.; Breda, J.J.; Cade, J.E. National nutrition surveys in Europe: A review on the current status in the 53 countries of the WHO European region. *Food Nutr. Res.* **2018**, *62*. [CrossRef]
7. WHO Regional Office for Europe. *European Food and Nutrition Action Plan 2015–2020*; World Health Organization: Copenhagen, Denmark, 2015.
8. Statistical Office of the Republic of Serbia; UNICEF. *Serbia Multiple Indicator Cluster Survey 2019 and Serbia-Roma settlements Multiple Indicator Cluster Survey 2019, Survey Findings Report*; Statistical Office of the Republic of Serbia and UNICEF: Belgrade, Serbia, 2020.

9. Gurinović, M.; Kadvan, A.; Vukotić, M. The quality of nutrition in schoolchildren and adult members of families. In *Monograph: Yugoslav Study of Atherosclerosis Precursors in Schoolchildren in Serbia: Twenty Years Follow-Up*; Nedeljkovic, S., Ed.; Medical Faculty University of Belgrade: Beograd, Serbia, 2011; pp. 343–368.
10. Breda, J.; Castro, L.S.N.; Whiting, S.; Williams, J.; Jewell, J.; Engesveen, K.; Wickramasinghe, K. Towards better nutrition in Europe: Evaluating progress and defining future directions. *Food Policy* **2020**, *96*, 101887. [CrossRef]
11. Zekovic, M.; Gurinovic, M.; Milesevic, J.; Kadvan, A.; Glibetic, M. National food consumption survey among children from 1 to 9 years old in Serbia. *EFSA Support. Publ.* **2021**, *18*, 6994E.
12. European Food Safety Authority (EFSA). Guidance on the EU Menu methodology. *EFSA J.* **2014**, *12*, 3944.
13. Gurinović, M.; Nikolić, M.; Zeković, M.; Milešević, J.; Kadvan, A.; Ranić, M.; Glibetić, M. Implementation of harmonized food consumption data collection in the Balkan Region according to the EFSA EU menu methodology standards. *Front. Nutr.* **2021**, *8*, 809328. [CrossRef] [PubMed]
14. Nikolić, M.; Milešević, J.; Zeković, M.; Gurinović, M.; Glibetić, M. The development and validation of food atlas for portion size estimation in the Balkan Region. *Front. Nutr.* **2018**, *5*, 78. [CrossRef]
15. Gurinović, M.; Milešević, J.; Kadvan, A.; Nikolić, M.; Zeković, M.; Djekić-Ivanković, M.; Dupouy, E.; Finglas, P.; Glibetić, M. Development, features and application of DIET ASSESS & PLAN (DAP) software in supporting public health nutrition research in Central Eastern European Countries (CEEC). *Food Chem.* **2018**, *238*, 186–194. [PubMed]
16. Gurinović, M.; Milešević, J.; Kadvan, A.; Djekić-Ivanković, M.; Debeljak-Martačić, J.; Takić, M.; Nikolić, M.; Ranković, S.; Finglas, P.; Glibetić, M. Establishment and advances in the online Serbian food and recipe data base harmonized with EuroFIR^{TM} standards. *Food Chem.* **2016**, *193*, 30–38. [CrossRef]
17. Gavrieli, A.; Naska, A.; Konstantinidi, C.; Berry, R.; Roe, M.; Harvey, L.; Finglas, P.; Glibetic, M.; Gurinovic, M.; Trichopoulou, A. Dietary monitoring tools for risk assessment. *EFSA Support. Publ.* **2014**, *11*, 607E. [CrossRef]
18. EFSA Food Classification, FoodEx2 Browsing Tool. Available online: https://github.com/openefsa/catalogue-browser/wiki (accessed on 15 December 2021).
19. State Gazette RS. *Pravilnik o Blizim Uslovima i Nacinu Ostvarivanja Ishrane Dece u Predskolskoj Ustanovi—Recommendations for Collective Nutrition of Children 39/2018*; Public Enterprise Official Gazette: Belgrade, Serbia, 2018.
20. European Food Safety Authority (EFSA). *DRV Finder—Dietary Reference Values for the EU*. Available online: https://multimedia.efsa.europa.eu/drvs/index.htm (accessed on 15 March 2022).
21. World Health Organization. *WHO Child Growth Standards: Length/Height-for-Age, Weight-for-Age, Weight-for-Length, Weight-for-Height and Body Mass Index-Forage: Methods and Development*; World Health Organization, Department of Nutrition for Health and Development: Geneva, Switzerland, 2006; pp. 1–312.
22. World Health Organization. *Regional Office for Europe The Minsk Declaration—The Life-Course Approach in the Context of Health 2020, WHO European Ministerial Conference on the Life-Course Approach in the Context of Health 2020*; World Health Organization: Copenhagen, Denmark, 2015.
23. Lamstein, S.; Pomeroy-Stevens, A.; Webb, P.; Kennedy, E. Optimizing the multisectoral nutrition policy cycle: A systems perspective. *Food Nutr. Bull.* **2016**, *37*, S107–S114. [CrossRef] [PubMed]
24. UNICEF. *Office of Research Innocenti Food Systems for Children and Adolescents: Working Together to Secure Nutritious Diets, Interim Summary Report*; UNICEF: Florence, Italy, 2018.
25. Raza, A.; Fox, E.L.; Morris, S.S.; Kupka, R.; Timmer, A.; Dalmiya, N.; Fanzo, J. Conceptual framework of food systems for children and adolescents. *Glob. Food Sec.* **2020**, *27*, 100436. [CrossRef]
26. Morales-Suárez-Varela, M.; Rubio-López, N.; Ruso, C.; Llopis-Gonzalez, A.; Ruiz-Rojo, E.; Redondo, M.; Pico, Y. Anthropometric status and nutritional intake in children (6–9 years) in Valencia (Spain): The ANIVA study. *Int. J. Environ. Res. Public Health* **2015**, *12*, 16082–16095. [CrossRef] [PubMed]
27. Smith, J.D.; Fu, E.; Kobayashi, M.A. Prevention and management of childhood obesity and its psychological and health comorbidities. *Annu. Rev. Clin. Psychol.* **2020**, *16*, 351. [CrossRef]
28. Simmonds, M.; Llewellyn, A.; Owen, C.G.; Woolacott, N. Predicting adult obesity from childhood obesity: A systematic review and meta-analysis. *Obes. Rev.* **2016**, *17*, 95–107. [CrossRef] [PubMed]
29. Sahoo, K.; Sahoo, B.; Choudhury, A.K.; Sofi, N.Y.; Kumar, R.; Bhadoria, A.S. Childhood obesity: Causes and consequences. *J. Fam. Med. Prim. Care* **2015**, *4*, 187.
30. Llewellyn, A.; Simmonds, M.; Owen, C.G.; Woolacott, N. Childhood obesity as a predictor of morbidity in adulthood: A systematic review and meta-analysis. *Obes. Rev.* **2016**, *17*, 56–67. [CrossRef]
31. Popkin, B.M. Nutrition in transition: The changing global nutrition challenge. *Asia Pac. J. Clin. Nutr.* **2001**, *10*, S13–S18. [CrossRef]
32. Jones, R.E.; Jewell, J.; Saksena, R.; Salas, X.R.; Breda, J. Overweight and obesity in children under 5 years: Surveillance opportunities and challenges for the WHO European Region. *Front. Public Health* **2017**, *5*, 58. [CrossRef]
33. Walls, H.L.; Johnston, D.; Mazalale, J.; Chirwa, E.W. Why we are still failing to measure the nutrition transition. *BMJ Glob. Health* **2018**, *3*, e000657. [CrossRef]
34. Singh, J.E.; Illner, A.K.; Dokova, K.; Usheva, N.; Kostadinova, T.; Aleksandrova, K. Mapping the global evidence on nutrition transition: A scoping review protocol. *BMJ Open* **2020**, *10*, e034730. [CrossRef] [PubMed]

35. Hebestreit, A.; Börnhorst, C.; Barba, G.; Siani, A.; Huybrechts, I.; Tognon, G.; Eiben, G.; Moreno, L.A.; Fernández Alvira, J.M.; Loit, H.M.; et al. Associations between energy intake, daily food intake and energy density of foods and BMI z-score in 2–9-year-old European children. *Eur. J. Nutr.* **2014**, *53*, 673–681. [CrossRef] [PubMed]
36. Afshin, A.; Sur, P.J.; Fay, K.A.; Cornaby, L.; Ferrara, G.; Salama, J.S.; Mullany, E.C.; Abate, K.H.; Abbafati, C.; Abebe, Z.; et al. Health effects of dietary risks in 195 countries, 1990–2017: A systematic analysis for the Global Burden of Disease Study 2017. *Lancet* **2019**, *393*, 1958–1972. [CrossRef]
37. Popkin, B.M.; Adair, L.S.; Ng, S.W. Now and then: The global nutrition transition: The pandemic of obesity in developing countries. *Nutr. Rev.* **2012**, *70*, 3. [CrossRef] [PubMed]
38. Zhao, J.; Zuo, L.; Sun, J.; Su, C.; Wang, H. Trends and urban-rural disparities of energy intake and macronutrient composition among Chinese children: Findings from the China health and nutrition survey (1991 to 2015). *Nutrients* **2021**, *13*, 1933. [CrossRef]
39. Cena, H.; Calder, P.C. Defining a healthy diet: Evidence for the role of contemporary dietary patterns in health and disease. *Nutrients* **2020**, *12*, 334. [CrossRef]
40. Cooke, L.J.; Wardle, J. Age and gender differences in children's food preferences. *Br. J. Nutr.* **2005**, *93*, 741–746. [CrossRef]
41. Rippin, H.L.; Hutchinson, J.; Jewell, J.; Breda, J.J.; Cade, J.E. Child and adolescent nutrient intakes from current national dietary surveys of European populations. *Nutr. Res. Rev.* **2019**, *32*, 38–69. [CrossRef]
42. Demory-Luce, D.; Morales, M.; Nicklas, T.; Baranowski, T.; Zakeri, I.; Berenson, G. Changes in food group consumption patterns from childhood to young adulthood: The Bogalusa Heart Study. *J. Am. Diet. Assoc.* **2004**, *104*, 1684–1691. [CrossRef]
43. Naude, C.E.; Visser, M.E.; Nguyen, K.A.; Durao, S.; Schoonees, A. Effects of total fat intake on bodyweight in children. *Cochrane Database Syst. Rev.* **2018**, *2*. [CrossRef]
44. State Gazette RS. *Pravilnik o Blizim Uslovima za Organizovanje, Ostvarivanje i Pracenje Ishrane Ucenika u Osnovnim Skolama—Recommendations for Collective Nutrition of School Children 68/2018*; Public Enterprise Official Gazette: Belgrade, Serbia, 2018.
45. Hojsak, I.; Benninga, M.A.; Hauser, B.; Kansu, A.; Kelly, V.B.; Stephen, A.M.; Morais Lopez, A.; Slavin, J.; Tuohy, K. Benefits of dietary fibre for children in health and disease. *Arch. Dis. Child.* **2022**. [CrossRef] [PubMed]
46. Anderson, J.W.; Baird, P.; Davis, R.H.; Ferreri, S.; Knudtson, M.; Koraym, A.; Waters, V.; Williams, C.L. Health benefits of dietary fiber. *Nutr. Rev.* **2009**, *67*, 188–205. [CrossRef] [PubMed]
47. Evans, C.E.L. Dietary fibre and cardiovascular health: A review of current evidence and policy. *Proc. Nutr. Soc.* **2020**, *79*, 61–67. [CrossRef]
48. European Food Safety Authority (EFSA). Scientific opinion on dietary reference values for carbohydrates and dietary fibre. *EFSA J.* **2010**, *8*, 1462.
49. Edwards, C.A.; Xie, C.; Garcia, A.L. Dietary fibre and health in children and adolescents. *Proc. Nutr. Soc.* **2015**, *74*, 292–302. [CrossRef]
50. Stephen, A.M.; Champ, M.M.J.; Cloran, S.J.; Fleith, M.; Van Lieshout, L.; Mejborn, H.; Burley, V.J. Dietary fibre in Europe: Current state of knowledge on definitions, sources, recommendations, intakes and relationships to health. *Nutr. Res. Rev.* **2017**, *30*, 149–190. [CrossRef] [PubMed]
51. Patrick, H.; Nicklas, T.A. A review of family and social determinants of children's eating patterns and diet quality. *J. Am. Coll. Nutr.* **2005**, *24*, 83–92. [CrossRef]
52. Herforth, A.; Arimond, M.; Álvarez-Sánchez, C.; Coates, J.; Christianson, K.; Muehlhoff, E. A Global review of food-based dietary guidelines. *Adv. Nutr.* **2019**, *10*, 590–605. [CrossRef]
53. Agostoni, C.; Bresson, J.-L.; Fairweather-Tait, S.; Flynn, A.; Golly, I.; Korhonen, H.; Lagiou, P.; Løvik, M.; Marchelli, R.; Martin, A.; et al. Scientific opinion on establishing food-based dietary guidelines. *EFSA J.* **2010**, *8*, 1460.
54. Cámara, M.; Giner, R.M.; González-Fandos, E.; López-García, E.; Mañes, J.; Portillo, M.P.; Rafecas, M.; Domínguez, L.; Martínez, J.A. Food-based dietary guidelines around the world: A comparative analysis to update AESAN scientific committee dietary recommendations. *Nutrients* **2021**, *13*, 3131. [CrossRef] [PubMed]
55. FAO/WHO. *Sustainable Healthy Diets: Guiding Principles*; FAO: Rome, Italy, 2019.
56. Hollis, J.L.; Collins, C.E.; DeClerck, F.; Chai, L.K.; McColl, K.; Demaio, A.R. Defining healthy and sustainable diets for infants, children and adolescents. *Glob. Food Sec.* **2020**, *27*, 100401. [CrossRef]

nutrients

MDPI

Article

Disparities in Nutritional Adequacy of Diets between Different Socioeconomic Groups of Finnish Adults

Liisa M. Valsta [1], Heli Tapanainen [1,*], Teea Kortetmäki [2], Laura Sares-Jäske [1], Laura Paalanen [1], Niina E. Kaartinen [1], Peppi Haario [1] and Minna Kaljonen [3]

[1] Department of Public Health and Welfare, Finnish Institute for Health and Welfare, 00271 Helsinki, Finland; liisa.valsta@thl.fi (L.M.V.); laura.sares-jaske@thl.fi (L.S.-J.); laura.paalanen@thl.fi (L.P.); niina.kaartinen@thl.fi (N.E.K.); peppi.haario@thl.fi (P.H.)
[2] Department of Social Sciences and Philosophy, University of Jyväskylä, 40014 Jyväskylä, Finland; teea.kortetmaki@jyu.fi
[3] Environmental Policy Centre, Finnish Environment Institute, 00790 Helsinki, Finland; minna.kaljonen@syke.fi
* Correspondence: heli.tapanainen@thl.fi; Tel.: +359-(0)29-524-8630

Abstract: Information on dietary adequacy is needed to assess food and nutrition security in a modern society, especially in the transition towards climate-friendly food systems. In this study, differences in the nutritional adequacy of diets among Finnish adults were evaluated in population groups of different education, income and urbanisation levels. The study used data from the FinDiet 2017 Survey (*n* = 1655, 18–74 years). Modelled usual intakes of foods and nutrients were evaluated relative to food-based dietary guidelines issued by the National Nutrition Council of Finland (FNNC) and with respect to nutrient adequacy following the Nordic Nutrition Recommendations and FNNC. For about half of the nutrients studied, intakes were found to be adequate. Intakes of protein, fat, saturated fatty acids and salt were estimated to be high. By contrast, inadequate intakes were seen in folate and vitamins A, D, B1, B2 and C in almost all groups studied. Groups with a higher education and income, groups that lived in urban areas and, in particular, women adhered more closely to recommended food consumption and nutrient intakes than others. However, major challenges posed by the Finnish diet are common to all groups studied, and only certain dietary features evaluated in view of nutritional adequacy are associated with socioeconomic differences.

Keywords: dietary intake; socioeconomic differences; urbanisation; 24 h dietary recall; usual intake modelling; dietary guidelines; dietary recommendations; macronutrients; micronutrients; climate-friendly

Citation: Valsta, L.M.; Tapanainen, H.; Kortetmäki, T.; Sares-Jäske, L.; Paalanen, L.; Kaartinen, N.E.; Haario, P.; Kaljonen, M. Disparities in Nutritional Adequacy of Diets between Different Socioeconomic Groups of Finnish Adults. *Nutrients* **2022**, *14*, 1347. https://doi.org/10.3390/nu14071347

Academic Editor: Megan A. McCrory

Received: 16 February 2022
Accepted: 20 March 2022
Published: 23 March 2022

1. Introduction

The increasing prevalence of diet-related non-communicable diseases (NCDs) is one important motivation to evaluate dietary adequacy as a baseline or a follow-up measure for health and nutrition policy actions [1–3]. Dietary intake data that record dietary habits and food consumption as well as nutrient intakes are needed to develop and evaluate health, nutrition and food policy actions at the national level [4,5] and in international coordinations [6–9]. Improvements in dietary assessment methods and greater harmonisation of surveillance activities have enhanced the accuracy of dietary assessments and the comparability of results internationally [10–12]. At this point, more insight is needed into the dietary differences between population groups that differ in their sociodemographic and socioeconomic status (SES) [13–15], in order to mitigate SES-based inequalities of health. This information is also increasingly important to predict the rate and the nutritional consequences of the transition towards more climate-friendly diets, i.e., the expected consumption increase in foods of vegetable origin and decrease in animal-based products [16–18]. Ensuring a socially just dietary transition requires attentiveness to existing SES-based inequalities in nutrition [15].

Efforts to harmonise dietary assessment methods and tools in Europe have made progress during the past decade [10,19–21]. Today, researchers also have available improved statistical modelling methods that permit the use of short-term individual food consumption data for estimations of usual intake [22,23] and, consequently, the evaluation of the dietary adequacy of whole population groups [24] using recent dietary recommendations and guidelines (e.g., [25,26]).

In Finland, gender differences in the quality of diets have been observed for several decades already. Women's diets have repeatedly been shown to be healthier than men´s [27,28]. SES-related health inequalities have received intense research interest during the past decades as well, and measures such as the National Action Plan to Reduce Health Inequalities 2008–2011 have been put forward [29]. However, no noticeable reduction in health inequalities was seen during the Action Plan period [30]. Correspondingly, a more recent study showed no narrowing of educational health inequalities in Finland between 2011 and 2017. Instead, according to some indicators, the disparities had widened [31]. SES differences between diets have been investigated for decades, but studies in Finland have been carried out based, for the most part, on information obtained by questionnaires on meal patterns, consumption frequencies of indicator foods, such as fresh vegetables, meat and meat products, butter and oil, or by comparing mean intakes at group level [7,32–37]. Although studies on SES-based differences in food consumption and nutrient intake exist, recent data from Finland are scarce as most quantitative studies only include data until 2007.

Results so far have shown generally higher vegetable and fruit consumption among higher education and income groups. Educational differences in nutrient intakes during the past decades have been most consistent for vitamin C intake, with a higher intake among high education groups, but differences have also been seen for other nutrients [28,38–40]. Results from other countries are in line with findings from Finland and point to poorer quality diets among individuals belonging to lower SES groups. In studies drawing on data from Europe or other high-income countries, the consumption of vegetables and fruit as well as the intake of vitamin C and vitamin D has tended to be lower for lower SES groups, whereas sodium intake has generally been higher for them [14,41–43].

Regional differences in food consumption have been less extensively studied in Finland, but the findings suggest more common use of vegetables and fruit in urban areas [44]. In addition, a recent study showed more common use of red and processed meat in rural areas [45]. Similar findings have been observed in other countries as well [46,47].

Evaluations of dietary adequacy based on usual intake and on an average requirement (AR) of micronutrients [25,48] are new in Europe [49] and have not been carried out for different SES groups in Finland before.

The present study aims to evaluate the nutritional adequacy of adult diets in different sociodemographic and socioeconomic population groups (education, income, urbanisation level), using as its baseline usual intake modelling [22] or, where this is not applicable, comparing mean intakes to recommended daily intake (RI) values by applying the most recent dietary reference values used in the Nordic Countries [25] and at the national level [26]. In addition, mean differences in food consumption, nutrient intakes and sources of nutrients in adult diets among these same population groups are estimated. To gain updated insights into the disparities between population groups, we used the most recent food consumption data of the FinDiet 2017 Survey [28], which represent the Finnish contribution to the EU Menu initiative of the European Food Safety Authority (EFSA) [50]. Based on previous findings [51,52], our hypothesis was that the proportion of individuals whose diet complies with current nutrient recommendations is highest among the highest educational and income groups.

2. Materials and Methods

2.1. Study Population and Data Collection

The FinDiet 2017 Survey data were collected as a sub-sample of and in collaboration with the FinHealth 2017 Study. The study population and data collection methods have been reported in detail before [50,53]. In brief, the FinHealth 2017 Study, a national health examination survey, was carried out in 50 study locations in mainland Finland between January and May 2017. The sampling design of the survey was based on the Health 2000 Survey [54]. For the FinHealth 2017 Study, a representative sample of adults aged 18 years and above (n = 10,247) was drawn from the Population Register using one- and two-stage stratified, random sampling. A 30% random sub-sample (n = 3099) among those aged 18–74 years of the FinHealth 2017 Study sample was invited to participate in the FinDiet 2017 Survey [28]. Their diet was assessed by two non-consecutive 24-h dietary recalls and recorded by dietary interviewers using the in-house dietary software Finessi (version 5.0.5, Finnish Institute for Health and Welfare, Helsinki, Finland) [55]. The software included the food list and descriptors of the national food composition database Fineli® [56]. Participants were first interviewed during the health examination part of the survey; the second 24-h dietary recall took place by telephone between February and October 2017, with a minimum interval of 8 days between recalls. A picture booklet of food portions was used to estimate portion sizes. The use of food supplements was recorded as well. The final food consumption data consisted of two accepted, non-consecutive 24-h recalls from 1655 participants, i.e., 53% of the original sub-sample. The under-reporting was evaluated following EU Menu methodology [10]; it was found to be on average 21% for the face-to-face interviews and 18% for the telephone interviews.

Background data, e.g., gender and age, were obtained from the sampling frame, information for the SES background variables (education and income) was obtained from the FinHealth 2017 Study questionnaires [57], and information on residential area was obtained from the Population Register Centre (coordinates of the residence of the participants) and Statistics Finland (categorisation of urbanisation level of residential area based on these coordinates). The three educational categories used here—"low", "middle" and "high" education level—were created by dividing self-reported number of years of fulltime studying (including primary school) in tertiles by sex and birth year. Questions on total household income before tax deductions during the previous year, and on number of adult and underage household members, were utilized to determine income group. The household income question contained ten categories—from "less than EUR 15,000" and "EUR 15,001–25,000" to "more than EUR 90,000" income per year. For this study, the upper limits of the bottom nine original response categories (e.g., EUR 15,000 for the lowest category and EUR 25,000 for the next category), and the lower limit of the highest category (i.e., EUR 90,000) multiplied by two, were divided by weighted sum of household members, assigning a weight of 1.0 to the first adult, 0.7 to additional adults and 0.5 to underage household members [58]. The resulting individual values were grouped into sex-specific quartiles (Qrt), which in turn were combined into three groups: "lowest Qrt", "middle Qrts" (2.–3. quartiles combined) and "highest Qrt". Urbanisation levels were categorized as follows: "urban" (urban areas), "semi-urban" (areas near urban areas and rural centres) and "rural" (remote rural areas). The study population and categorisation into these groups is described in Table 1.

Table 1. The participants of the FinDiet 2017 Survey by gender, education, income and urbanisation level.

	Men					Women				
	n	%	Mean Age, Years	% Under-Reporters	% BMI ≥30 kg/m²	*n*	%	Mean Age, Years	% Under-Reporters	% BMI ≥30 kg/m²
Education										
Low	259	33	50.7	26	26	269	31	50.3	29	33
Middle	258	33	50.1	30	23	305	35	51.7	30	26
High	256	33	54.0	24	22	285	33	52.7	15	22
Missing	7	1				16	2			
Income										
Lowest Qrt	200	26	45.4	26	28	187	21	45.5	33	28
Middle (2.–3. Qrt)	389	50	54.8	27	22	419	48	54.7	25	31
Highest Qrt	175	22	51.6	26	22	235	27	50.6	15	18
Missing	16	2				34	4			
Urbanisation level										
Urban	451	58	50.1	29	22	535	61	50.6	24	24
Semi-urban	207	27	53.3	26	26	192	22	52.7	28	28
Rural	122	16	54.9	19	29	147	17	53.8	25	40
Missing	0	0				1	0			
All	780	100	51.7	27	24	875	100	51.6	25	27

BMI, body mass index; Qrt, quartile.

2.2. Food Consumption and Compatibility with Food-Based Dietary Guidelines

Consumption of foods and dishes was compared between the different SES groups at the ingredient level after disaggregating the consumed foods according to the recipies of the National Food Composition Database, Fineli® [56]. Exceptions to this principle were processed meat products (e.g., sausages) and processed fish products (e.g., canned fish), which were not disaggregated into ingredients but rather quantified as purchased. Food consumption was classified according to the Fineli® food grouping system, of which results are shown for 13 food groups. These food groups include groups that are highlighted in the Finnish food-based dietary guidelines, namely vegetables and fruit, legumes, nuts and seeds, potatoes, red and processed meat (further broken down into beef, pork and sausages), fish and seafood, liquid milk products (including yogurts) and cheese, and cereals. These are also food groups that are expected to change markedly during the dietary transition towards more climate-friendly diets, i.e., the expected consumption increase in foods of vegetable origin and decrease in animal-based products [16,59].

Consumption of four main food groups was compared to the food-based dietary guidelines, which are as follows: vegetables and fruit consumption (500 g/day excluding juices) and red and processed meat intake (no more than 500 g as cooked/week). For milk products, the guideline (which is 5–6 dL of milk and 2–3 slices of cheese) was summed up and expressed as raw milk needed to produce these amounts, taking into account the higher energy intake of men compared to women; this resulted in an approximate guideline value of 900 g/day of raw milk for men and 700 g/day for women. The amount of raw milk was calculated for men as 6 dL (600 g) of liquid milk plus 3 slices, 30 g each, of cheese multiplied by 10 as a commonly used conversion factor from milk to cheese; for women, this was 5 dL (500 g) of liquid milk plus 2 slices, 20 g each, of cheese. The guidelines for cereal products (9 portions and 6 portions for men and women, respectively) were multiplied by an average amount of cereal per portion, i.e., 27 g, for bread and porridge, and other cereal product portions commonly used in Finland, resulting in an approximate rounded guideline of 245 g/day and 160 g/day for men and women, respectively [26].

2.3. Nutrient Intake and Evaluation of Adequacy Relative to Reference Values

The mean intakes of 20 macro- and micronutrients from food alone (i.e., excluding food supplements) were analysed and intake differences between the SES groups evaluated. The adequacy evaluation followed a modification of an evaluation protocol previously reported by Steenbergen et al. [49]. The AR of nutrients was used to estimate the proportion of Finnish adults in different SES groups with inadequate intake, using modelled usual intake distributions and the AR cut-point method [48]. If the proportion of a population group reaching the AR level was ≥90% (i.e., the proportion below the AR level was <10%), the nutrient intake was considered "probably adequate". If <90% of the population group met the AR level, the intake was judged "not adequate". When the AR was not available, the RI was used as advised by the Nordic Nutrition Recommendations (NNR2012) [25]. According to the NNR2012, if the mean intake of a group is at or above the RI, there is probably a "low prevalence of inadequacy", and if the mean intake is below the RI, "no firm conclusions can be drawn regarding the prevalence of inadequacy at the group level". With respect to macronutrients, the RI ranges as % of total energy (E%) were considered. The macronutrient intake of a SES group was considered to be adequate if inside the RI range. Sodium intake was evaluated as salt intake and compared to the Finnish population reference intake of 5 g/day [26]. For iron in pre-menopausal women, RI was used for intake evaluation since one of the underlying assumptions of the AR cut-point method, symmetrical requirement distribution, is not met by this group [48].

Upper limits were evaluated as well. If >2.5% of a population group exceeded the upper limit of the RI range, the intake was judged to be "high". For micronutrients, the upper intake level (UL) reference values were adopted from the nutrient recommendations of the National Nutrition Council of Finland [26] and were used to estimate the proportion of Finnish adults that may potentially be at risk of adverse effects due to high intake of a certain nutrient. If the proportion of a population group exceeding the UL was larger than 2.5% of the population, the intake was considered "high"; if below, it was considered to be "safe".

In addition, the usual intake distributions of macronutrients and micronutrients from food only and from food and dietary supplements combined, for both men and women, were evaluated for the whole sample by comparison with the AR values or RI values, as described above.

2.4. Statistical Analyses

All analyses were performed for men and women separately. Non-participation bias was corrected using weighting factors, which improves the representativeness of the results with respect to the Finnish adult population overall [60]. The energy under-reporters were identified following the instructions of EU Menu methodology [10].

Mean consumption or intake with 95% confidence intervals (CIs) was calculated using the mean of the data for two days for each subject. Regression analysis was used to test the mean differences between sociodemographic groups. Age was included as a covariate in the regression models. Consumption or intake data were transformed prior to regression analysis using either log or cube root transformation to achieve normality. For pair-wise tests, multiple comparisons were taken into account using the Tukey–Kramer adjustment. For some episodically consumed food groups, it was not possible to transform the consumption data into a normal distribution. For these, the Kruskal–Wallis non-parametric test was used. Pair-wise comparisons were not performed for non-parametric tests. For all food groups, consumption adjusted for energy intake (g/MJ) was used in statistical tests. Usual intake and the proportion of the population below or above the reference value were estimated with statistical program SPADE (R package SPADE.RIVMNwCore 4.0.92, RIVM, Bilthoven, The Netherlands) [22]. The 95% CIs of the proportions were generated by the bootstrap function available in SPADE, with 500 iterations. Significant differences in proportions between sociodemographic groups were evaluated by non-overlapping 95% CIs.

3. Results

3.1. Population Characteristics

The characteristics of the whole population, including gender, SES groups and urban–rural categorisation, are shown in Table 1. The sample was evenly divided among educational groups, but the proportions of subjects in the middle income and the urban groups were larger compared to other groups within the income and urbanisation sets, respectively.

3.2. Food Consumption by Gender, SES Groups and Urbanisation

The consumption of different foods at ingredient level by gender, SES groups and urbanisation level is shown in Table 2. The highest educated group consumed more vegetables and fruit but less red and processed meat compared to the two lower educated groups. The same pattern, i.e., the highest educated group consuming more foods of vegetable origin and fewer of animal origin, could also be seen in the consumption of nuts and seeds and pork (general test statistically significant). Similar findings with respect to education level were seen within both genders. In addition, the lowest educated group of women consumed less cheese compared to the middle education group.

The highest income group consumed more vegetables and fruit, whereas the consumption of cereals was more likely to be lower among men and women in the highest income group when compared to the lower income groups. Among men, potatoes were consumed less in the highest income group compared to the middle income group (Table 2).

Urban men consumed more vegetables and fruit compared to semi-urban or rural men. Rural men consumed more potatoes and liquid milk compared to urban men. Semi-urban men consumed more red and processed meat than urban men. Rural women consumed more potatoes, red and processed meat and milk fats compared to urban women. Semi-urban women did not differ from urban women regarding potato and milk fat consumption, but consumed more red and processed meat and liquid milk than urban women (Table 2).

3.3. Food Consumption in Relation to Food-Based Dietary Guidelines

The usual food consumption distribution obtained in this study for vegetables and fruit, red and processed meat, milk products and cereals were compared to the reference consumption levels given by the Finnish food-based dietary guidelines. The comparisons for vegetables and fruit and for red and processed meat are shown in Figures 1 and 2, respectively.

The vegetable and fruit consumption guideline (500 g/day) was met by 20–29% of the highest educational and income groups of men and women as well as by 24% of the middle educational group of women (Figure 1). In these SES groups, the proportion of those who reached the guideline was significantly higher compared to other SES groups. On average, 8–15% of men in different urbanisation groups, and 16–23% of women in these groups, met the vegetables and fruit consumption guideline, but differences between the urbanisation groups were not statistically significant within genders.

Overall, 18–22% of men met the guideline value for red and processed meat consumption (no more than 500 g/week as cooked). Among men, the groups adhering more closely to the guideline than others were the highest educated and urban men (34% and 24%, respectively) (Figure 2). There was no significant difference by income level. Among women, adherence to the meat guideline was generally higher; the overall proportion not exceeding the recommended intake was highest among the highest educated of these (83%) and also higher among urban women (77%) compared to other urbanization levels (Figure 2, bottom). Again, income was not a significant factor.

Table 2. Average consumption of food groups (g/day) in men and women by education, income and urbanisation level.

	Education								Income								Urbanisation Level							
	Low (1)		Middle (2)		High (3)		General Test	Pair-Wise Comparison	Lowest Qrt (1)		Middle (2.–3. Qrt) (2)		Highest Qrt (3)		General Test	Pair-Wise Comparison	Urban (1)		Semi-Urban (2)		Rural (3)		General Test	Pair-Wise Comparison
Food Group	Mean	95% CI	Mean	95% CI	Mean	95% CI	p-Value	Sign. Diff. [1]	Mean	95% CI	Mean	95% CI	Mean	95% CI	p-Value	Sign. Diff. [1]	Mean	95% CI	Mean	95% CI	Mean	95% CI	p-Value	Sign. Diff. [1]
Men																								
Veg. and fruit	289	265–314	283	257–310	371	340–402	<0.001	1, 2 < 3	280	246–313	313	289–337	360	329–391	<0.001	1, 2 < 3	327	304–349	298	267–329	271	229–312	0.001	1 > 2, 3
Potatoes	89	79–99	86	76–95	79	68–89	NS	NS	84	73–95	91	82–100	71	60–83	0.040	2 > 3	80	72–87	85	77–93	107	91–124	0.021	1 < 3
Legumes	8	5–10	13	7–20	16	10–22	NS	[2]	14	7–21	12	8–15	12	8–16	NS	[2]	15	11–20	8	5–11	7	4–10	NS	[2]
Nuts and seeds	5	3–6	8	5–10	9	6–11	<0.001	[2]	7	4–10	6	5–7	8	5–11	0.001	[2]	8	6–10	6	4–8	4	2–5	0.017	[2]
Red and proc. Meat	146	131–161	149	136–163	117	102–131	<0.001	1, 2 > 3	141	124–158	135	125–145	144	125–163	NS	NS	133	122–144	150	133–167	139	119–160	0.007	1 < 2
Beef	27	21–33	41	32–51	31	23–39	0.008	[2]	37	27–47	28	23–34	39	29–48	NS	[2]	35	29–42	33	26–40	22	17–28	NS	[2]
Pork	42	32–51	42	35–50	24	18–30	<0.001	[2]	43	32–54	37	31–43	28	20–36	NS	[2]	35	29–41	43	32–54	32	24–41	NS	[2]
Sausages	40	32–47	31	26–37	35	25–45	NS	[2]	31	23–40	36	31–42	40	30–49	NS	[2]	31	25–36	41	32–49	46	33–59	0.001	[2]
Fish and seafood	34	27–41	32	16–47	43	35–51	0.032	[2]	34	17–51	35	29–41	39	30–48	0.020	[2]	40	30–49	27	22–32	34	21–48	NS	[2]
Liquid milk	407	358–457	389	329–450	342	301–384	NS	NS	444	385–503	355	320–389	358	301–416	NS	NS	361	320–402	400	339–462	444	376–512	0.016	1 < 3
Milk fats	19	16–22	18	15–21	19	16–22	NS	NS	18	14–22	19	17–21	18	15–21	NS	NS	18	16–20	19	17–22	20	16–23	NS	NS
Cheese	32	28–35	36	30–42	33	29–37	NS	NS	33	27–40	35	31–38	32	27–38	NS	NS	32	28–35	36	30–43	35	29–41	NS	NS
Cereals	146	137–156	154	137–171	150	141–158	NS	NS	157	138–176	152	144–159	139	128–150	0.035	2 > 3	151	140–162	144	133–154	156	144–169	NS	NS
Women																								
Veg. and fruit	326	294–357	394	365–423	422	390–453	<0.001	1 < 2, 3	342	307–377	385	358–412	418	387–448	0.009	1 < 3	393	370–417	375	321–429	340	293–387	NS	NS
Potatoes	64	57–71	64	55–73	58	52–65	NS	NS	61	51–70	65	58–72	58	51–65	NS	NS	59	53–65	60	53–67	76	66–87	0.004	1, 2 < 3
Legumes	11	8–14	14	8–19	15	11–20	0.038	[2]	14	10–19	14	10–19	12	9–16	NS	[2]	15	11–19	11	6–16	9	5–12	NS	[2]
Nuts and seeds	6	4–8	8	6–10	11	9–13	<0.001	[2]	7	5–9	8	6–10	12	9–14	<0.001	[2]	10	8–11	7	5–9	7	4–9	0.005	[2]
Red and proc. meat	79	72–86	72	64–80	61	55–68	<0.001	1, 2 > 3	76	66–85	71	66–77	65	57–73	NS	NS	66	60–71	78	70–86	83	73–94	<0.001	1 < 2, 3
Beef	17	13–22	21	17–26	17	13–21	NS	[2]	18	13–24	18	15–21	20	16–25	NS	[2]	19	16–23	17	14–21	17	14–20	NS	[2]
Pork	20	16–24	19	15–23	14	11–18	0.015	[2]	22	16–28	18	15–22	14	11–17	0.016	[2]	15	13–18	21	16–26	25	17–33	0.006	[2]
Sausages	20	16–25	14	10–17	13	10–16	NS	[2]	19	14–24	16	13–19	11	8–14	NS	[2]	13	11–16	20	13–27	19	14–25	NS	[2]
Fish and seafood	24	19–29	27	22–32	34	28–39	0.011	[2]	22	17–27	27	23–31	35	29–41	0.003	[2]	30	26–34	25	19–30	23	16–30	NS	[2]
Liquid milk	333	295–370	290	260–320	286	260–313	NS	NS	297	257–338	306	282–331	307	274–341	NS	NS	280	258–303	341	303–380	332	278–386	0.043	1 < 2
Milk fats	12	11–14	14	12–15	14	12–15	NS	NS	11	9–14	14	12–15	14	12–16	NS	NS	13	11–14	13	10–15	16	13–19	0.001	1, 2 < 3
Cheese	19	17–21	24	21–26	24	21–27	0.004	1 < 2	20	17–23	23	21–25	24	22–27	NS	NS	22	21–24	22	19–25	22	19–26	NS	NS
Cereals	110	103–117	108	101–116	116	109–122	NS	NS	119	108–129	110	105–115	109	103–115	0.015	1 > 2, 3	110	105–116	114	109–120	109	100–119	NS	NS

[1] Considered significantly different with group rankings as indicated, if for the general test $p < 0.05$ and for pair-wise comparison $p < 0.05$. [2] Pair-wise comparisons not produced for food groups where the non-parametric general test had to be used. CI, confidence interval; Sign. Diff., significant difference; NS, not statistically significant; veg., vegetables; proc., processed.

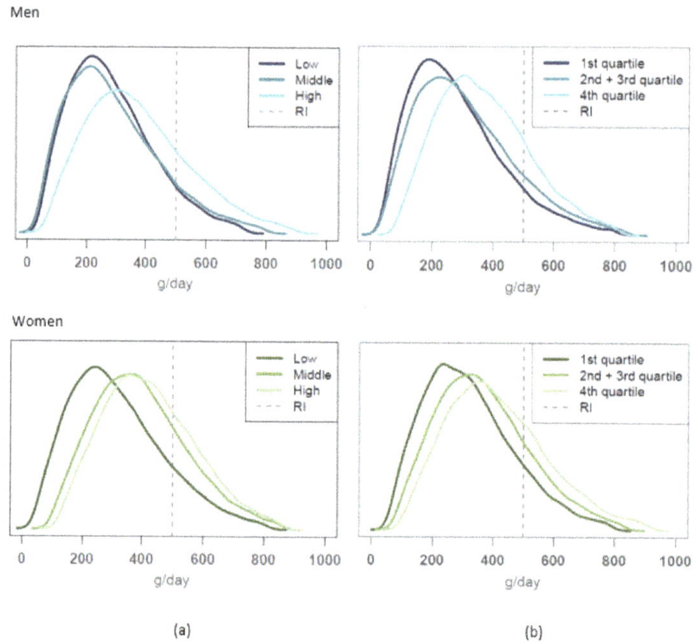

Figure 1. Usual intake distribution of vegetables and fruit consumption compared to the dietary guideline (recommended daily intake (RI); marked as dashed line) of minimally 500 g/day among men (upper figures) and women (lower figures) according to (**a**) educational group, (**b**) income level.

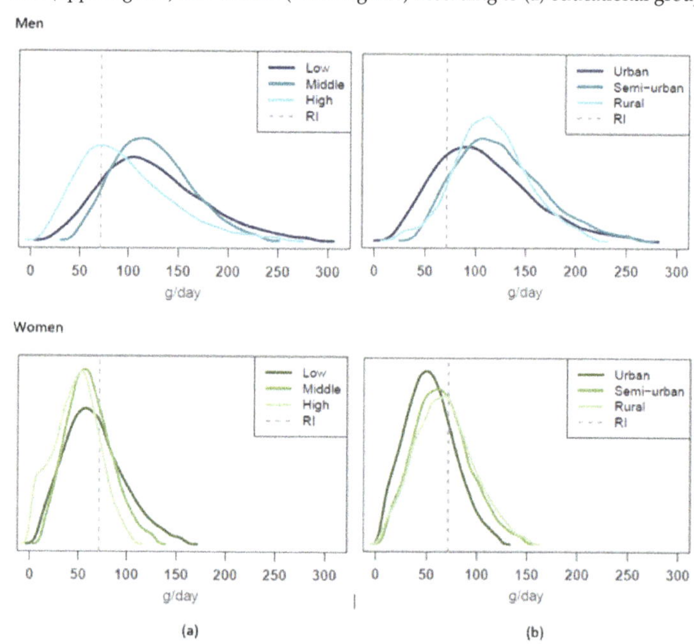

Figure 2. Usual intake distribution of red and processed meat consumption compared to the dietary guideline (RI; marked as dashed line) of a maximum of 500 g/week (= 71 g/day as cooked meat) among men (upper figures) and women (lower figures) according to (**a**) educational group and (**b**) urbanisation level.

About half the men and women met the milk guideline, but no differences were seen between SES groups. Among men, only 1–6% consumed enough cereals to reach the consumption guideline. No differences were found among men between the SES groups. Among women, the cereal consumption guideline was met best by the lowest income group (15%) and worst among the highest income group (4%). There were no differences between educational or urbanisation levels.

3.4. Nutrient Intake Differences and Intake Adequacy Evaluation Based on RI

The evaluation of nutrient intake differences and the adequacy evaluation for nutrients that did not have AR values available and for which the evaluation was based on RI values instead (i.e., macronutrients and salt) are shown in Table 3. In addition, RI was also used for iron in pre-menopausal women. Proportions of population groups meeting the RI reference intakes according to our data on intake distributions in men and women by education, income and urbanisation level are presented in the Supplementary Material, Table S1.

Table 3. Average nutrient intakes and adequacy evaluation based on recommended daily intake (RI) values in men and women by education, income and urbanisation level.

A. Men	**Reference Value**							**General Test**	**Pair-Wise Comparison**	**Overall Adequacy Evaluation** [2]
Nutrient	**RI**	**Mean**	**95% CI**	**Mean**	**95% CI**	**Mean**	**95% CI**	***p*-Value**	**Sign. Diff.** [1]	
Education		Low (1)		Middle (2)		High (3)				
Energy (MJ)	-	9.4	9.0–9.8	9.5	8.9–10.2	9.5	9.1–9.8	NS	NS	-
Protein (E%)	10–20	17.8	17.4–18.3	18.1	17.6–18.6	18.1	17.5–18.7	NS	NS	High intake
Total Carbohydrates (E%)	45–60	44.2	43.2–45.2	42.6	41.7–43.5	43.2	42.1–44.4	0.036	1 > 2	No firm conclusions can be drawn
Fibre (g)	>35	21.9	20.7–23.2	21.1	20–22.1	24.4	23–25.8	0.001	3 > 1, 2	No firm conclusions can be drawn
Fat (E%)	25–40	38.0	37.1–38.9	39.3	38.5–40.1	38.6	37.6–39.7	0.036	1 < 2	Low prevalence of inadequacy
Saturated f.a. (SFA) (E%)	<10	15.1	14.6–15.5	15.4	14.8–15.9	14.7	14.2–15.2	NS	NS	High intake
Polyunsaturated f.a. (PUFA) (E%)	5–10	6.6	6.4–6.8	6.7	6.5–7	7.1	6.8–7.3	0.023	3 > 1, 2	Low prevalence of inadequacy
N-3 PUFA (E%)	1	1.5	1.5–1.6	1.5	1.4–1.6	1.7	1.6–1.8	0.0163	3 > 2	Low prevalence of inadequacy
Salt (g)	≤5	8.8	8.4–9.1	9.0	8.3–9.7	8.4	8–8.8	NS	NS	High intake
Income		Lowest Qrt (1)		Middle (2)		Highest Qrt (3)				
Energy (MJ)	-	9.9	9.2–10.5	9.3	9.0–9.7	9.4	9.0–9.8	NS	NS	-
Protein (E%)	10–20	18.0	17.4–18.6	17.6	17.2–18	18.8	18.1–19.5	0.041	2 < 3	High intake
Total Carbohydrates (E%)	45–60	43.0	42–44	44.1	43.4–44.9	42.4	41–43.8	NS	NS	No firm conclusions can be drawn
Fibre (g)	>35	21.3	19.8–22.8	23.0	21.9–24.1	23.1	21.6–24.7	0.036	1 < 3	No firm conclusions can be drawn
Fat (E%)	25–40	39.0	38–39.9	38.2	37.5–39	38.8	37.6–40	NS	NS	Low prevalence of inadequacy
Saturated f.a. (SFA) (E%)	<10	15.2	14.6–15.8	15.1	14.8–15.5	14.8	14.2–15.4	NS	NS	High intake
Polyunsaturated f.a. (PUFA) (E%)	5–10	6.8	6.4–7.1	6.7	6.5–6.9	7.0	6.7–7.3	NS	NS	Low prevalence of inadequacy
N-3 PUFA (E%)	1	1.5	1.4–1.6	1.5	1.5–1.6	1.6	1.5–1.8	NS	NS	Low prevalence of inadequacy
Salt (g)	≤5	9.0	8.3–9.7	8.7	8.4–9	8.7	8.3–9.2	NS	NS	High intake
Urbanisation level		Urban (1)		Semi-urban (2)		Rural (3)				
Energy (MJ)	-	9.4	9.0–9.8	9.6	9.1–10.0	9.6	9.1–10.0	NS	NS	-
Protein (E%)	10–20	18.2	17.8–18.6	18.2	17.5–18.9	16.8	16.2–17.3	0.003	1,2 > 3	High intake
Total Carbohydrates (E%)	45–60	43.1	42.2–43.9	43.0	41.9–44.2	45.5	43.5–47.6	NS	NS	No firm conclusions can be drawn
Fibre (g)	>35	22.4	21.4–23.4	22.0	20.5–23.5	23.0	20.9–25	NS	NS	No firm conclusions can be drawn
Fat (E%)	25–40	38.8	37.9–39.6	38.7	37.9–39.5	37.7	35.9–39.5	NS	NS	Low prevalence of inadequacy
Saturated f.a. (SFA) (E%)	<10	14.8	14.4–15.3	15.3	14.8–15.7	15.7	14.8–16.6	NS	NS	High intake
Polyunsaturated f.a. (PUFA) (E%)	5–10	7.0	6.7–7.2	6.7	6.5–6.9	6.2	5.8–6.6	0.006	1 > 3	Low prevalence of inadequacy
N-3 PUFA (E%)	1	1.6	1.5–1.7	1.5	1.4–1.6	1.4	1.3–1.6	0.026	1 > 3	Low prevalence of inadequacy
Salt (g)	≤5	8.7	8.3–9.2	8.8	8.4–9.3	8.7	8.2–9.2	NS	NS	High intake

Table 3. *Cont.*

B. Women	Reference value							General test	Pair-wise comparison	Overall adequacy evaluation[2]
Nutrient	**RI**	**Mean**	**95% CI**	**Mean**	**95% CI**	**Mean**	**95% CI**	***p*-Value**	**Sign. Diff.**[1]	
Education		Low (1)		Middle (2)		High (3)				
Energy (MJ)	-	7.1	6.8–7.4	7.2	6.9–7.5	7.9	7.6–8.1	0,0005	3 > 1, 2	-
Protein (E%)	10–20	17.6	17.1–18.1	17.7	17.2–18.2	17.1	16.5–17.6	NS	NS	High intake
Total Carbohydrates (E%)	45–60	44.8	43.8–45.8	44.9	43.9–45.9	44.5	43.6–45.4	NS	NS	No firm conclusions can be drawn
Fibre (g)	>25	18.5	17.5–19.6	20.8	19.7–22	22.2	21.1–23.3	0.000	3,2 > 1	No firm conclusions can be drawn
Fat (E%)	25–40	37.6	36.7–38.6	37.4	36.5–38.3	38.4	37.6–39.2	NS	NS	Low prevalence of inadequacy
Saturated f.a. (SFA) (E%)	<10	14.6	14–15.2	14.4	13.9–14.9	14.0	13.6–14.4	NS	NS	High intake
Polyunsaturated f.a. (PUFA) (E%)	5–10	6.6	6.3–7	6.7	6.4–7	7.4	7.1–7.7	0.000	3 > 1, 2	Low prevalence of inadequacy
N-3 PUFA (E%)	1	1.6	1.5–1.7	1.6	1.5–1.7	1.8	1.7–1.9	0.0090	3 > 1, 2	Low prevalence of inadequacy
Salt (g)	≤5	6.3	6.1–6.6	6.2	5.9–6.5	6.7	6.4–6.9	NS	NS	High intake
Iron (18–50 years) (mg)	15	9.0	8.5–9.5	10.2	9.6–10.8	10.9	10.3–11.5	0.0000	3, 2 > 1	No firm conclusions can be drawn
Income		Lowest Qrt (1)		Middle (2)		Highest Qrt (3)				
Energy (MJ)	-	7.2	6.8–7.6	7.3	7.1–7.6	7.7	7.4–8.0	NS	NS	-
Protein (E%)	10–20	17.3	16.7–17.9	17.3	17–17.7	17.7	17.1–18.2	NS	NS	High intake
Total Carbohydrates (E%)	45–60	45.8	44.6–46.9	45.1	44.3–45.8	43.5	42.3–44.6	0.021	3 < 1	1,2 low prevalence of inadequacy, 3 no firm conclusions can be drawn
Fibre (g)	>25	20.0	18.7–21.3	20.6	19.5–21.6	21.4	20–22.7	NS	NS	No firm conclusions can be drawn
Fat (E%)	25–40	36.9	35.6–38.2	37.6	36.9–38.3	38.9	37.7–40	NS	1 < 3	Low prevalence of inadequacy
Saturated f.a. (SFA) (E%)	<10	14.0	13.3–14.7	14.5	14.1–14.9	14.6	13.8–15.3	NS	NS	High intake
Polyunsaturated f.a. (PUFA) (E%)	5–10	6.7	6.4–7.1	6.8	6.5–7	7.2	6.9–7.5	0.020	2 < 3	Low prevalence of inadequacy
N-3 PUFA (E%)	1	1.6	1.5–1.7	1.7	1.6–1.8	1.8	1.7–1.9	0.031	1 < 3	Low prevalence of inadequacy
Salt (g)	≤5	6.4	6.1–6.7	6.3	6.1–6.6	6.5	6.2–6.8	NS	NS	High intake
Iron (18–50 years) (mg)	15	9.5	8.9–10.1	10.1	9.5–10.7	10.3	9.6–11.1	NS	NS	No firm conclusions can be drawn
Urbanisation level		Urban (1)		Semi-urban (2)		Rural (3)				
Energy (MJ)	-	7.4	7.2–7.6	7.3	7.0–7.7	7.2	6.9–7.5	NS	NS	-
Protein (E%)	10–20	17.6	17.2–17.9	17.5	16.9–18.1	17.2	16.6–17.7	NS	NS	High intake
Total Carbohydrates (E%)	45–60	44.3	43.6–45	45.6	44.1–47.2	45.2	43.9–46.5	NS	NS	1 no firm conclusions can be drawn, 2,3 low prevalence of inadequacy
Fibre (g)	>25	20.7	19.9–21.6	20.3	18.6–22	19.9	18.2–21.6	NS	NS	No firm conclusions can be drawn
Fat (E%)	25–40	38.2	37.5–38.8	36.9	35.4–38.4	37.6	36.5–38.7	NS	NS	Low prevalence of inadequacy
Saturated f.a. (SFA) (E%)	<10	14.3	13.9–14.6	14.3	13.3–15.3	14.9	14.3–15.4	NS	NS	High intake
Polyunsaturated f.a. (PUFA) (E%)	5–10	7.1	6.8–7.4	6.6	6.3–6.8	6.5	6.2–6.9	0.011	1 > 2, 3	Low prevalence of inadequacy
N-3 PUFA (E%)	1	1.7	1.6–1.8	1.6	1.5–1.7	1.6	1.5–1.7	NS	NS	Low prevalence of inadequacy
Salt (g)	≤5	6.4	6.2–6.6	6.5	6.2–6.8	6.3	5.9–6.6	NS	NS	High intake
Iron (18–50 years) (mg)	15	10.2	9.8–10.6	9.8	8.8–10.8	9.1	8.5–9.8	NS	NS	No firm conclusions can be drawn

[1] Considered significantly different with group rankings as indicated, if for the general test $p < 0.05$ and for pair-wise comparison $p < 0.05$. [2] Using RI reference values provided by Nordic Nutrition Recommendations (NNR2012) [25]. If the mean intake of a group is at or above the RI, there is probably a "low prevalence of inadequacy" and if it is below the RI, "no firm conclusions can be drawn regarding the prevalence of inadequacy at the group level", according to the NNR2012 [25]. RI, recommended daily intake; E%, % of total energy; f.a., fatty acids; N-3, omega-3.

3.4.1. Education

Energy intake varied between 9.4 MJ/day and 9.5 MJ/day among men across educational groups (NS). Energy intake varied between 7.1 MJ/day and 7.9 MJ/day among women across educational groups, being highest for the highest educational group. In men, fat intake was higher in the middle education group than in the lowest education group. The highest educational group had the highest intakes of fibre and total polyunsaturated fatty acid (PUFA). It also had higherintake of omega-3 (n-3) PUFA, compared to the middle education group. Similar differences were seen in women. The intake of fibre was higher in the highest and middle education groups compared to the lowest education group, and the intake of PUFA and n-3 PUFA was highest among those in the highest education group (Table 3).

The mean intakes of total fat, protein, total PUFA and n-3 PUFA met the lower bound of the RI reference values. However, total fat, protein, saturated fatty acid (SFA) and salt intakes were found to be high both in men and women. The higher level of recommended protein intake (20 E%) was exceeded by 18–25% of men and by 4–19% of women; there were no differences between educational levels in this regard (Supplementary Material, Table S2). Mean total carbohydrate and fibre intakes were both below the RI reference values, but based on this fact alone, no firm conclusions can be drawn about the adequacy of intakes either in men or in women (Table 3) [25]. The intake of salt in both men and women exceeded the population goal (5 g/day) in over 95% of men and in about 85–90% of women in different educational groups. The mean iron intake of pre-menopausal women fell below the RI reference value; thus, no firm conclusions can be drawn about the adequacy of their intake [25] (Table 3).

3.4.2. Income

Energy intake by income group ranged between 9.3 and 9.9 MJ/day among men and between 7.2 MJ/day and 7.7 MJ/day among women; differences between income groups within genders were not statistically significant. Intake by income did differ notably, however, for several important nutrients. Thus, with respect to fibre, the lowest income group of men had the lowest intake. For protein, the intake was lower in the middle income group compared to the highest income group of men (Table 3). Intakes of fat, PUFA and n-3 PUFA were highest in the highest income group of women, while intake of carbohydrates was highest in the lowest income group of women (Table 3).

The intakes of total fat, protein, total PUFA and n-3 PUFA met the minimum recommendations in all income groups of men and women. The intakes of total fat and protein were sufficient and indeed exceeded recommended levels in all income groups of men and women. The higher level of recommended protein intake was exceeded by 16–33% of men, with the greatest excess recorded in the highest income group, and by 11–21% of women in the income groups, but without differences between groups (Supplementary Material, Table S2). SFA and salt intakes were found to be high throughout: salt intake in both men and women exceeded the population goal (5 g/day) in over 95% of men and in about 85% of women. The mean iron intake of pre-menopausal women in all income groups was below the RI reference value; thus, no firm conclusions can be drawn about the adequacy of their intake (Table 3).

3.4.3. Urbanisation Level

Energy intake was between 9.4 MJ/day and 9.6 MJ/day among men and between 7.2 MJ/day and 7.4 MJ/day among women based on urbanisation level and did not differ by urbanisation level for either gender. In men, the intakes of PUFA and n-3 PUFA were lower in rural areas compared to urban areas. The protein intake of rural men was also lower in comparison to urban and semi-urban men. In a similar trend, in women, urban residents had higher PUFA intakes compared to semi-urban or rural residents (Table 3).

Intakes of total fat, protein, total PUFA and n-3 PUFA met the recommendations in all urbanisation groups. Practically all men and women met the lower limit of the protein intake recommendation range, i.e., 10 E%. Except for semi-urban women, in all population groups evaluated, far below 10% met the saturated fatty acid recommendation, i.e., <10 E%. Thus, total fat, protein and saturated fatty acid intakes were evaluated to be high in both men and women. The intake of salt exceeded the population goal (5 g/day) in more than 95% of all men and in about 80–85% of women in all urbanisation groups (Supplementary Material, Table S2). The mean iron intake of pre-menopausal women at all urbanisation levels was below the RI reference value, so no firm conclusions can be drawn about intake adequacy (Table 3).

3.5. Nutrient Intake Differences and Adequacy Evaluation Based on the AR Cut-Point Method

The nutrient intake differences for micronutrients are shown in Supplementary Material, Table S3. The adequacy evaluations based on the AR cut-point method [48] are shown in Tables 4–6.

3.5.1. Education

In men, intakes of vitamin E, folate, vitamin C and iron were higher in the highest educational group compared to the two other groups. In women, intakes of vitamin A, folate and vitamin C were higher in the highest and middle education groups compared to the lowest education group. In addition, the intake of vitamin E was highest in the highest education group (Supplementary Material, Table S3).

The intakes of vitamin E, vitamin B12, calcium, iodine and zinc met the AR reference values in all educational groups of men and women. In addition, both men and post-menopausal women met the iron requirement and women the B2 requirement. Intakes of vitamins A, D, B1 and folate were not adequate (Table 4). In addition, among men, the intake of vitamin B2 was not adequate.

3.5.2. Income

The lowest income group of men had the lowest vitamin C intake. For iron, the intake was lower in the lowest income group compared to the highest, whereas for B12, the intake was lower in the middle income group compared to the highest. In women, the intakes of vitamin D, vitamin E, vitamin B2, folate, vitamin B12 and vitamin C were highest in the highest income group (Supplementary Material, Table S3).

The intakes of vitamin E, vitamin B12, calcium, iodine and zinc met the adequacy criteria in all income groups of men and women. In addition, the adequacy criteria for iron were met in all income groups of men and in post-menopausal women. Vitamins A, D, B1, B2, folate and vitamin C levels were evaluated as not adequate, with the exception of the highest income group of men (vitamin C and D), the highest and middle income groups of women (vitamin B2) and vitamin C in all income groups of women (Table 5).

3.5.3. Urbanisation Level

Only a few differences in nutrient intakes for micronutrients by urbanisation level were seen. Calcium intake was higher among rural men compared to urban men. By contrast, vitamin C intake was higher among urban compared to semi-urban men. In women, urban women had higher folate intakes compared to rural women (Supplementary Material, Table S3).

The intakes of vitamin E, vitamin B12, calcium, iodine and zinc met the adequacy criteria in all urbanisation level groups. In addition, all urbanisation level groups of men and post-menopausal women met the iron recommendations. Under 90% of the population met the AR reference values of vitamin A, D, B1, B2, folate and vitamin C. These intakes are thus considered "not adequate". The exceptions are the intake of vitamin C among urban women and of vitamin B2 among urban and semi-urban women (Table 6).

3.5.4. Case Vitamin C

Of all nutrients evaluated, vitamin C intake distributions differed the most between different SES groups. Vitamin C intakes were not adequate for any of the educational groups of men or for the lowest educational group of women. Vitamin C intakes were also not adequate for the two lowest income groups of men or for men at any urbanisation level; semi-urban and rural women had inadequate vitamin C intakes as well (Figure 3). In contrast, in other population groups, over 90% of participants exceeded the AR reference value of vitamin C.

3.6. Nutrient Sources of the Different Population Groups Studied

The highest educated group obtained more nutrients from vegetables, fruit and berries, from legumes and nuts, and from fish compared to the lowest educated group, for whom fats, meats, cereals, potatoes and beverages ranked as more important nutrient sources (data only shown for folate in Supplementary Material, Figure S1a–c).

Across income groups, the picture was more mixed. For many nutrients, both the lowest and highest income groups or else two adjacent groups shared the same important food sources; e.g., meat, eggs and legumes served as an important source of folate, and vegetables as an important source of vitamin C, among both the lowest and the highest income groups of men). In this inspection of nutrient sources, meats or legumes were not divided into detailed sub-categories. Among women, the important food groups of the highest income group were more often vegetables, fruit and berries, fish, legumes and nuts, milk, sugars and sweets, whereas with the lower income groups, cereals, meat, potatoes, eggs and fats served more often as important nutrient sources.

Some important food sources of nutrients among urban men and women were fish, vegetables, legumes and nuts, fats, and fruit and berries; by contrast, in semi-urban and rural population groups, meat, eggs, and sugary ingredients and confectionary were seen to be more common. In addition, milk, cereals and potatoes were especially common nutrient sources in the rural population groups.

Table 4. Proportion of population groups reaching the average requirement (AR) values, and adequacy evaluation based on usual intake distributions in men and women by education.

	Reference Value	Low (1)		Middle (2)		High (3)		Sign. Diff. [1]	≥90% of Population Group > AR	Overall Adequacy Evaluation [2]
Nutrient	AR	%	95% CI	%	95% CI	%	95% CI		Yes/No	
Men										
Vitamin A (μg RE)	600	81	72–90	71	67–77	77	69–85	NS	No	Not adequate
Vitamin D (μg)	7.5	89	85–95	86	79–93	86	80–92	NS	No	Not adequate
Vitamin E (mg)	6	96	94–99	96	95–98	99	97–100	NS	Yes	Adequate
Vitamin B1 (mg)	1.2	65	59–72	65	59–72	64	58–70	NS	No	Not adequate
Vitamin B2 (mg)	1.4	85	80–89	82	78–87	80	76–85	NS	No	Not adequate
Folate (μg)	200	65	59–71	67	61–75	80	73–87	3 > 1	No	Not adequate
Vitamin B12 (μg)	1.4	100	100–100	100	100–100	100	100–100	NS	Yes	Adequate
Vitamin C (mg)	60	70	63–77	72	66–80	82	76–88	NS	No	Not adequate
Calcium (mg)	500	97	95–99	97	95–99	98	96–100	NS	Yes	Adequate
Iron (mg)	7	95	92–99	93	90–97	95	93–98	NS	Yes	Adequate
Iodine (μg)	100	100	99–100	99	98–100	100	99–100	NS	Yes	Adequate
Zinc (mg)	6	99	98–100	99	98–100	99	98–100	NS	Yes	Adequate
Women										
Vitamin A (μg RE)	500	84	74–94	86	80–93	89	82–100	NS	No	Not adequate [3]
Vitamin D (μg)	7.5	69	63–77	70	63–78	69	63–75	NS	No	Not adequate
Vitamin E (mg)	5	97	94–99	96	94–99	100	99–100	3 > 2	Yes	Adequate
Vitamin B1 (mg)	0.9	72	66–77	76	70–83	79	73–86	NS	No	Not adequate
Vitamin B2 (mg)	1.1	90	86–95	91	88–95	91	88–95	NS	Yes	Adequate
Folate (μg)	200	45	38–52	60	55–67	74	68–81	2 > 1, 3 > 1, 2	No	Not adequate
Vitamin B12 (μg)	1.4	100	100–100	100	100–100	100	100–100	NS	Yes	Adequate
Vitamin C (mg)	50	88	82–93	94	91–98	96	94–99	3 > 1	No (1), Yes (2, 3)	1 not adequate, 2,3 adequate
Calcium (mg)	500	96	94–99	98	97–100	98	97–99	NS	Yes	Adequate
Iron (51–74 years) (mg)	6	98	94–100	94	91–98	96	93–100	NS	Yes	Adequate
Iodine (μg)	100	99	97–100	99	98–100	99	98–100	NS	Yes	Adequate
Zinc (mg)	5	100	99–100	100	99–100	99	98–100	NS	Yes	Adequate

[1] Significant differences in proportions between educational groups were evaluated by non-overlapping 95% CI. [2] If the proportion of the group reaching the average requirement (AR) level was ≥90%, the intake was considered "adequate". If <90% of the group met the AR level, the intake was considered "not adequate". If over 2.5% of the group exceeded the upper limit of the RI range of macronutrients as E% or exceeded the UL level of micronutrients, the intake was considered "high". [3] Based on the confidence interval, the highest educational group is close to adequate vitamin A intake. RE, retinol equivalents.

Table 5. Proportion of population groups reaching the average requirement (AR) values, and adequacy evaluation according to usual intake distributions in men and women by income.

	Reference Value	Lowest Qrt (1)		Middle (2.–3. Qrt) (2)		Highest Qrt (3)		Sign. Diff. [1]	≥90% of Population Group > AR	Overall Adequacy Evaluation [2]
Nutrient	AR	%	95% CI	%	95% CI	%	95% CI		Yes/No	
Men										
Vitamin A (μg RE)	600	77	69–88	72	67–80	81	74–90	NS	No	Not adequate
Vitamin D (μg)	7.5	88	80–94	83	77–89	90	84–98	NS	Yes (3), No (1, 2)	3 adequate, 1 and 2 not adequate
Vitamin E (mg)	6	98	96–99	96	94–98	99	99–100	3 > 2	Yes	Adequate
Vitamin B1 (mg)	1.2	63	56–70	63	58–70	67	60–73	NS	No	Not adequate
Vitamin B2 (mg)	1.4	85	80–89	80	75–85	88	83–92	NS	No	Not adequate
Folate (μg)	200	70	62–77	67	62–74	85	76–92	3 > 2	No	Not adequate
Vitamin B12 (μg)	1.4	100	100–100	100	100–100	100	100–100	NS	Yes	Adequate
Vitamin C (mg)	60	72	65–79	70	65–75	93	87–98	3 > 1, 2	Yes (3), No (1, 2)	3 adequate, 1,2 not adequate
Calcium (mg)	500	98	96–99	97	95–99	97	95–99	NS	Yes	Adequate
Iron (mg)	7	94	90–98	95	92–97	97	95–99	NS	Yes	Adequate
Iodine (μg)	100	99	99–100	100	99–100	100	99–100	NS	Yes	Adequate
Zinc (mg)	6	99	98–100	99	98–100	99	99–100	NS	Yes	Adequate
Women										
Vitamin A (μg RE)	500	88	80–100	89	83–96	85	78–94	NS	No	Not adequate
Vitamin D (μg)	7.5	64	58–73	71	64–78	69	61–80	NS	No	Not adequate
Vitamin E (mg)	5	96	94–99	97	95–99	100	99–100	3 > 2	Yes	Adequate
Vitamin B1 (mg)	0.9	72	65–81	76	70–81	78	72–86	NS	No	Not adequate
Vitamin B2 (mg)	1.1	87	82–92	92	88–95	95	93–98	3 > 1	No (1), Yes (2, 3)	1 not adequate, 2,3 adequate
Folate (μg)	200	51	45–60	57	52–62	78	71–87	3 > 1, 2	No	Not adequate
Vitamin B12 (μg)	1.4	100	100–100	100	100–100	100	100–100	NS	Yes	Adequate
Vitamin C (mg)	50	91	86–97	93	89–96	95	93–98	NS	Yes	Adequate
Calcium (mg)	500	95	92–99	98	96–99	99	98–100	NS	Yes	Adequate
Iron (51–74 years) (mg)	6	94	90–99	94	91–97	99	98–100	3 > 2	Yes	Adequate
Iodine (μg)	100	99	97–100	98	96–99	100	99–100	3 > 2	Yes	Adequate
Zinc (mg)	5	99	98–100	100	99–100	100	100–100	NS	Yes	Adequate

[1] Significant differences in proportions between income groups were evaluated by non-overlapping 95% CI. [2] If the proportion of the group reaching the AR level was ≥90%, the intake was considered "adequate". If <90% of the group met the AR level, the intake was evaluated to be "not adequate". If over 2.5% of the group exceeded the upper limit of the RI range of macronutrients as E% or exceeded the UL level of micronutrients, the intake was evaluated to be "high".

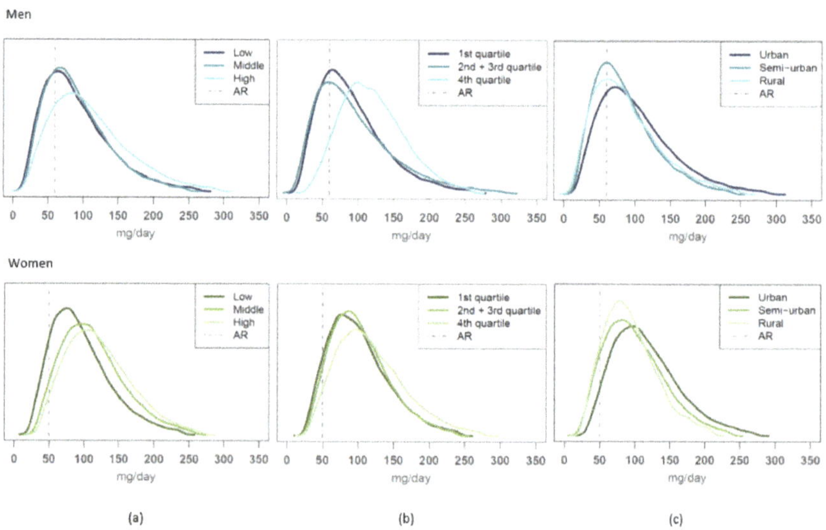

Figure 3. Usual intake distribution of vitamin C intakes compared to the average requirement (AR) value among men (average requirement (AR) = 60 mg/day, upper figures) and women (AR = 50 mg/day, lower figures) according to (**a**) educational group, (**b**) income level and (**c**) urbanization level.

Table 6. Proportion of population groups reaching the average requirement (AR) values, and adequacy evaluation according to usual intake distributions in men and women by urbanisation level.

	Reference Value	Urban (1)		Semi-Urban (2)		Rural (3)		Sign. Diff. [1]	≥90% of Population Group > AR	Overall Adequacy Evaluation [2]
Nutrient	AR	%	95% CI	%	95% CI	%	95% CI		Yes/No	
Men										
Vitamin A (μg RE)	600	74	68–79	83	70–97	72	66–79	NS	No	Not adequate
Vitamin D (μg)	7.5	86	80–91	87	81–93	86	80–93	NS	No	Not adequate
Vitamin E (mg)	6	98	96–99	95	93–98	96	94–100	NS	Yes	Adequate
Vitamin B1 (mg)	1.2	59	53–64	67	61–75	65	60–74	NS	No	Not adequate
Vitamin B2 (mg)	1.4	81	77–86	82	77–87	86	81–95	NS	No	Not adequate
Folate (μg)	200	73	67–78	62	56–69	68	62–77	NS	No	Not adequate
Vitamin B12 (μg)	1.4	100	100–100	100	100–100	100	100–100	NS	Yes	Adequate
Vitamin C (mg)	60	78	72–83	67	59–74	68	63–77	NS	No	Not adequate
Calcium (mg)	500	97	95–98	97	95–99	99	99–100	3 > 1	Yes	Adequate
Iron (mg)	7	94	91–96	97	94–100	92	88–97	NS	Yes	Adequate
Iodine (μg)	100	99	99–100	100	99–100	100	100–100	NS	Yes	Adequate
Zinc (mg)	6	99	98–100	99	98–100	100	99–100	NS	Yes	Adequate
Women										
Vitamin A (μg RE)	500	86	81–92	86	77–96	84	72–96	NS	No	Not adequate
Vitamin D (μg)	7.5	67	61–74	66	61–73	72	64–80	NS	No	Not adequate
Vitamin E (mg)	5	98	96–99	98	96–100	96	93–98	NS	Yes	Adequate
Vitamin B1 (mg)	0.9	75	70–81	71	65–77	76	68–83	NS	No	Not adequate
Vitamin B2 (mg)	1.1	90	87–94	94	90–97	87	84–91	NS	Yes (1, 2), No (3)	1, 2 adequate, 3 not adequate
Folate (μg)	200	65	61–71	56	49–64	44	38–49	3 < 1, 2	No	Not adequate
Vitamin B12 (μg)	1.4	100	100–100	100	99–100	100	100–100	NS	Yes	Adequate
Vitamin C (mg)	50	95	93–98	89	85–94	88	84–93	NS	Yes (1), No (2, 3)	1 adequate, 2, 3 not adequate
Calcium (mg)	500	98	97–99	98	97–99	94	92–97	3 < 1, 2	Yes	Adequate
Iron (51–74 years) (mg)	6	95	92–99	94	90–98	98	96–100	NS	Yes	Adequate
Iodine (μg)	100	99	97–100	99	98–100	99	97–100	NS	Yes	Adequate
Zinc (mg)	5	100	99–100	99	99–100	100	100–100	NS	Yes	Adequate

[1] Significant differences in proportions between urbanisation level groups were evaluated by non-overlapping 95% CI. [2] If the proportion of the group reaching the AR level was ≥90%, the intake was considered "adequate". If <90% of the group met the AR level, the intake was considered "not adequate". If over 2.5% of the group exceeded the upper limit of the RI range of macronutrients as E% or exceeded the UL level of micronutrients, the intake was considered "high".

4. Discussion

In this study, the adequacy of the Finnish diet among adults in different SES groups was evaluated to be close to adequate—or the prevalence of inadequacy to be low—in the case of total fat, PUFA, n-3 PUFA, vitamin E, vitamin B12, calcium, iodine and zinc intakes of the whole population and of iron among men and post-menopausal women. On the other hand, improvements are clearly needed to address the levels of high saturated fatty acid and salt intake in all population groups studied. Inadequate intakes were seen for folate, vitamin A, vitamin D, vitamin B1, vitamin B2 as well as vitamin C in almost all SES groups studied. Additionally, protein intake was unnecessarily high, and total carbohydrate and fibre intakes were prone to being below the recommended level across groups.

In the case of most nutrients, either all or none of the studied groups exceeded or did not reach a recommended upper or lower reference intake. This shows that the major challenges in the Finnish diet cover all the groups studied here, and that only a few dietary features evaluated for nutritional adequacy are associated with SES differences. One such exception was the highest income group of men, which had adequate vitamin C and D intakes (>90% of the group reaching the AR value), while the lower income groups did not. Among women, such exceptions were seen in vitamin B2 and C intakes. For vitamin B2, differences in nutritional adequacy were seen by income groups and urbanisation level and for vitamin C by education and urbanisation levels. It was also seen that even when none of the population groups met the threshold for nutritional adequacy set in the evaluation for a certain nutrient, there were differences between the proportions reaching the reference intake. For example, PUFA intakes, which were evaluated to have low prevalence of inadequacy in all studied groups, were nevertheless higher among the highest educational and urban groups compared to the lower educational and rural population groups. Thus, this study shows that attention needs to be paid to nutrition policy actions ensuring the availability of nutritious food for the whole population, but especially for lower SES groups and those living in semi-urban or rural areas. In this study, we covered education, income and urbanisation levels as SES indicators. It may be, though, that age should be a factor of concern as well, insofar as the elderly may need special attention [61].

In general, nutrient intakes in Finland do not differ very much from general European levels [62]. Vitamin D is an exception. In the European context, Finland has one of the highest vitamin D intakes [7]. This is due to the Finnish vitamin D fortification program that was started in the 1940s and was upgraded in 2002 [5,63]. Despite this effort, we estimate vitamin D intake to be adequate only in the highest income group of men (≥90% of the group reaching the AR reference value). However, the differences in the within-group proportion of SES population groups reaching the average vitamin D requirement were shown to be small in our study, insofar as a proportion of over 80% was reached in all groups studied. This is the situation when only food sources of vitamin D are taken into account (excluding food supplement sources). The main sources of vitamin D in the Finnish diet are fortified foods and fish [28,64]. Similarly, there were very few differences between population groups in the proportion of subjects reaching the AR of iodine intake. Iodine has been shown before to be an important nutrient showing lower intakes in lower SES groups in Europe [14]. Again, the small differences in iodine uptake between SES groups in Finland today are due to a salt fortification program in place since the 1940s and upgraded in 2015 [65]. These examples show that food fortification is an effective tool of nutrition policy when it comes to mitigating dietary disparities between different SES or other population groups.

The differences observed in vitamin C intakes here resemble those seen in earlier years in Europe [14]. Vegetables and fruit are the best sources of vitamin C, providing 63% and 71% of vitamin C for men and women, respectively, in the Finnish diet [28]. In this study, men in the highest educational and income groups and urban men and women in the two highest educational groups, women in all income groups and urban women consumed more of these foods compared to other groups within the same gender. Significantly higher proportions in these population groups also met the AR reference value for vitamin C intake. Moreover, although none of the studied groups exceeded the 90% proportion for reaching the AR of folate, a significantly higher proportion of higher-educated men and women and in higher income groups met the AR reference intake for this nutrient. An increase in vegetable and fruit consumption is important, therefore, from the point of view of nutritional equity, health and climate-friendly diets that are rich in foods of vegetable origin.

Protein intake in the Finnish diet stems mainly from animal-based foods (close to 70%) [28]. The fact that red and processed meat intakes among 70–90% of men and among 20–40% of women are above food-based dietary guideline levels in most population groups studied, and that vegetable and fruit consumption is still below guideline levels for about 80% of these groups, provides evidence and motivation for the need to move further towards a more vegetable-based, healthier diet.

We see in this study that women´s food consumption and dietary intakes are closer to dietary guidelines and nutrition recommendations in the case of certain foods (e.g., red and processed meat) and nutrients (e.g., sodium or salt). These are foods and nutrients whose recommended intake represents an upper limit; the same reference criteria (maximum amounts) are given for both genders [25,26]. In this study, the average energy intake of women was 7.3 MJ/day and of men 9.5 MJ/day [28]. This means that staying at or below the maximum recommended intake would require greater dietary adjustments in men, e.g., consumption of less meat/MJ or less salty food compared to women, to ensure similar adequacy outcomes in both genders. In the case of the vegetables and fruit guideline (500 g/day) and certain micronutrients (e.g., vitamin D, folate and calcium) that have the same minimum requirement set for men and women, men have an advantage over women in adequacy evaluation simply because they should and do eat more food. The same applies with respect to different adult age groups, when the same absolute reference values are applied over a broad age range with varying energy needs. In the case of macronutrients (e.g., protein and fatty acids) different energy intakes are taken into account since recommendations are expressed relative to total energy intakes [25,26]. It might be useful in the future to consider whether differences in energy intake between genders should be taken into account more comprehensively in setting dietary guidelines or nutrient recommendations than is done today.

One limitation of this study is that it covered nutrient intake from foods only, excluding food supplements. We address this in the Supplementary Materials, where information on combined intakes of foods and food supplements, and an evaluation of proportions reaching the AR reference values of men and women based on these additional data, are provided (Supplementary Material, Table S4). These data show that some inadequate intakes, e.g., in the case of vitamin D, are resolved by the use of food supplements; but for most nutrients this is not the case.

A known limitation of dietary studies based on self-reporting is the possibility of misreporting, especially an under-reporting of energy consumed. In this study, under-reporting was estimated to be on average about 25–27% for men and about 24–26% for women. These figures are lower than those of national FinDiet Surveys in the past [66]. In certain previous studies, high educational level has shown to predict under-reporting [66], but in other settings, results have been reported to the effect that under-reporting is more prevalent among individuals in lower SES groups [67]. In this study, the reason why higher energy intake among women with high education compared to other educational groups is probably due to differences in reporting of dietary intake. Indeed, women with low or middle educational groups were found to under-report their dietary intake more often than women with high education. This is supported also by the observation of obesity (BMI $\geq$ 30 kg/m^2) prevalence being lower among the highest educated women and not the other way around, which is concordant with earlier findings [68]. The fact that our data included approximately 25% of energy under-reporters may have also affected the nutrient adequacy evaluation. It may be possible that the proportion of inadequate diets according to this study is an overestimation of the real situation. This would mean that among women, the lower SES groups would have adequate diets more commonly than was found by this study. If that was the case, this would mean that the disparities between SES groups would be even less among Finnish women than evaluated here. Among men, under-reporting did not differ between the SES groups.

The strengths of this study were the very careful sampling design based on population register data; the data collection methods, involving a selection of supporting tools for data collection; extensive quality controls; standardization interviews for the dietary interviewers throughout the data collection period using the method reported by Gavrieli and co-workers [21]; availability of a broad set of background data substantiating SES variables; data management and weighting methods used to tackle the non-response bias [50,53]. The data collection methods used were in line with the European guidance on methodology for harmonised food consumption data collection in EU member states put forward by EFSA [10]. Advanced modelling methods were used to obtain usual food consumption and nutrient intake estimates and population distributions based on short-term data collection [22], which enabled more accurate dietary adequacy evaluations, using also the AR-based cut-point method [48].

While addressing nutritional disparities and the related health inequalities has been acknowledged as an important societal goal in itself, the increasing pressure to transition to more climate-friendly diets will make this issue even more urgent in the future. Active interventions are needed to achieve these goals in tandem. According to the present study, such interventions should certainly aim for an increase in vegetable and fruit consumption across the entire population; initiatives such as the inclusive school meal program in Finland [69–71] still deserve to be fostered. Moreover, the successful fortification programs implemented since the 1940s (Vitamin D, iodine) may become an important way to make sure lower SES classes are quickly brought on board. However, policy makers should also consider more targeted nutritional interventions with respect to meat and dairy consumption. Meanwhile, our results underline the need to update nutritional guidelines while taking into account a more nuanced understanding of SES-based differences. Finally, any initiatives taxing foods on the basis of climate or health impacts should consider their impacts on different SES groups.

5. Conclusions

This study shows that the major challenges in the Finnish diet apply to all population groups studied; only certain dietary features affecting nutritional adequacy are associated with SES differences. Urban, higher educated population groups with a higher income –and especially women among these groups—adhere more closely to dietary guidelines and recommended nutrient intakes than other population groups, but not even these groups reach the desired reference intakes of all nutrients evaluated. The dietary transition towards healthier and more climate-friendly diets is more advanced among these trailblazer population groups. Meanwhile, those groups whose protein intake is, for the most part, still based on red and processed meat, i.e., men in general and less educated and non-urban men in particular, will need to make the greatest adjustments.

Means to mitigate nutritional disparities must be applied broadly and pragmatically to safeguard equal opportunities to achieve adequate nutrition and health regardless of a person's income, education or place of residence. The transition to climate-friendly diets needs to ensure the right to good nutrition for all.

Supplementary Materials: The following supporting information can be downloaded at: https://www.mdpi.com/article/10.3390/nu14071347/s1. Figure S1: Food groups as sources of folate in the diet of men and women according to (a) educational group, (b) income level and (c) urbanisation level (% of daily intake); Table S1: Proportions of population groups reaching the recommended daily intake (RI) values according to usual intake distributions in men and women by education, income and urbanisation level; Table S2: Proportions of population groups exceeding the upper value of macronutrient RI range (E%) or the UL value [26], and evaluation of the intakes; Table S3: Average nutrient intakes in men and women by education, income and urbanisation level; Table S4: Nutrient intakes from food and combined intakes from food and food supplements, and proportion of men and women reaching the dietary reference intakes, modified from [28].

Author Contributions: Conceptualization, L.M.V., H.T., T.K. and M.K.; methodology, H.T., L.M.V. and N.E.K.; formal analysis, H.T.; investigation, L.M.V., P.H., L.S.-J. and L.P.; resources, N.E.K., H.T., L.S.-J. and L.M.V.; data curation, H.T., L.S.-J. and L.M.V.; writing—original draft preparation, L.M.V.; writing—review and editing, P.H., H.T., L.P., L.S.-J., T.K., M.K. and N.E.K.; visualization, H.T. and L.S.-J.; supervision, L.M.V.; project administration L.M.V. and M.K.; funding acquisition, L.M.V., M.K. and T.K. All authors have read and agreed to the published version of the manuscript.

Funding: The data collection was funded in addition to THL, partially by the European Food Safety Authority (EFSA), contract OC/EFSA/DATA/2015/03 CT 01 (EU Menu, Lot 2, Finland/Adults) * and by the Strategic Research Council (SRC) of the Academy of Finland, grant number 327370. * Disclaimer: The publication is produced by the Finnish Institute for Health and Welfare (THL) and authors and not by EFSA and only represents the views of THL and the authors and not EFSA's position.

Institutional Review Board Statement: The FinHealth 2017 Study (including the FinDiet 2017 Survey) was conducted in accordance with the Declaration of Helsinki and approved by the Coordinating Ethics Committee at the Hospital District of Helsinki and Uusimaa (Reference 37/13/03/00/2016, approved on 22 March 2016) for studies involving humans.

Informed Consent Statement: Informed consent was obtained from all subjects involved in the study before the first interview.

Data Availability Statement: The data presented in this study (The FinDiet 2017 Survey data) are available on request from THL Biobank at: https://thl.fi/en/web/thl-biobank/for-researchers (accessed on 14 February 2022). The individual-level data are not publicly available due to privacy restrictions.

Acknowledgments: We acknowledge the expert work carried out by the FinHealth 2017 Study team and the FinDiet 2017 Survey team throughout the data collection process, Anne Juolevi for the data compilation, Tommi Härkänen for the expert support in statistical methods, the Fineli team for the challenging task of keeping the food and nutrient composition data up-to-date, and Maria Vaalavuo and Tuija Martelin for expert support in compiling the socioeconomic background data. Christine Bartels is gratefully acknowledged for the language editing of the manuscript. We thank the participants of the FinDiet 2017 Survey.

Conflicts of Interest: The authors declare no conflict of interest.

References

1. GBD 2015 Obesity Collaborators; Afshin, A.; Forouzanfar, M.H.; Reitsma, M.B.; Sur, P.; Estep, K.; Lee, A.; Marczak, L.; Mokdad, A.H.; Moradi-Lakeh, M.; et al. Health Effects of Overweight and Obesity in 195 Countries over 25 Years. *N. Engl. J. Med.* **2017**, *377*, 13–27. [PubMed]
2. GBD 2017 Risk Factor Collaborators. Global, Regional, and National Comparative Risk Assessment of 84 Behavioural, Environmental and Occupational, and Metabolic Risks or Clusters of Risks for 195 Countries and Territories, 1990–2017: A Systematic Analysis for the Global Burden of Disease Study 2017. *Lancet* **2018**, *392*, 1923–1994.
3. Roth, G.A.; Mensah, G.A.; Johnson, C.O.; Addolorato, G.; Ammirati, E.; Baddour, L.M.; Barengo, N.C.; Beaton, A.Z.; Benjamin, E.J.; Benziger, C.P.; et al. Global Burden of Cardiovascular Diseases and Risk Factors, 1990–2019: Update from the GBD 2019 Study. *J. Am. Coll. Cardiol.* **2020**, *76*, 2982–3021. [CrossRef] [PubMed]
4. Pietinen, P.; Männistö, S.; Valsta, L.M.; Sarlio-Lahteenkorva, S. Nutrition Policy in Finland. *Public Health Nutr.* **2010**, *13*, 901–906. [CrossRef] [PubMed]
5. Raulio, S.; Erlund, I.; Männistö, S.; Sarlio-Lahteenkorva, S.; Sundvall, J.; Tapanainen, H.; Vartiainen, E.; Virtanen, S.M. Successful Nutrition Policy: Improvement of Vitamin D Intake and Status in Finnish Adults over the Last Decade. *Eur. J. Public Health* **2017**, *27*, 268–273. [CrossRef] [PubMed]
6. Roos, G.; Prättälä, R. *Disparities in Food Habits—Review of Research in 15 European Countries*; Publications of the National Public Health Institute B24/1999: Helsinki, Finland, 1999.
7. Rippin, H.L.; Hutchinson, J.; Greenwood, D.C.; Jewell, J.; Breda, J.J.; Martin, A.; Rippin, D.M.; Schindler, K.; Rust, P.; Fagt, S.; et al. Inequalities in Education and National Income Are Associated with Poorer Diet: Pooled Analysis of Individual Participant Data across 12 European Countries. *PLoS ONE* **2020**, *15*, e0232447. [CrossRef]
8. Powles, J.; Fahimi, S.; Micha, R.; Khatibzadeh, S.; Shi, P.; Ezzati, M.; Engell, R.E.; Lim, S.S.; Danaei, G.; Mozaffarian, D.; et al. Global, Regional and National Sodium Intakes in 1990 and 2010: A Systematic Analysis of 24 H Urinary Sodium Excretion and Dietary Surveys Worldwide. *BMJ Open* **2013**, *3*, e003733. [CrossRef]
9. Wang, Q.; Afshin, A.; Yakoob, M.Y.; Singh, G.M.; Rehm, C.D.; Khatibzadeh, S.; Micha, R.; Shi, P.; Mozaffarian, D.; Global Burden of Diseases Nutrition and Chronic Diseases Expert Group (NutriCoDE). Impact of Nonoptimal Intakes of Saturated, Polyunsaturated, and Trans Fat on Global Burdens of Coronary Heart Disease. *J. Am. Heart Assoc.* **2016**, *5*, 2891. [CrossRef]
10. EFSA (European Food Safety Authority). Guidance on the EU Menu Methodology. *EFSA J.* **2014**, *12*, 3944. [CrossRef]
11. Micha, R.; Khatibzadeh, S.; Shi, P.; Andrews, K.G.; Engell, R.E.; Mozaffarian, D.; Global Burden of Diseases Nutrition and Chronic Diseases Expert Group (NutriCoDE). Global, Regional and National Consumption of Major Food Groups in 1990 and 2010: A Systematic Analysis including 266 Country-Specific Nutrition Surveys Worldwide. *BMJ Open* **2015**, *5*, e008705. [CrossRef]
12. Burrows, T.L.; Ho, Y.Y.; Rollo, M.E.; Collins, C.E. Validity of Dietary Assessment Methods When Compared to the Method of Doubly Labeled Water: A Systematic Review in Adults. *Front. Endocrinol.* **2019**, *10*, 850. [CrossRef] [PubMed]
13. Darmon, N.; Drewnowski, A. Does Social Class Predict Diet Quality? *Am. J. Clin. Nutr.* **2008**, *87*, 1107–1117. [CrossRef] [PubMed]
14. Novakovic, R.; Cavelaars, A.; Geelen, A.; Nikolic, M.; Altaba, I.I.; Vinas, B.R.; Ngo, J.; Golsorkhi, M.; Medina, M.W.; Brzozowska, A.; et al. Socio-Economic Determinants of Micronutrient Intake and Status in Europe: A Systematic Review. *Public Health Nutr.* **2014**, *17*, 1031–1045. [CrossRef] [PubMed]
15. Kaljonen, M.; Kortetmäki, T.; Tribaldos, T.; Huttunen, S.; Karttunen, K.; Maluf, R.S.; Niemi, J.; Saarinen, M.; Salminen, J.; Vaalavuo, M.; et al. Justice in Transitions: Widening Considerations of Justice in Dietary Transition. *Environ. Innov. Soc. Transit.* **2021**, *40*, 474–485. [CrossRef]
16. Willett, W.; Rockstrom, J.; Loken, B.; Springmann, M.; Lang, T.; Vermeulen, S.; Garnett, T.; Tilman, D.; DeClerck, F.; Wood, A.; et al. Food in the Anthropocene: The EAT-Lancet Commission on Healthy Diets from Sustainable Food Systems. *Lancet* **2019**, *393*, 447–492. [CrossRef]
17. Hemler, E.C.; Hu, F.B. Plant-Based Diets for Personal, Population, and Planetary Health. *Adv. Nutr.* **2019**, *10*, S275–S283. [CrossRef]
18. Bock, A.K.; Bontoux, L.; Rudkin, J. *Concepts for a Sustainable EU Food System, EUR 30894 EN*; Publications Office of the European Union: Luxembourg, 2022.

19. Ocké, M.; de Boer, E.; Brants, H.; van der Laan, J.; Niekerk, M.; van Rossum, C.; Temme, L.; Freisling, H.; Nicolas, G.; Casagrande, C.; et al. PANCAKE—Pilot Study for the Assessment of Nutrient Intake and Food Consumption among Kids in Europe. *EFSA Support. Publ.* **2012**, *9*, 339E. Available online: www.efsa.europa.eu/publications (accessed on 14 February 2022). [CrossRef]
20. Ambrus, Á; Horváth, Z.; Farkas, Z.; Cseh, J.; Petrova, S.; Dimitrov, P.; Duleva, V.; Rangelova, L.; Chikovalscener, E.; Ovaskainen, M.; et al. Pilot Study in the View of a Pan-European Dietary Survey—Adolescents, Adults and Elderly. *EFSA Support. Publ.* **2013**, *10*, 508E. Available online: http://www.efsa.europa.eu/en/supporting/pub/508e.htm (accessed on 14 February 2022).
21. Gavrieli, A.; Naska, A.; Berry, R.; Roe, M.; Harvey, L.; Finglas, P.; Glibetic, M.; Gurinovic, M.; Trichopoulou, A. Dietary Monitoring Tools for Risk Assessment. *EFSA Support. Publ.* **2014**, *11*, 607E. Available online: www.efsa.europa.eu/publications (accessed on 14 February 2022). [CrossRef]
22. Dekkers, A.L.; Verkaik-Kloosterman, J.; van Rossum, C.T.; Ocke, M.C. SPADE, a New Statistical Program to Estimate Habitual Dietary Intake from Multiple Food Sources and Dietary Supplements. *J. Nutr.* **2014**, *144*, 2083–2091. [CrossRef]
23. Tooze, J.A.; Midthune, D.; Dodd, K.W.; Freedman, L.S.; Krebs-Smith, S.M.; Subar, A.F.; Guenther, P.M.; Carroll, R.J.; Kipnis, V. A New Statistical Method for Estimating the Usual Intake of Episodically Consumed Foods with Application to Their Distribution. *J. Am. Diet. Assoc.* **2006**, *106*, 1575–1587. [CrossRef] [PubMed]
24. Souverein, O.W.; Dekkers, A.L.; Geelen, A.; Haubrock, J.; de Vries, J.H.; Ocke, M.C.; Harttig, U.; Boeing, H.; van't Veer, P.; EFCOVAL Consortium. Comparing Four Methods to Estimate Usual Intake Distributions. *Eur. J. Clin. Nutr.* **2011**, *65* (Suppl. 1), 92. [CrossRef] [PubMed]
25. Nordic Council of Ministers. *Nordic Nutrition Recommendations 2012. Integrating Nutrition and Physical Activity*; Norden, Nord 2014:002; Nordic Council of Ministers: Copenhagen, Denmark, 2014.
26. National Nutrition Council of Finland. *Terveyttä Ruoasta: Suomalaiset Ravitsemussuositukset 2014*; Valtion Ravitsemusneuvottelukunta: Helsinki, Finland, 2014. (In Finnish)
27. Prättälä, R.; Paalanen, L.; Grinberga, D.; Helasoja, V.; Kasmel, A.; Petkeviciene, J. Gender Differences in the Consumption of Meat, Fruit and Vegetables are Similar in Finland and the Baltic Countries. *Eur. J. Public Health* **2007**, *17*, 520–525. [CrossRef] [PubMed]
28. Valsta, L.; Kaartinen, N.; Tapanainen, H.; Männistö, S.; Sääksjärvi, K. *Nutrition in Finland—The National FinDiet 2017 Survey*; Report 12/2018; Finnish Institute for Health and Welfare (THL): Helsinki, Finland, 2018; (In Finnish, Abstract and Table and Chart Titles in English).
29. Ministry of Social Affairs and Health. *National Action Plan to Reduce Health Inequalities 2008–2011*; 2008:16; Publications of the Ministry of Social Affairs and Health: Helsinki, Finland, 2008; (In Finnish, Abstract in English).
30. Rotko, T.; Kauppinen, T.; Mustonen, N.; Linnanmäki, E. *National Action Plan to Reduce Health Inequalities 2008–2011, Final Report*; Report 41; National Institute for Health and Welfare (THL): Helsinki, Finland, 2012; (In Finnish, Abstract in English).
31. Jokela, S.; Kilpeläinen, K.; Parikka, S.; Sares-Jäske, L.; Koskela, T.; Lumme, S.; Martelin, T.; Koponen, P.; Koskinen, S.; Rotko, T. *Health Inequalities in Finland—Proposal for the Development of a Health Inequality Monitoring System in Finland*; Report 5/2021; Finnish Institute for Health and Welfare (THL): Helsinki, Finland, 2021; (In Finnish, Abstract in English).
32. Roos, E.; Talala, K.; Laaksonen, M.; Helakorpi, S.; Rahkonen, O.; Uutela, A.; Prättälä, R. Trends of Socioeconomic Differences in Daily Vegetable Consumption, 1979–2002. *Eur. J. Clin. Nutr.* **2008**, *62*, 823–833. [CrossRef]
33. Lallukka, T.; Pitkäniemi, J.; Rahkonen, O.; Roos, E.; Laaksonen, M.; Lahelma, E. The Association of Income with Fresh Fruit and Vegetable Consumption at Different Levels of Education. *Eur. J. Clin. Nutr.* **2010**, *64*, 324–327. [CrossRef]
34. Ovaskainen, M.L.; Paturi, M.; Tapanainen, H.; Harald, K. Educational Differences in the Diet of Finnish Adults and the Associations between Education and the Determinants and Facilitators of Dietary Fat Quality. *Public Health Nutr.* **2010**, *13*, 925–931. [CrossRef]
35. Petkeviciene, J.; Klumbiene, J.; Prattala, R.; Paalanen, L.; Pudule, I.; Kasmel, A. Educational Variations in the Consumption of Foods Containing Fat in Finland and the Baltic Countries. *Public Health Nutr.* **2007**, *10*, 518–523. [CrossRef]
36. Puolakka, E.; Pahkala, K.; Laitinen, T.T.; Magnussen, C.G.; Hutri-Kahonen, N.; Männistö, S.; Palve, K.S.; Tammelin, T.; Tossavainen, P.; Jokinen, E.; et al. Childhood Socioeconomic Status and Lifetime Health Behaviors: The Young Finns Study. *Int. J. Cardiol.* **2018**, *258*, 289–294. [CrossRef]
37. Paalanen, L.; Prättälä, R.; Palosuo, H.; Laatikainen, T. Socio-Economic Differences in the Consumption of Vegetables, Fruit and Berries in Russian and Finnish Karelia: 1992–2007. *Eur. J. Public Health* **2011**, *21*, 35–42. [CrossRef]
38. Männistö, S.; Ovaskainen, M.; Valsta, L. *The National Findiet 2002 Study*; National Public Health Institute: Helsinki, Finland, 2003; (In Finnish, Summary in English).
39. Paturi, M.; Tapanainen, H.; Reinivuo, H.; Pietinen, P. *The National FINDIET 2007 Survey*; B23/2008; Publications of the National Public Health Institute: Helsinki, Finland, 2008; (In Finnish, Abstract in English).
40. Helldán, A.; Kosola, M.; Ovaskainen, M.; Raulio, S.; Tapanainen, H.; Virtanen, S. *The National FINDIET 2012 Survey*; National Institute for Health and Welfare (THL): Helsinki, Finland, 2013; (In Finnish, Abstract in English).
41. Irala-Estevez, J.D.; Groth, M.; Johansson, L.; Oltersdorf, U.; Prattala, R.; Martinez-Gonzalez, M.A. A Systematic Review of Socio-Economic Differences in Food Habits in Europe: Consumption of Fruit and Vegetables. *Eur. J. Clin. Nutr.* **2000**, *54*, 706–714. [CrossRef]

42. Nikolic, M.; Glibetic, M.; Gurinovic, M.; Milesevic, J.; Khokhar, S.; Chillo, S.; Abaravicius, J.A.; Bordoni, A.; Capozzi, F. Identifying Critical Nutrient Intake in Groups at Risk of Poverty in Europe: The CHANCE Project Approach. *Nutrients* **2014**, *6*, 1374–1393. [CrossRef] [PubMed]
43. De Mestral, C.; Mayen, A.L.; Petrovic, D.; Marques-Vidal, P.; Bochud, M.; Stringhini, S. Socioeconomic Determinants of Sodium Intake in Adult Populations of High-Income Countries: A Systematic Review and Meta-Analysis. *Am. J. Public Health* **2017**, *107*, e1–e12. [CrossRef] [PubMed]
44. Similä, M.; Taskinen, O.; Männistö, S.; Lahti-Koski, M.; Karvonen, M.; Laatikainen, T.; Valsta, T. *Maps on Health Promoting Diet, Obesity and Serum Cholesterol in Finland*; B20/2005; Publications of the National Public Health Institute: Helsinki, Finland, 2005. (In Finnish)
45. Sares-Jäske, L.; Valsta, L.; Haario, P.; Martelin, T. Population Group Differences in Subjective Importance of Meat in Diet and Red and Processed Meat Consumption. *Appetite* **2022**, *169*, 105836. [CrossRef] [PubMed]
46. Guenther, P.M.; Jensen, H.H.; Batres-Marquez, S.P.; Chen, C.F. Sociodemographic, Knowledge, and Attitudinal Factors Related to Meat Consumption in the United States. *J. Am. Diet. Assoc.* **2005**, *105*, 1266–1274. [CrossRef]
47. Martin, J.C.; Moran, L.J.; Teede, H.J.; Ranasinha, S.; Lombard, C.B.; Harrison, C.L. Exploring Diet Quality between Urban and Rural Dwelling Women of Reproductive Age. *Nutrients* **2017**, *9*, 586. [CrossRef]
48. Institute of Medicine. *Dietary Reference Intakes: Applications in Dietary Assessment*; National Academies Press: Washington, DC, USA, 2000.
49. Steenbergen, E.; Krijger, A.; Verkaik-Kloosterman, J.; Elstgeest, L.E.M.; Ter Borg, S.; Joosten, K.F.M.; van Rossum, C.T.M. Evaluation of Nutrient Intake and Food Consumption among Dutch Toddlers. *Nutrients* **2021**, *13*, 1531. [CrossRef]
50. Kaartinen, N.; Tapanainen, H.; Reinivuo, H.; Pakkala, H.; Aalto, S.; Raulio, S.; Männistö, S.; Korhonen, T.; Virtanen, S.; Borodlin, K.; et al. The Finnish National Dietary Survey in Adults and Elderly (FinDiet 2017). *EFSA Support. Publ.* **2020**, *17*, 1914E. [CrossRef]
51. Roos, E.; Ovaskainen, M.; Raulio, S.; Pietikäinen, M.; Sulander, T.; Prättälä, R. Dietary habits. In *Trends in Socioeconomic Health Differences 1980–2005*; Palosuo, H., Koskinen, S., Lahelma, E., Kostiainen, E., Prättälä, R., Martelin, T., Ostamo, A., Keskimäki, I., Sihto, M., Linnanmäki, E., Eds.; 2009:9; Publications of the Ministry of Social Affairs and Health: Helsinki, Finland, 2009.
52. Raulio, S.; Tapanainen, H.; Männistö, S.; Valsta, L.; Kuusipalo, H.; Vartiainen, E.; Virtanen, S. *Ravitsemuksessa Eroja Koulutusryhmien Välillä—Finravinto-Tutkimuksen Tuloksia. Tutkimuksesta Tiiviisti 26*; Finnish Institute for Health and Welfare: Helsinki, Finland, 2016. (In Finnish)
53. Borodulin, K.; Sääksjärvi, K. *FinHealth 2017 Study—Methods*; Report 17/2019; Finnish Institute for Health and Welfare: Helsinki, Finland, 2019.
54. Heistaro, S. *Methodology Report: Health 2000 Survey*; National Public Health Institute: Helsinki, Finland, 2008.
55. Reinivuo, H.; Hirvonen, T.; Ovaskainen, M.L.; Korhonen, T.; Valsta, L.M. Dietary Survey Methodology of FINDIET 2007 with a Risk Assessment Perspective. *Public Health Nutr.* **2010**, *13*, 915–919. [CrossRef]
56. Finnish Institute for Health and Welfare. National Food Composition Database FINELI®, Release 19. Available online: https://fineli.fi/fineli/en/index? (accessed on 14 February 2022).
57. Finnish Institute for Health and Welfare. FinHealth 2017 Questionnaires. Available online: https://thl.fi/en/web/thlfi-en/research-and-expertwork/population-studies/national-finhealthstudy/questionnaires/finhealth2017 (accessed on 14 February 2022).
58. OECD Project on Income Distribution and Poverty. What Are Equivalence Scales? Available online: https://www.oecd.org/economy/growth/OECD-Note-EquivalenceScales.pdf (accessed on 14 February 2022).
59. Saarinen, M.; Kaljonen, M.; Niemi, J.; Antikainen, R.; Hakala, K.; Hartikainen, H.; Heikkinen, J.; Joensuu, K.; Lehtonen, H.; Mattila, T.; et al. *Effects of Dietary Change and Policy Mix Supporting the Change: End Report of the FoodMin Project*; Publications of the Government's Analysis, Assessment and Research Activities: Helsinki, Finland, 2019; p. 47, (In Finnish, Abstract in English).
60. Härkänen, T.; Karvanen, J.; Tolonen, H.; Lehtonen, R.; Djerf, K.; Juntunen, S.; Koskinen, S. Systematic Handling of Missing Data in Complex Study Designs: Experiences from the Health 2000 and 2011 Surveys. *J. Appl. Stat.* **2016**, *43*, 2772–2790. [CrossRef]
61. Hengeveld, L.M.; Boer, J.M.A.; Gaudreau, P.; Heymans, M.W.; Jagger, C.; Mendonca, N.; Ocke, M.C.; Presse, N.; Sette, S.; Simonsick, E.M.; et al. Prevalence of Protein Intake below Recommended in Community-Dwelling Older Adults: A Meta-Analysis across Cohorts from the PROMISS Consortium. *J. Cachexia Sarcopenia Muscle* **2020**, *11*, 1212–1222. [CrossRef]
62. Rippin, H.L.; Hutchinson, J.; Jewell, J.; Breda, J.J.; Cade, J.E. Adult Nutrient Intakes from Current National Dietary Surveys of European Populations. *Nutrients* **2017**, *9*, 1288. [CrossRef] [PubMed]
63. Itkonen, S.T.; Andersen, R.; Bjork, A.K.; Brugard Konde, A.; Eneroth, H.; Erkkola, M.; Holvik, K.; Madar, A.A.; Meyer, H.E.; Tetens, I.; et al. Vitamin D Status and Current Policies to Achieve Adequate Vitamin D Intake in the Nordic Countries. *Scand. J. Public Health* **2021**, *49*, 616–627. [CrossRef] [PubMed]
64. Raulio, S.; Tapanainen, H.; Valsta, L.; Jääskeläinen, T.; Virtanen, S.; Erlund, I. Vitamin D Intake and Serum Levels in the Finnish Adult Population—Results from the National FinDiet 2017 Survey. *Finn. Med. J.* **2021**, *76*, 2578–2584, (In Finnish, Abstract in English).
65. Nyström, H.F.; Brantsaeter, A.L.; Erlund, I.; Gunnarsdottir, I.; Hulthen, L.; Laurberg, P.; Mattisson, I.; Rasmussen, L.B.; Virtanen, S.; Meltzer, H.M. Iodine Status in the Nordic Countries-Past and Present. *Food Nutr. Res.* **2016**, *60*, 31969. [CrossRef]

66. Hirvonen, T.; Männistö, S.; Roos, E.; Pietinen, P. Increasing Prevalence of Underreporting Does Not Necessarily Distort Dietary Surveys. *Eur. J. Clin. Nutr.* **1997**, *51*, 297–301. [CrossRef] [PubMed]
67. Grech, A.; Hasick, M.; Gemming, L.; Rangan, A. Energy Misreporting Is More Prevalent for Those of Lower Socio-Economic Status and Is Associated with Lower Reported Intake of Discretionary Foods. *Br. J. Nutr.* **2021**, *125*, 1291–1298. [CrossRef]
68. Cohen, A.K.; Rai, M.; Rehkopf, D.H.; Abrams, B. Educational attainment and obesity: A systematic review. *Obes. Rev.* **2013**, *14*, 989–1005. [CrossRef]
69. Kaljonen, M.; Peltola, T.; Kettunen, M.; Salo, M.; Furman, E. Vegs to School—Experimentative Research for Sustainability Transition. *Alue Ympäristö* (In Finnish, Abstract in English). **2018**, *47*, 32–47. [CrossRef]
70. Kaljonen, M.; Peltola, T.; Salo, M.; Furman, E. Attentive, Speculative Experimental Research for Sustainability Transitions: An Exploration in Sustainable Eating. *J. Clean. Prod.* **2019**, *206C*, 365–373. [CrossRef]
71. Kortetmäki, T.; Pudas, T.; Saralahti, I. School Meals 2030—How to Halve the Climate Impact; School Meals 2030 Experiment. 2021. Available online: https://www.aka.fi/globalassets/3-stn/1-strateginen-tutkimus/tiedon-kayttajalle/politiikkasuositukset/politiikkasuositukset/21_03_school_meals_2030.pdf (accessed on 14 February 2022).

nutrients

MDPI

Article

Dietary Intake and Status of Vitamin B12 in Slovenian Population

Živa Lavriša [1], Hristo Hristov [1], Maša Hribar [1,2], Katja Žmitek [1,3], Anita Kušar [1], Barbara Koroušić Seljak [4], Matej Gregorič [5], Urška Blaznik [5], Nadan Gregorič [6,7], Katja Zaletel [6,7], Adrijana Oblak [6], Joško Osredkar [6,8] and Igor Pravst [1,2,3,*]

1 Nutrition Institute, Tržaška Cesta 40, SI-1000 Ljubljana, Slovenia; ziva.lavrisa@nutris.org (Ž.L.); hristo.hristov@nutris.org (H.H.); masa.hribar@nutris.org (M.H.); katja.zmitek@vist.si (K.Ž.); anita.kusar@nutris.org (A.K.)
2 Biotechnical Faculty, University of Ljubljana, Jamnikarjeva 101, SI-1000 Ljubljana, Slovenia
3 VIST–Faculty of Applied Sciences, Gerbičeva Cesta 51A, SI-1000 Ljubljana, Slovenia
4 Computer Systems Department, Jozef Stefan Institute, SI-1000 Ljubljana, Slovenia; barbara.korousic@ijs.si
5 National Institute of Public Health, Trubarjeva 2, SI-1000 Ljubljana, Slovenia; matej.gregoric@nijz.si (M.G.); urska.blaznik@nijz.si (U.B.)
6 University Medical Centre Ljubljana, Zaloška Cesta 7, SI-1000 Ljubljana, Slovenia; nadan.gregoric@kclj.si (N.G.); katja.zaletel@kclj.si (K.Z.); adrijana.oblak@kclj.si (A.O.); josko.osredkar@kclj.si (J.O.)
7 Faculty of Medicine, University of Ljubljana, Vrazov trg 2, SI-1000 Ljubljana, Slovenia
8 Faculty of Pharmacy, University of Ljubljana, Aškerčeva Cesta 7, SI-1000 Ljubljana, Slovenia
* Correspondence: igor.pravst@nutris.org; Tel.: +386-590-68871; Fax: +386-310-07981

Abstract: Vitamin B12 deficiency poses a health concern, especially in vulnerable populations. Dietary vitamin B12 intake was obtained by two 24 h dietary recalls and food propensity questionnaires in a representative Slovenian cross-sectional food consumption survey, SI.Menu (n = 1248 subjects; 10–74 years). For a subgroup of 280 participants, data on serum vitamin B12 were available through the Nutrihealth study. The estimated usual population-weighted mean daily vitamin B12 intakes were 6.2 µg (adults), 5.4 µg (adolescents), and 5.0 µg (elderly). Lower intakes were observed in females. Inadequate daily vitamin B12 intake (<4 µg) was detected in 37.3% of adolescents, 31.7% of adults, and 58.3% elderlies. The significant predictors for inadequate daily vitamin B12 intake were physical activity score in all age groups, sex in adolescents and adults, financial status and smoking in elderly, and employment in adults. Meat (products), followed by milk (products), made the highest vitamin B12 contribution in all age groups. In adolescents, another important vitamin B12 contributor was cereals. The mean population-weighted serum vitamin B12 levels were 322.1 pmol/L (adults) and 287.3 pmol/L (elderly). Low serum vitamin B12 concentration (<148 nmol/L) and high serum homocysteine (>15 µmol/L) were used as criteria for vitamin B12 deficiency. The highest deficiency prevalence was found in elderlies (7.0%), particularly in males (7.9%). Factors associated with high serum homocysteine were also investigated. In conclusion, although vitamin B12 status was generally not critical, additional attention should be focused particularly to the elderly.

Keywords: vitamin B12; deficiency; homocysteine; folate; Slovenia; EU Menu

Citation: Lavriša, Ž.; Hristov, H.; Hribar, M.; Žmitek, K.; Kušar, A.; Koroušić Seljak, B.; Gregorič, M.; Blaznik, U.; Gregorič, N.; Zaletel, K.; et al. Dietary Intake and Status of Vitamin B12 in Slovenian Population. *Nutrients* **2022**, *14*, 334. https://doi.org/10.3390/nu14020334

Academic Editor: Ebba Nexø

Received: 21 December 2021
Accepted: 12 January 2022
Published: 13 January 2022

1. Introduction

Vitamin B12, also known as cobalamin, is a water-soluble vitamin that plays a significant role in cellular metabolism. It acts as a cofactor in metabolic processes that are important, especially in DNA methylation and synthesis, as well as in mitochondrial metabolism [1]. Consequently, it is required for the normal functioning of the nervous system and red blood cell formation. Vitamin B12 deficiency can cause megaloblastic anemia, where, due to its incapacity for DNA synthesis, the bone marrow produces large blood cells with immature nuclei [2]. Other prominent manifestations of B12 deficiency are neuropsychiatric, with a wide range of symptoms, whereby the elderly are particularly at risk [3,4]. The metabolism of vitamin B12 is closely intertwined with folate [5]; insufficient vitamin B12 status can also lead to a functional folate deficiency, as folate becomes trapped and cannot be properly utilized [6]. Disruption of these metabolic pathways can lead to the accumulation of homocysteine, which is a known risk factor for cardiovascular diseases, such as atherosclerosis, endothelial dysfunction, and thromboembolism [7–10].

Naturally, vitamin B12 is produced by bacterial synthesis [11]. Its main foods sources include foods of animal origin, such as meat, fish, eggs, and dairy [12]. Foods of plant origin are typically not sources of this vitamin; however, certain fermented foods, such as fermented soybean (tempeh), can contain considerable amounts of vitamin B12, depending on the production procedure [13]. It should be also noted that blue-green algae (cyanobacteria), which can be found in some dietary supplements, usually contain pseudovitamin B12, which is inactive in humans [12]. A valuable source of vitamin B12 for those who avoid foods of animal origin can be fortified foods, i.e., breakfast cereals. In the European Union (EU), food fortification with vitamin B12 is voluntary, with cyanocobalamin and hydroxocobalamin as authorized sources [14]. Vitamin B12 naturally present in foods is bound to peptides, proteins, and glycoproteins, and must be released during the digestion process to be utilized by the body. The bioavailability of vitamin B12 is the greatest in fish, meat, and milk (42–89%) [12,15] and lower in eggs (<9%) [16]. It should be acknowledged that at high single doses of vitamin B12, the proportion of absorbed vitamin B12 declines, so dispersing the intake among daily meals could result in a higher amount being absorbed [17,18]. Losses in vitamin B12 in foods can predominantly occur during storage, and with light and heat exposure, with losses between 30% and 50% [12,19]. The presence of other vitamins can also affect its stability [20].

The major causes of vitamin B12 deficiency are inadequate intake, inadequate bioavailability, and malabsorption. An inadequate intake of vitamin B12 can occur especially in subjects who exclude foods of animal origin from their diet, such as vegans [21,22], but is present in the general population as well. Low dietary intake usually does not result in symptomatic cobalamin deficiency until hepatic reserves are exhausted [23]. In the elderly, the main risk for vitamin B12 deficiency is usually due to impaired absorption, and less due to inadequate dietary intake [24–26]. Food-cobalamin malabsorption, which is primarily caused by gastric atrophy, is a common cause of vitamin B12 deficiency, particularly in the elderly, and is often also caused by infection with *Helicobacter pylori* [23]. Another common risk factor for developing B12 deficiency due to malabsorption is pernicious anemia, in which a lack of gastric intrinsic factor impairs the intestinal B12 absorption [27]. Vitamin B12 deficiency and anemia can also occur in some long-term drug treatments, such as metformin in diabetic patients, so for these patients, routine testing for deficiency and/or supplementation is advised [28]. Additionally, other risk factors for deficiency include several gastric diseases and/or surgeries that compromise vitamin B12 absorption [1]. Several diagnostic approaches are available for the diagnosis of vitamin B12 malabsorption and deficiency [29], and a food-cobalamin malabsorption test can be used to determine the organism's ability to release vitamin B12 from foods [30].

To evaluate serum vitamin B12 status, several biomarkers exist. For a general estimation of vitamin B12 status, total serum vitamin B12 concentration is typically used, which measures both holotranscobalamin and transcobalamin [31]. Additionally, holotranscobalamin (bioactive protein-bound form of vitamin B12) can be used as a marker of vitamin B12 status, together with methylmalonic acid (MMA) and homocysteine (HC) to test for deficiency [32]. These auxiliary markers are used because the serum vitamin B12 alone has limited diagnostic value. While currently there is no consensus on the best marker or best combination of markers to be applied for the assessment of vitamin B12 status, it is advised to use at least one metabolic biomarker together with serum vitamin B12 levels [33]. In the literature, vitamin B12 serum status in adults is commonly defined with the following cut-offs: >221 pmol/L—vitamin "B12 adequacy", 148–221 pmol/L—"low B12", and <148 pmol/L—"B12 deficiency" [33–36]. Homocysteine concentrations above 15 µmol/L are typical for vitamin B12 deficiency [35,37]. In the present study, homocysteine concentrations between 10 and 15 µmol/L were considered as marginally elevated. As a criterion for high-risk vitamin B12 deficiency, serum vitamin B12 concentration <148 pmol/L, together with homocysteine concentration >15 µmol/L, was applied.

Internationally, there are quite considerable differences in reference intake values for vitamin B12. The World Health Organization (WHO) and the United States Institute of Medicine (IoM) reference value for vitamin B12 intake is set at 2.4 µg/day for adolescents and adults [38]. Nordic Council of Ministers' reference value is even lower, at 2 µg for adolescents and adults [39]. The current reference daily vitamin B12 intake in Slovenia [40], which was implemented according to German–Austrian–Switzerland (D-A-CH) recommendations [41], is 3.5 µg for adolescents of 10–13 years and 4 µg for adolescents older than 13 years and adults, including the elderly [40]. It should be noted that the D-A-CH reference intake was previously set at 3 µg but was recently revised considering newer vitamin B12-related biomarker studies [42]. The same reference values are also set by the European Food Safety Authority (EFSA) [43].

According to the WHO, in some populations worldwide, vitamin B12 status is a public health concern [34]. However, more country representative data are lacking for assessment [44]. Vitamin B12 deficiency is more prevalent in less developed countries, which is closely related to the diet. In Latin America and some parts of Africa, around 40% of the population is deficient [45], and these proportions are even higher in parts of India, where deficiency prevalence can reach up to 70% of the population. In a recent study from Latin America, particularly elderly women (33%) were found to be vitamin B12 deficient [46]. In developed countries, such as the United States of America and the United Kingdom, deficiency is around 6% in the general population and around 20% in the elderly population [47]. A study on European populations reported serum vitamin B12 concentrations mostly around 350 pmol/L, with few countries reporting marginal vitamin B12 status [48]. Data from dietary surveys in nine European Union countries showed that average vitamin B12 intakes across countries range between 3.3 and 6.6 µg/day in adolescents (10–18 years) and 4.2 and 8.6 µg/day in adults [43]. In Slovenia, comprehensive, nationally representative data on B12 status or intake in the population have not been available; this topic has only been investigated in a few studies on smaller population subgroups or patients [49–52].

The objective of the present study was to estimate the dietary vitamin B12 intake and prevalence of vitamin B12 deficiency in the Slovenian population. The secondary objective was to determine the main sources of vitamin B12 in the daily diet and the determinants of low vitamin B12 intake and deficiency. Vitamin B12 intake was investigated for adolescents, adults, and the elderly using the data collected in the nationally representative food consumption SI.Menu study, while vitamin B12 status was assessed using Nutrihealth study data on adults and the elderly.

2. Material and Methods

2.1. Study Design and Participants

Data for the present study were obtained from the nationally representative cross-sectional food consumption study SI.Menu. Data were collected between March 2017 and April 2018 in Slovenia (population ca. 2 million). The study was performed based on the European Food Safety Agency (EFSA) Guidance on EU Menu Methodology [53]. The SI.Menu study methodology details have previously been published [54]. Participants were randomly selected from the Central Register of Population of Slovenia, with consideration of place of residency, age, and sex. Participants were selected only from private households; people living in institutions were not included. Statistical units were individuals. The study sample included 2280 participants, stratified into three age groups: adolescent (10–17 years), adult (18–64 years), and elderly population (65–74 years). The final response rate was 62% (1319 participants); the lowest participation rates were observed in adults (57%), and these were higher in adolescents (69%) and the elderly (65%). In these two groups, participation rates were very similar between males and females, while in the adults, the participation rate of males was notably lower (50% vs. 61%). The sampling was performed according to the EFSA guidelines on minimum sample sizes for each of the age groups [53], with consideration of sample sizes of previous Slovenian dietary surveys [55,56]. The population of Slovenia at the time of the sampling was 2.064 million people. The study protocol was approved by the National Medical Ethics Committee (KME 53/07/16; approval No. 0120-337/2016 issued on 19 July 2016). Prior to inclusion in the study, all subjects were informed about the study and signed an informed consent form. In the case of adolescents, informed consent was also obtained from the parent or legal guardian.

As an extension of the SI.Menu study, data on serum biomarkers were collected via the Nutrihealth study for a subsample of adults and elderly people who participated in a SI.Menu study. Altogether, 34% of adults and 37% of elderly people from the SI.Menu study were further included in the Nutrihealth study. Blood and urine samples were collected for $n = 280$ participants (125 adults and 155 elderly). The sample size is comparable with other previously published studies on this topic [57,58]. In the blood samples, serum vitamin B12 and homocysteine were determined. The detailed methodology of the Nutrihealth study has previously been published [59].

2.2. Data Collection

In the SI.Menu study, data were collected using a general questionnaire, a food propensity questionnaire (FPQ), and two 24 h dietary recalls. During the first interview, data from the general questionnaire, anthropometric data, and FPQ data were collected, together with the first 24 h dietary recall. During the second interview, data for the second 24 h dietary recall were collected.

Via the general questionnaire, sociodemographic and socioeconomic data were obtained, such as place of residency, education, employment, financial status, smoking, and the following of specific dietary patterns. A cut-off point for below/above average self-reported financial status was at a monthly income of EUR 1300. Participants also self-reported their level of physical activity, which was transformed into an International Physical Activity Questionnaire (IPAQ) score [60]. Anthropometric data were collected, including body height and body weight for body mass index (BMI) calculation. A cut-off point for overweight was established at 25 kg/m^2 for adults; those below this cut-off point were grouped into "normal", even if their BMI was lower than 18.5 kg/m^2. For adolescents, sex-/age-adjusted cut-off points (>1 standard deviation (SD)) were applied as described in the literature [61,62].

Dietary habits of participants were collected by a trained interviewer performing 24 h dietary recalls for two nonconsecutive days, which were 7–21 days apart, covering workdays and weekends (71% and 29%, respectively). For the estimation of portion sizes, a national, validated picture book was used, which depicted commonly consumed foods and simple dishes in six portion sizes [63]. Further, the usual consumption frequency of specific foods in the past 12 months was collected using FPQ with the following frequency response options: never, 1–3 times per month or less, once per week, 2–3 times per week, 4–6 times per week, 1–2 times per day or more [54].

2.3. Dietary Vitamin B12 Intake

Data on food consumption from 24 h recalls were inserted into the Open Platform for Clinical Nutrition (OPEN), which is a web application based on the Slovenian food composition database [64]. The OPEN platform consists of nutritional composition data of branded and generic foods, as well as commonly used recipes in Slovenia. The sample of extracted foods from dietary recalls consisted of 2377 foods. For foods with missing vitamin B12 content in the OPEN, this information was searched for in other food composition databases, particularly the National Food Composition Database in Finland (Fineli) [65] and the United States Department of Agriculture Food Composition Database (USDA) [66]. Altogether, 37.1% of the foods from the sample were found to be sources of vitamin B12.

The Multiple Source Method (MSM) was used for estimating usual daily intakes, with consideration of FPQ [67], in line with the previously reported approach [68].

2.4. Vitamin B12 and Homocysteine Status

A subsample of adult and elderly participants from the SI.Menu study was included in the Nutrihealth study, where they provided fasting blood samples. Serum vitamin B12 and homocysteine concentrations were measured in human serum at the Department of Nuclear Medicine (University Medical Center, Ljubljana).

Total serum vitamin B12 concentration was analyzed with the chemiluminescence immunoassay, determined on an Immulite 2000 XPi analyzer (Siemens Healthineers, Gwynedd, UK). The performance characteristics of the assay are as follows: the vitamin B12 concentration limit of detection is 92 pmol/L; the linearity of the assay is in the range of 111 to 738 pmol/L with a recovery range of 92% to 123%. The intra-assay and interassay coefficients of variation range from 6.7% to 13.0% and from 6.0% to 15.0%, respectively. The cross-reactivity of the vitamin B12 assay was shown to be nondetectable for cobinamide.

In the assessment of vitamin B12 status, we also considered serum homocysteine levels, already reported in our previous study [68]. A chemiluminescence immunoassay method carried out on an IDS-iSYS analyzer (Immunodiagnostic Systems, Boldon, UK) was used for the determination of homocysteine. A cut-off point for high homocysteine was set at 15 μmol/L [37,69]. In the assessment of factors influencing serum homocysteine levels, we also considered serum folate concentrations, which were previously reported in the above-mentioned paper [68]. Serum folate was measured with the chemiluminescence immunoassay on an Immulite 2000 XPi analyzer (Siemens Healthineers, Gwynedd, UK), and a cut-off of 7 nmol/L was used to identify subjects with low serum folate level.

2.5. Data and Statistical Analysis

Data cleaning was conducted separately for each of the two dietary recalls. To examine under- and over-reporting, the adapted Goldberg et al. method was used, as previously described in the study of Black et al. [70]. The method is based on the ratio of reported daily energy intake and basic metabolic rate (BMR). The BMR was estimated using sex, age, body height, and body weight, based on the method developed by Harris et al. [71] and adapted by Roza and Shizgal [72]. Additionally, a low energy intake cut-off value was introduced to exclude participants reporting energy intakes lower than 500 kcal/day. This procedure was presented in detail in our previous paper [73]. After the exclusion of 97 subjects (incomplete or missing anthropometric data: $n = 12$; missing one set of the 24 h recall data: $n = 36$; under-/over-reporting: $n = 49$), our study sample included $n = 1248$ subjects: 468 adolescents, 364 adults, and 416 elderly people.

To calculate the usual daily dietary vitamin B12 intake, two 24 h recalls and FPQs were used. Day-to-day inter- and intraindividual variations in the vitamin B12 intake distribution within age groups were modeled using the Multiple Source Method (MSM) [67], with participants' age, sex, and BMI used as covariates. The MSM modeling approach using FPQ data corrects for within-individual variation in vitamin B12 intake and provides data on usual dietary intake on an individual level [74].

To assure national representativeness, the descriptive analysis considered weighting for each of the three cohorts (age/sex) using the iterative proportional fitting method [75]. Census data from the year 2017 were used for population weighting [54]. For the assessment of the proportion of each population meeting recommended daily intake, we used the nationally accepted threshold of 4 μg/day vitamin B12 [40]. For the reporting of the relative contributions of different food categories in daily vitamin B12 intake, foods were also categorized using a modified categorization system, developed by the Global Food Monitoring Initiative [76]. Population-weighted serum vitamin B12 and homocysteine levels were also calculated for all age groups and sexes. The population-weighted proportions of low vitamin B12 status and high homocysteine were calculated using the above-mentioned cut-off levels. Prevalence of risk for vitamin B12 deficiency was determined using a combination of two criteria: serum vitamin B12 concentration <148 pmol/L and homocysteine concentration >15 μmol/L.

Multiple linear and logistic regression analyses were performed to assess mean usual vitamin B12 intake and odds ratios (ORs) for inadequate daily vitamin B12 intake (<4 μg/day) in adolescents, adults, and the elderly and for selected subpopulations within these three age cohorts. In the dietary intake model analysis, we used sex, residential area, financial status, education level, BMI, IPAQ, employment status, smoking status, and following a specific medical or behavioral diet as predictor variables. Logistic regression analysis was also used to investigate risk for vitamin B12 deficiency; analysis was performed on a combined sample of adult and elderly participants (n = 271); 9 participants were excluded due to incomplete vitamin B12 intake data. Analyses were conducted using the following factors: age cohort, sex, residential area, financial status, smoking status, BMI, IPAQ, usual daily vitamin B12 intake, and energy intake. Furthermore, the same sample was used for linear and logistic regression analyses to assess adjusted serum homocysteine concentrations and ORs for high homocysteine concentrations (>15 μmol/L). In addition to the above-mentioned parameters, both models also included serum concentrations of vitamin B12 and folate.

All statistical analyses were performed using STATA (version 17.0; StataCorp LLC, College Station, TX, USA). The outcomes were statistically significant at $p < 0.05$, except in the logistic regression analysis for adequate daily dietary intake of vitamin B12, where marginal statistical significance is reported for $p < 0.1$.

3. Results

The sociodemographic characteristics of the SI.Menu sample are described in Table 1. The sample included a total of 1248 participants, divided into three population groups: adolescents (n = 468), adults (n = 364), and elderly (n = 416). About a quarter of these participants reported the use of multivitamin supplements, but specific data about supplementation with vitamin B12 were not available. Altogether, 34% of adults and 37% of elderly people from the SI.Menu study were further included in the Nutrihealth study. For this subsample, fasting blood samples were collected, so data on serum vitamin B12 and homocysteine levels were available and included in the analysis.

The estimated usual population-weighted mean daily vitamin B12 intake was above the recommended 4 μg in all age groups (Table 2). Generally, mean intake was the highest in adults (6.2 μg/day) and lower in adolescents and the elderly, where quite similar values were observed (5.4. and 5.0 μg/day, respectively). The distribution of the usual daily dietary intake of vitamin B12 among age groups is presented in Figure 1.

Despite the mean intakes being above 4 μg/day, a notable proportion of the population does not meet this threshold for recommended vitamin B12 intake (Table 2). This proportion is particularly high in the elderlies, where the recommended daily vitamin B12 intake is not met in 58.3% of the population. In adolescents and adults, the proportions of the population with inadequate vitamin B12 intakes are 37.3% and 31.7%, respectively. Lower intakes of vitamin B12 were observed in females, particularly among adolescents and adults. This trend was confirmed in the linear regression analysis, where sex was the only parameter with significant association with vitamin B12 intake (Supplementary Table S1). Interestingly, the population-weighted daily vitamin B12 intake, calculated per 1000 kcal, was also lower in females in the adult and elderly populations, but not in adolescents.

Table 1. Demographic characteristics of the SI.Menu study sample for all three age cohorts (adolescents: 10–17 years; adults: 18–64 years; elderly: 65–74 years).

Variable		Adolescents n (%)	Adults n (%)	Elderly n (%)
Overall		n = 468 (100)	n = 364 (100)	n = 416 (100)
Age (mean ± SD)		13.4 (2.37)	43.6 (13.81)	68.7 (2.7)
Residential area	rural	270 (57.7)	202 (55.5)	229 (55.1)
	intermediate	76 (16.2)	56 (15.4)	71 (17.1)
	urban	122 (26.1)	106 (29.1)	116 (27.9)
Sex	male	238 (50.9)	173 (47.5)	213 (51.2)
	female	230 (49.1)	191 (52.5)	203 (48.8)
Education	no university degree	n.a.	249 (68.4)	342 (82.2)
	university degree	n.a.	115 (31.6)	74 (17.8)
Financial status	below average	n.a.	118 (38.4)	269 (71.5)
	above average	n.a.	189 (61.6)	107 (28.5)
Employment	employed	n.a.	226 (62.1)	n.a.
	unemployed	n.a.	42 (11.5)	n.a.
	student	n.a.	32 (8.8)	n.a.
	retired	n.a.	64 (17.6)	n.a.
BMI (mean ± SD)		21.0 (4.2)	26.7 (5.2)	28.4 (5.0)
BMI	normal	301 (64.6)	148 (40.7)	108 (26.0)
	overweight and obese	167 (35.7)	216 (59.3)	308 (74.0)
Smoking status	current, occasional, ex-smoker	30 (6.4)	165 (45.3)	185 (44.5)
	nonsmoker	438 (93.6)	199 (54.7)	231 (55.5)
IPAQ	low	108 (23.3)	127 (35.3)	137 (33.4)
	moderate	141 (30.5)	108 (30.0)	133 (32.4)
	high	214 (46.2)	125 (34.7)	140 (34.2)
Supplement use	multivitamins	129 (27.6)	140 (38.4)	95 (22.8)
	use not reported	339 (72.4)	224 (61.5)	321 (77.2)
Diet type	vegetarian/vegan	12 (2.6)	8 (2.2)	3 (0.7)
	no diet	456 (97.4)	356 (97.8)	413 (99.3)
	medical and/or weight loss	13 (2.8)	32 (8.8)	51 (12.3)
	no diet	455 (97.2)	332 (91.2)	465 (87.7)
Participation in the Nutrihealth study *			125 (34.3)	155 (37.3)

Notes: SD = standard deviation; BMI = body mass index; for adults and elderly people, normal BMI was considered below 25 kg/m^2, while sex-/age-adjusted cut-off points were used for adolescents [61,62]; IPAQ = physical activity according to International Physical Activity Questionnaire; * serum vitamin B12 and homocysteine levels available for the subgroup participating in the Nutrihealth study. n.a.: not applicable.

Table 2. Population-weighted usual daily vitamin B12 intake and prevalence of inadequate vitamin B12 intake (<4 µg/day) and serum markers for deficiency.

	Adolescents *n* (%)			Adults *n* (%)			Elderly *n* (%)		
	All	**Male**	**Female**	**All**	**Male**	**Female**	**All**	**Male**	**Female**
SI.Menu study *n* (%)	468 (100)	238 (50.9)	230 (49.2)	364 (100)	173 (47.5)	191 (52.5)	416 (100)	213 (51.2)	203 (48.8)
Weighted * *n* (%)	150,674 (78.2)	75,580 (50.2)	73,094 (49.8)	1,302,132 (78.2)	670,464 (51.5)	631,668 (48.5)	212,793 (12.8)	100,247.5 (47.1)	112,545.5 (52.9)
Usual daily vitamin B12 intake									
Mean (95%CI) (µg/day)	5.4 (5.0–5.8)	6.0 (5.4–6.5)	4.7 (4.1–5.4)	6.2 (5.7–6.8)	6.9 (6.1–7.8)	5.5 (4.8–6.2)	5.0 (4.4–5.6)	5.7 (4.5–6.8)	4.4 (3.8–5.0)
Q25 (µg/day)	3.4	4.4	2.8	3.6	4.3	3.3	2.9	3.2	2.7
Median (µg/day)	4.7	5.3	3.9	5.0	5.4	4.4	3.6	3.8	3.4
Q75 (µg/day)	5.9	6.7	5.4	7.1	8.1	6.1	4.9	5.5	4.5
Mean (95%CI) (µg/per 1000 kcal/day)	2.0 (1.9–2.2)	2.0 (1.8–2.3)	2.0 (1.8–2.3)	2.5 (2.3–2.7)	2.5 (2.2–2.9)	2.4 (2.1–2.6)	2.1 (1.9–2.4)	2.3 (1.8–2.7)	2.0 (1.7–2.2)
Prevalence of inadequate daily vitamin B12 intake (< 4µg) (%) (95% CI)									
<4 µg/day	37.3 (30.6–44.6)	24.0 (17.7–31.8)	51.7 (42.2–61.2)	31.7 (26.5–37.3)	20.6 (14.9–27.8)	42.9 (35.1–51.0)	58.3 (49.5–66.7)	53.2 (37.7–68.0)	63.1 (52.7–72.3)
Nutrihealth study *n* (%)				125 (100)	52 (41.6)	73 (58.4)	155 (100)	76 (49.0)	79 (51.0)
Serum vitamin B12 (pmol/L) (95% CI)									
Mean (95%CI)				322.1 (294.2–350.1)	329.8 (284.7–374.9)	314.0 (282.2–345.7)	287.3 (258.2–316.2)	250.2 (219.1–281.2)	320.3 (274.1–366.5)
Std. Err.				14.1	22.8	16.1	14.7	15.8	23.4
Median				283	280	283	232	204	232
Prevalence of low serum vitamin B12 (%) (95% CI)									
<148 pmol/L				3.7 (1.6–8.3)	2.4 (0.6–9.2)	5.3 (1.9–13.9)	10.4 (6.4–16.4)	11.8 (6.2–21.4)	9.1 (4.4–18.0)
<221 pmol/L				21.1 (14.5–29.7)	18.9 (10.2–32.2)	23.6 (14.8–35.5)	46.0 (38.2–54.1)	57.9 (46.5–68.5)	35.1 (25.2–46.4)
Serum homocysteine (µmol/L) (95% CI)									
Mean (95% CI)				12.6 (11.9–13.3)	13.6 (12.6–14.6)	14.6 (13.9–15.2)	14.6 (13.9–15.2)	16.1 (15.2–17.0)	13.2 (12.5–13.9)
Std. Err.				0.35	0.50	0.42	0.32	0.47	0.36
Median				12.1	12.6	15.7	14.2	10.9	13.0
Prevalence of high serum homocysteine (µmol/L) (%) (95% CI)									
>10 µmol/L				75.3 (66.4–82.4)	88.7 (76.5–95.0)	61.0 (48.6–72.2)	88.9 (82.7–93.0)	96.0 (88.3–98.7)	82.5 (72.5–89.4)
>15 µmol/L				20.5 (13.9–29.1)	26.4 (15.8–40.8)	14.2 (8.0–23.9)	39.9 (32.4–47.8	56.6 (45.2–67.3)	25.0 (16.7–35.7)
Prevalence of vitamin B12 deficiency (%) (95% CI) using criteria for low serum vitamin B12 (<148 nmol/L) and high serum homocysteine (>15µmol/L)									
				1.2 (0.2–4.7)	2.3 (0.5–9.1)	/	7.0 (3.9–12.3)	7.9 (3.6–16.6)	6.3 (2.6–14.3)

Notes: CI—confidence interval. Dietary intake of Vitamin B12 is estimated with consideration of regular foods (without food supplements). * Number of people and their respective population share regarding age and sex cohorts (census data in 2017). Serum homocysteine levels from [68].

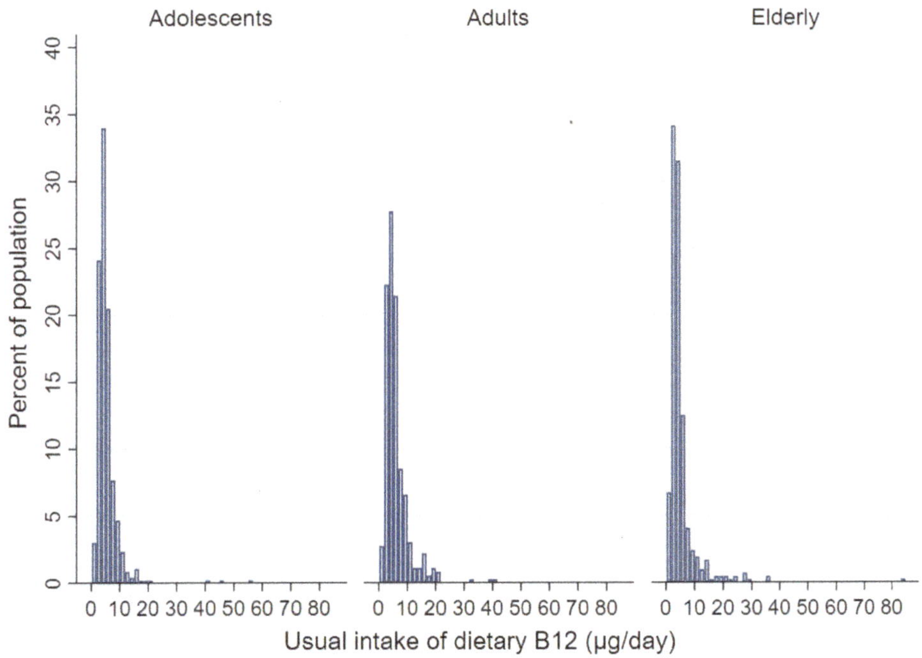

Figure 1. Histograms of the sample population distribution of estimated usual daily intake of vitamin B12 for different age groups (adolescents: 10–17 years; adults: 18–64 years; elderly: 65–74 years).

Logistic regression analysis was used to determine predictors for the prevalence of inadequate daily vitamin B12 intake below 4 μg (Table 3). This analysis also showed significantly higher odds ratios (ORs) for inadequate vitamin B12 intakes in female adolescents (OR 2.70; CI: 1.80–4.03; $p < 0.001$) and adults (OR 2.51; CI: 1.45–4.36, $p < 0.01$). While a similar trend was observed in elderly women, the difference was not significant. Particularly in adolescents and the elderly, we also observed a significant association with physical activity IPAQ score. Interestingly, a higher level of physical activity was associated with a higher prevalence of inadequate vitamin B12 intake. Marginally significant associations were found for employment (adults) and for smoking status and financial status (elderly). Compared to employed participants, those who were retired had a higher risk for inadequate daily vitamin B12 intake (OR = 2.80; CI = 1.32–5.94, $p < 0.1$). We also observed a trend of lower risk for inadequate vitamin B12 intake (OR = 0.65; CI = 0.40–1.08, $p < 0.1$) in elderly people with above-average financial status.

The relative contributions of different food categories to daily vitamin B12 intake are presented in Figure 2 and Supplementary Table S2. The most important vitamin B12 contributors were meat and meat products (including both unprocessed and processed meat), fish and fish products, and milk and milk products. In meat and meat products, which made the highest vitamin B12 contributions among all food categories in all age groups, unprocessed meat made the greatest contribution in the elderly population. In this food category, the most notable differences were observed between adolescents and the elderly, where the contribution was 32% vs. 55%, respectively. In adolescents, an important contributor to vitamin B12 intake was milk and milk products (23.4%); interestingly, another important contributor in adolescents was cereals and cereal products (18.7%), which on the other hand did not make any contribution in the elderly.

Table 3. Association between the prevalence of inadequate intake of vitamin B12 (<4 μg/day) and sex, place of living, education, income, employment, smoking status, BMI, IPAQ, vegetarian/vegan diet, and diet restrictions for different age groups.

Variable		Adolescents (10–17 Years)			Adults (18–64 Years)			Elderly (65–74 Years)		
		Prevalence (%)	Crude OR	Adjusted OR	Prevalence (%)	Crude OR	Adjusted OR	Prevalence (%)	Crude OR	Adjusted OR
Overall		175 (37.4)			118 (32.4)			215 (51.7)		
Sex	male	62 (26.1)	1	1	40 (23.1)	1	1	94 (44.1)	1	1
	female	113 (49.1)	2.74 (1.83–4.12)	2.70 (1.80–4.03)	78 (40.8)	2.29 (1.42–3.73)	2.51 (1.45–4.36)	121 (59.6)	1.87 (1.24–2.81)	1.44 (0.90–2.31)
Place of living	rural	99 (36.7)	1	1	68 (33.7)	1	1	115 (50.2)	1	1
	intermediate	24 (31.6)	0.80 (0.44–1.41)	0.78 (0.44–1.38)	22 (39.3)	1.27 (0.66–2.44)	1.27 (0.63–2.59)	41 (57.8)	1.35 (0.77–2.41)	1.34 (0.74–2.43)
	urban	52 (42.6)	1.28 (0.81–2.03)	1.20 (0.76–1.92)	28 (26.4)	0.71 (0.40–1.22)	0.73 (0.40–1.33)	59 (50.9)	1.03 (0.64–1.64)	1.02 (0.61–1.70)
Education	no university degree		n.a.	n.a.	78 (31.3)	1	1	180 (52.6)	1	1
	university degree				40 (34.8)	1.17 (0.71–1.91)	1.62 (0.88–3.00)	35 (47.3)	0.81 (0.47–1.37)	0.94 (0.51–1.72)
Financial status	below average		n.a.	n.a.	40 (33.9)	1	1	148 (55.0)	1	1
	above average				57 (30.2)	0.84 (0.50–1.42)	0.94 (0.51–1.71)	48 (44.9)	0.67 (0.41–1.07)	0.65 (0.40–1.08)
BMI	normal	110 (35.5)	1	1	48 (32.4)	1	1	57 (52.8)	1	1
	overweight and obese	65 (38.9)	1.10 (0.73–1.66)	1.13 (0.75–1.72)	70 (32.4)	1.00 (0.62–1.60)	1–03 (0.59–1.80)	158 (51.3)	0.94 (0.59–1.50)	1.00 (0.61–1.65)
IPAQ	low intensity	27 (25.0)	1	1	36 (28.4)	1	1	64 (46.7)	1	1
	moderate	68 (48.2)	2.79 (1.57–5.03)	2.48 (1.41–4.38)	33 (30.6)	1.11 (0.61–2.02)	1.28 (0.66–2.46)	83 (62.4)	1.89 (1.13–3.17)	1.86 (1.10–3.14)
	high intensity	76 (35.5)	1.65 (0.96–2.89)	1.64 (0.96–2.80)	47 (37.6)	1.52 (0.87–2.68)	2.13 (1.12–4.03)	64 (45.7)	0.96 (0.58–1.58)	1.02 (0.61–1.71)
Employment	employed		n.a.	n.a.	65 (28.8)	1	1		n.a.	n.a.
	unemployed				15 (35.7)	1.37 (0.64–2.88)	1.71 (0.72–4.05)			
	student				8 (25.0)	0.83 (0.30–2.02)	1.13 (0.38–3.34)			
	retired				30 (46.9)	2.19 (1.18–4.01)	2.80 (1.32–5.94)			
Smoking status	nonsmoker	164 (37.4)	1	1	64 (32.2)	1	1	133 (57.6)	1	1
	current/ ex-smoker	11 (36.7)	0.97 (0.41–2.20)	0.96 (0.42–2.20)	54 (32.7)	1.03 (0.64–1.63)	1.02 (0.59–1.76)	82 (44.3)	0.59 (0.39–0.88)	0.64 (0.40–1.02)
Medical diet	no special diet	169 (37.1)	1	1	105 (31.6)	1	1	184 (50.4)	1	1
	medical/weight loss	6 (46.2)	1.45 (0.40–5.13)	1.61 (0.49–5.26)	13 (40.6)	1.48 (0.64–3.29)	1.28 (0.53–3.08)	31 (60.8)	1.52 (0.81–2.93)	1.34 (0.70–2.56)
Behavioral diet	no diet	166 (36.4)	1	n.a.	113 (31.7)	1	n.a.	212 (51.3)	n.a.	n.a.
	vegetarian/vegan	9 (75.0)	5.24 (1.28–30.40)	n.a.	5 (62.5)	3.58 (0.68–23.39)	n.a.	3 (100.0)	n.a.	n.a.

Notes: n.a.—not applicable; CI—confidence interval; body mass index (BMI) was considered as normal below 25 kg/m^2, except for adolescents, where sex-/age-adjusted cut-off points [61,62] were used. Logistic regression analysis was conducted on samples with excluded missing values (financial status: n = 57 (adults) and 40 (elderly); IPAQ (International Physical Activity Questionnaire): n = 5 (adolescents), 4 (adults), 6 (elderly)). Cut-off odds ratios calculated with threshold of vitamin B12 intake < 4 μg/day; association was significant ($p < 0.05$) or marginally significant ($p < 0.1$) for the following variables: $p < 0.001$ sex (adolescents), $p < 0.01$ IPAQ (adolescents); $p < 0.01$ sex (adults), $p < 0.1$ IPAQ (adults), $p < 0.1$ employment (adults); $p < 0.05$ IPAQ (elderly), $p < 0.1$ smoking status (elderly), $p < 0.1$ financial status (elderly).

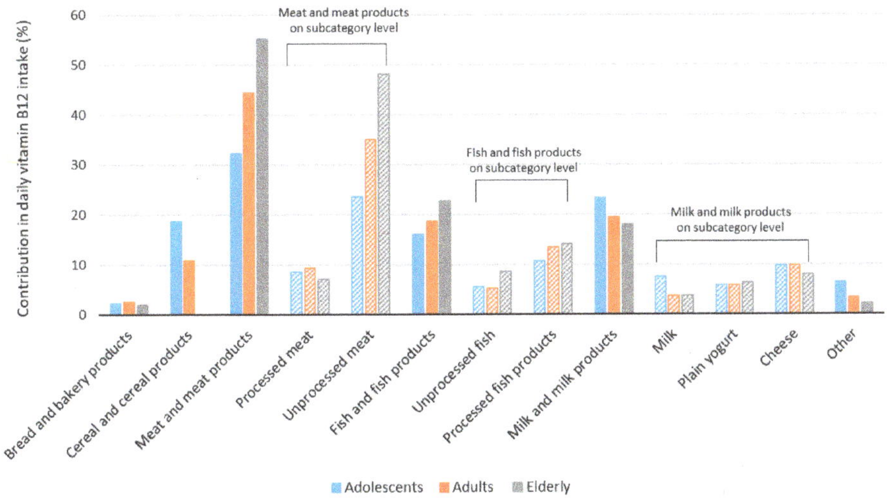

Figure 2. Relative contributions of selected food categories to usual daily vitamin B12 intake among different age groups (% of total vitamin B12 intake).

As an extension of the SI.Menu study, a subsample of adults and elderly also participated in the Nutrihealth study, where blood samples were collected. Vitamin B12 status was determined using two biomarkers—serum vitamin B12 and homocysteine concentrations. The mean population-weighted serum vitamin B12 levels were 322.1 pmol/L in adults and 287.3 pmol/L in the elderly (Table 2), which are above the cut-off point for low vitamin B12 status, at 221 pmol/L. However, a low vitamin B12 status was observed in 21% of adults and in almost half of the elderly population (46%). Vitamin B12 deficiency (<148 pmol/L) was detected in 3.7% of adults and 10.4% of the elderly. The highest prevalence of low vitamin B12 status and deficiency was observed in elderly males (57.9% and 11.8%, respectively). Furthermore, when vitamin B12 deficiency was assessed together with high homocysteine (serum B12 < 148 nmol/L, serum homocysteine > 15 µmol/L), deficiency was observed in 1.2% of adults (2.3% of males and no women) and 7.0% of elderly people (7.9% of males and 6.3% of women). We should note a strong negative linear correlation between log-transformed serum vitamin B12 and homocysteine concentration in adults (r = −0.27, $p = 0.003$) and the elderly (r = −0.29, $p < 0.001$), as well as in males (r = −0.24, $p = 0.005$) and females (r = −0.32, $p < 0.001$) (Figure 3). It should be noted that serum homocysteine concentration is also affected by several other factors, including folate status. Results of the linear regression analysis model (Supplementary Table S3) identified age, sex, serum vitamin B12, and serum folate as significant parameters associated with serum homocysteine concentrations. In additional logistic regression analyses, the same parameters were also found as significant predictors of high homocysteine concentrations (>15 µmol/L): age ($p < 0.01$; higher prevalence in elderly: OR 2.42; CI: 1.23–4.73), sex ($p < 0.001$; lower prevalence in females: OR 0.21; CI: 0.1–0.41), low serum vitamin B12 concentration ($p < 0.05$; lower prevalence in those above 221 pmol/L: OR 0.52; CI: 0.28–0.99), and low serum folate ($p < 0.01$; lower prevalence in those above 7 nmol/L: OR 0.33; CI: 0.15–0.73).

The results of logistic regression analyses on the combined sample of adults and the elderly identified age group and sex as the only significant predictors of vitamin B12 deficiency (Figure 4). A significantly higher risk for deficiency was observed in the elderly (OR: 4.42; CI: 1.72–13.37), and lower risk was observed in females (OR: 0.24; CI: 0.09–0.51). Additionally, we tested a model which considered the use of multivitamin food supplements (in addition to the above-mentioned parameters), but this was not found as a significant predictor for vitamin B12 deficiency ($p = 0.49$).

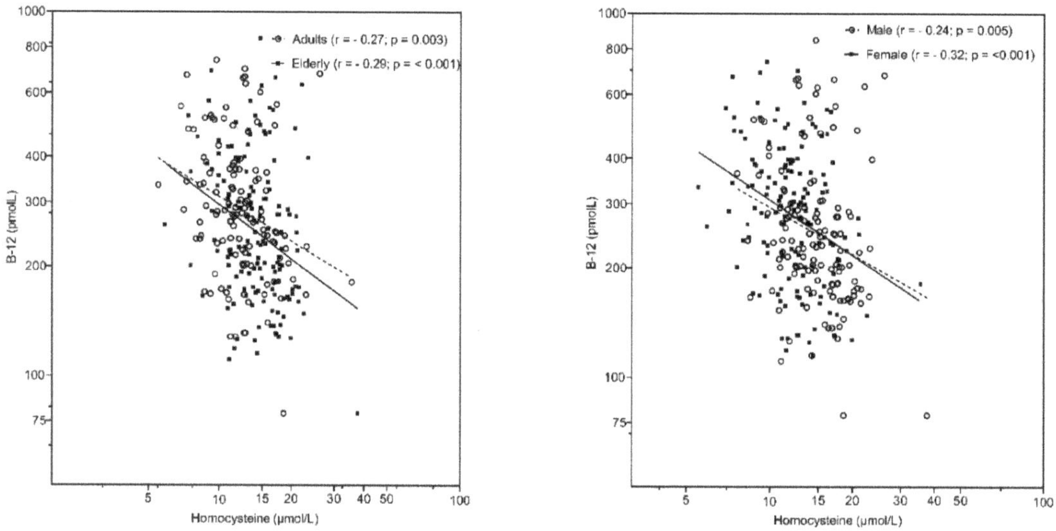

Figure 3. Association of serum vitamin B12 with serum homocysteine concentration in different age cohorts (**left**) and sex groups (**right**) using log scale.

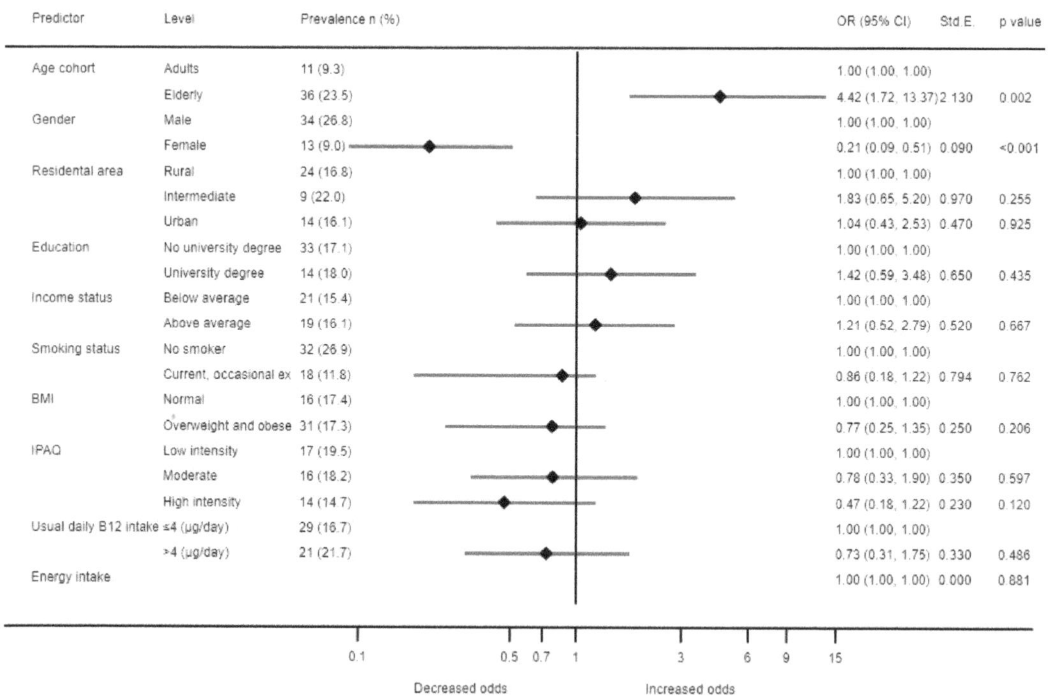

Figure 4. Association between prevalence of vitamin B12 deficiency (serum vitamin B12 concentration < 221 pmol/L and serum homocysteine > 15 µmol/L) and age, sex, residential area, education, financial status, body mass index (BMI), IPAQ (International Physical Activity Questionnaire) score, usual intake of vitamin B12, and usual daily energy intake.

Analyses of differences in the proportion of the prevalence of vitamin B12 deficiency showed significantly higher deficiency prevalence in males in the elderly population ($p < 0.001$) (Figure 5).

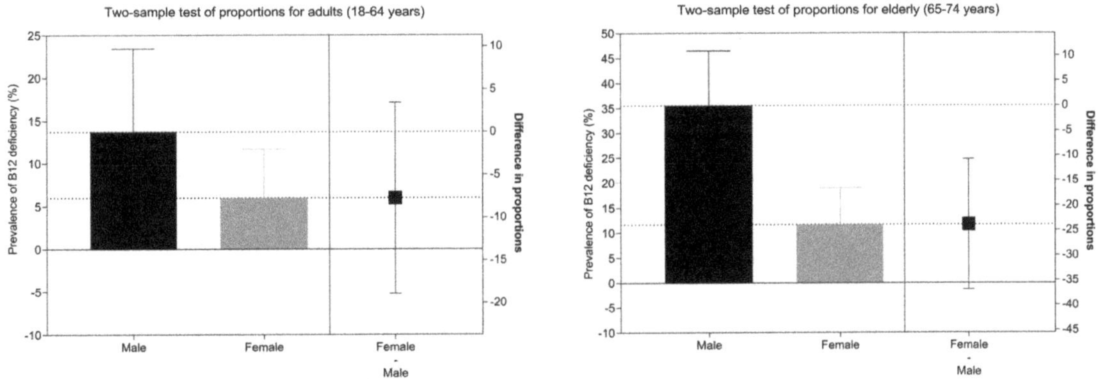

Figure 5. Analysis of differences in prevalence of vitamin B12 deficiency (serum vitamin B12 concentration < 221 pmol/L and homocysteine > 15 μmol/L) between males and females in different age cohorts.

4. Discussion

Although clinically manifested vitamin B12 deficiency is relatively rare in developed countries, low intakes and statuses of vitamin B12 are much more common and can pose a public health concern, with certain population groups being particularly at risk [34]. The present study is the first one evaluating the intake and status of vitamin B12 in a nationally representative sample in Slovenia. In all population groups, the population-weighted mean vitamin B12 intakes are above the national recommendations of 4 μg/day, making the situation in Slovenia comparable to the findings from other EU countries [43,77]. In all population groups, males had higher population-weighted mean vitamin B12 intakes than women. The highest population-weighted vitamin B12 mean intake was found in the adult male population (6.9 μg/day), and the lowest was found in elderly females (4.4 μg/day). In adults and the elderly, lower mean vitamin B12 intakes in females were also found when calculating the daily vitamin B12 intake per 1000 kcal/day, while male and female adolescents had equal daily vitamin B12 intakes per 1000 kcal/day.

Even though population-weighted mean vitamin B12 intakes were generally sufficient in all age groups, more than 30% of the adolescent and adult populations did not meet the recommended daily intake of vitamin B12. Among all age groups, the situation was particularly concerning among the elderly, where 58.3% had inadequate vitamin B12 daily intakes. Looking into the distribution of vitamin B12 intake (Figure 1) especially in the elderly population, besides certain individuals with particularly low vitamin B12 intakes, there is a proportion of individuals with considerably higher vitamin B12 intakes. This influences the overall mean intake in the population, which appears higher due to the higher interindividual variability, although a notable proportion of the population did not meet the recommended daily vitamin B12 intake. Lower vitamin B12 intakes in females could be concerning, as especially in adolescents and elderly females, the median population-weighted vitamin B12 intake was lower than 4 μg/day, while consistent evidence shows that vitamin B12 intake above 4 μg/day is associated with an adequate serum vitamin B12 status [43].

Our findings are in line with the results of studies conducted in other European countries, where the reported mean intakes were usually above the national recommendations, but with notable variations across countries. In a review of studies from 15 European countries, the highest vitamin B12 intakes were seen in Finland (>7 μg/day), followed by Sweden, Germany, France, and Spain, with mean vitamin B12 intakes above 5 μg/day [48]. Denmark, Ireland, and the Netherlands had moderate vitamin intakes between 2.5 and 5 μg/day. The lowest daily vitamin B12 intake was observed in Greece (<2.5 μg/day), which is also below their national recommendation for vitamin B12 intake [78].

Dietary intakes of vitamin B12 are highly dependent on the type of diet. Diets that are low in meat and foods of animal origin are often low in vitamin B12; therefore, in countries where more foods of animal origin are consumed, vitamin B12 intakes are usually higher [1]. According to the latest study on dietary habits in Slovenia, meat is usually consumed several times weekly in all age groups, with adult males being the largest consumers of meat [79]. This is reflected in the mean vitamin B12 intakes in our study, which were also the highest in this population. As seen in Figure 2, where the relative contributions of food categories to the usual daily intake of vitamin B12 among different age groups are presented, the highest contribution of daily vitamin B12 comes from meat and meat products, with the elderly deriving notably higher levels from unprocessed meat, compared to adolescents. For adolescents, important vitamin B12 contributors were milk and milk products and, interestingly, cereal and cereal products, which made no dietary vitamin B12 contribution among the elderly. This is because fortified cereal products (i.e., breakfast cereals) can be a source of vitamin B12, but such products are more popular among adolescents than in the elderly population. We should mention that in our previous study, this food category was also found to be an important contributor to folate intake in adolescents [68]. To derive the optimal benefits from the daily diet, adolescents should be encouraged to eat more unprocessed and minimally processed foods. Although they tend to intake certain micronutrients from fortified products, such as fortified breakfast cereals, the frequent consumption of such processed foods can have detrimental effects on health. Processed foods, particularly processed breakfast cereals, can often have a less favorable nutritional composition [80] and can be high in sugar and refined grains; this was particularly observed in foods specifically marketed to children [81,82]. Considering the fact that the majority of the general population meets the daily vitamin B12 requirement, and that meats and foods of animal origin are abundantly consumed in Slovenia [79], additional encouragement of the consumption of such foods is not relevant. Currently, around 20% of daily vitamin B12 intake originates in processed meat and fish products; therefore, switching to unprocessed meat and fish should be supported.

A logistic regression analysis of adequate daily dietary intake of vitamin B12 (Table 3) showed that sex, IPAQ, smoking status, employment, and financial status were (marginally) significant predictors of adequate vitamin B12 intake, exposing adult and adolescent females, elderly nonsmokers, retired adults, elderly people with poor financial status, and all of those with higher IPAQ scores as more susceptible to inadequate daily vitamin B12 intake. This could be linked to the lower consumption of meat in females and those with lower financial status, which was also reported in the latest Slovenian dietary survey [79]. It is established that meat consumption is associated with economic status [83], and those with poor financial status usually consume less meat, especially unprocessed meat, probably also due to its higher price compared to some other foods. It was previously reported that financial status impacts food choices, which further manifests in micronutrient intake [84,85]. Regarding IPAQ score, it is possible that those who are more physically active will more commonly follow dietary restrictions, which could result in lower vitamin B12 intake. It should be mentioned that increased physical activity may raise homocysteine levels; therefore, such individuals could benefit from a higher intake of B vitamins, including vitamin B12, so as to balance homocysteine levels and reduce the risk of cardiovascular disease later in life [86].

The investigation of serum biomarkers in adults and the elderly showed that in both population groups, the weighted mean serum vitamin B12 levels were above our cut-off point for low vitamin B12 status (221 pmol/L), which is comparable to other European countries [48]. The serum vitamin B12 status was adequate in 79% of adults and 54% of the elderly. Serum vitamin B12 levels were lower in the elderly than in adults, with a higher prevalence of vitamin B12 deficiency observed in this population group, compared to adults. Higher homocysteine levels were observed in the elderly and in males, in line with previous studies [87]. The prevalence of vitamin B12 deficiency was also higher in men than in women, especially in the elderly, which is consistent with previous reports [30,88]. It is well established that certain vitamins, including vitamin B12, influence serum homocysteine status and that increased homocysteine is related to a higher risk for cardiovascular disease [89]. In the elderly population, we observed a strong negative linear association between high homocysteine and low serum vitamin B12 within age groups (Figure 3). Taking into account both low serum vitamin B12 (<148 nmol/L) and high serum homocysteine (>15 μmol/L), vitamin B12 deficiency was observed in 7% of the Slovenian elderly population, which is similar to the rates observed in the United States and the United Kingdom [90]. A much lower prevalence of deficiency (1.2%) was observed in adults. The logistic regression analysis of the prevalence of vitamin B12 deficiency (Figure 4) confirmed that elderly people and males had a higher likelihood of deficiency, while other demographic factors were not shown to be statistically significant.

Vitamin B12 intake and status are not always associated, particularly due to different causes of vitamin B12 absorption; moreover, because vitamin B12 is one of the vitamins that are best stored in the body, it can take a few years for an inadequate status to develop. This was observed, for example, in the Netherlands and Germany, where mean vitamin B12 intakes were above the recommendations, but status appeared to be inadequate [48]. We identified the elderly as a particularly vulnerable population in terms of vitamin B12 intake and status; this problem was highlighted also in numerous studies in other regions [25,91–99]. Depending on the diagnostic method and cut-offs, international deficiency rates among the elderly population were reported somewhere between 6% and 43%, with higher deficiency prevalence observed in developing countries [26,90,100,101]. In the elderly, the predominant factor for a high prevalence of low vitamin B12 status is malabsorption. Due to the complex absorption process of vitamin B12, several factors contribute to malabsorption, such as gastric abnormalities, small bowel diseases, pancreatic insufficiency, and several medications, most commonly protein pump inhibitors and metformin; as the age advances, the probability of such conditions increases, together with malabsorption [102]. Atrophic gastritis, often accompanied by *H. pylori* infection, is one of the more common causes of vitamin B12 deficiency, which affects 20–50% of the elderly and is typically asymptomatic [103,104]. Here, reduced gastric acid and pepsinogen secretion results in a decreased intestinal absorption of vitamin B12–protein complexes from food. Considering the high probability of malabsorption, along with the fact that 58.3% of this population had inadequate vitamin B12 intakes, this would explain our findings related to low vitamin B12 status in the elderly, as the mean population vitamin B12 intakes did not significantly deviate in comparison with other population groups. Although elderly males had higher daily vitamin B12 intakes compared to women, they also had a higher prevalence of vitamin B12 deficiency. This is seen also in Figure 5, where the data show a significantly higher deficiency prevalence in elderly males in comparison to females, while this difference was not significant for adults. As the literature's data show that atrophic gastritis is more common in men and in the elderly [105–107], we could speculate that absorption impairment could be the reason for the higher vitamin B12 deficiency rates among the Slovenian elderly, especially males. The capacity for the efficient absorption of vitamin B12 from foods decreases with age; however, evidence shows that synthetic vitamin B12 from fortified foods and supplements could be more efficiently absorbed, as the synthetic form of vitamin B12 is not usually protein-bound [34]. Due to the considerable prevalence of atrophic gastritis and lower stomach acid excretion in the elderly, in some countries, certain institutions already propose that the elderly meet their daily requirements with vitamin B12 supplements and/or fortified foods [25]. Furthermore, research data show a good metabolic response to vitamin B12 supplementation in elderly people with malabsorption issues [103].

In Slovenia, testing for vitamin B12 status is currently mostly indicated for those with clinically relevant symptoms of vitamin B12 deficiency. In view of the higher prevalence of inadequate vitamin B12 intake and deficiency, as well as vitamin B12 absorption issues, in the elderly population, routine screening may be advantageous in this population. The use of vitamin B12 supplements in the elderly should also be considered in Slovenia, as the additional intake of foods naturally rich in vitamin B12 would probably not be sufficient due to absorption issues [91]. For example, because of the specific metabolic characteristics of the elderly, the American dietary guidelines have already been modified for people over 70 years of age. Dietary supplementation with vitamins B12 and D and calcium was recommended, as was the use of folic acid fortified foods, such as flour, which has been subjected to mandatory folic acid fortification in the USA since 1998 [108]. In addition, some studies suggest that besides fortifying flour with folic acid, vitamin B12 should also be added, due to the metabolic associations of these two vitamins and related public health issues [109,110]. Interventions in the elderly population are therefore much needed, as the timely detection of vitamin B12 deficiency and appropriate diagnosis could prevent severe hematological and irreversible neurological complications [111,112].

Another critical population group, for which vitamin B12 supplementation is already advised, includes those who do not consume foods of animal origin, such as vegans. In the present study, suboptimal intakes of vitamin B12 were observed for those following such diets (Supplementary Table S1), although this should be considered with caution due to the small sample size. Crude odds ratios (Table 3) also showed a higher risk for vitamin B12 deficiency in this population. Vitamin B12 intake was observed as critical in a recent systematic review of the adequacy of a vegan diet [21]. Default vitamin B12 supplementation should therefore be additionally encouraged in such individuals, and regular monitoring of vitamin B12 status should be advised.

A major strength of the present SI.Menu study was that it was performed using an internationally acknowledged EU Menu methodology for dietary intake assessment, published by the EFSA [53]. We were able to analyze nationally representative data on food consumption, using 24 h dietary recalls and FPQ. Data were collected for a representative sample of Slovenian adolescents, adults, and elderly. In a Nutrihealth study, we also investigated serum biomarkers for vitamin B12 deficiency in a subsample of adults and elderly. One of the study's strengths was that, besides data on serum vitamin B12 concentration, the homocysteine status was also available, which supported the assessment of vitamin B12 deficiency prevalence. However, the study results should be interpreted with consideration of some limitations. The SI.Menu study only included data on the general use of specific food supplements, for example, multivitamins, but not specifically vitamin B12. Therefore, the dietary intake of vitamin B12 was only calculated from food sources. Furthermore, while a combination of serum vitamin B12 and homocysteine concentration is considered a much better indicator for the assessment of vitamin B12 deficiency than serum vitamin B12 alone, we were not able to analyze methylmalonic acid, which is also considered a very valuable biomarker for the assessment of vitamin B12 status [33]. We should also note that adults in the Nutrihealth study were slightly overrepresented by those above 50 years of age. Another limitation is that serum biomarker data were not available for adolescents. Furthermore, although the standard chemiluminescence immunoassay medical diagnostic method was used to determine vitamin B12 in serum samples, we should note that this method is sensitive to very high biotin concentrations. In serum samples with biotin concentration above 1500 μg/L, the measurement error was less than 10%, and therefore serum biotin concentration was not controlled.

5. Conclusions

In Slovenia, the estimated population-weighted mean daily vitamin B12 intake was above the recommended 4 μg in all age groups. However, inadequate daily vitamin B12 intake was observed in 37.3% of adolescents, 31.7% of adults, and 58.3% of elderly. The mean population-weighted daily vitamin B12 intake was lower in females in all population groups. Besides sex, other notable predictors for inadequate vitamin B12 intake were physical activity IPAQ score, financial status, and smoking and employment status. The study also showed that in all population groups, the most important contributors to vitamin B12 intake were meat and meat products, followed by milk and milk products. The mean population-weighted serum vitamin B12 concentration was 322.1 pmol/L in adults and 287.3 pmol/L in the elderly population. Vitamin B12 deficiency (<148 pmol/L, homocysteine >15 µmol/L) was observed in 1.2% of adults and 7.0% of the elderly. In the elderly, vitamin B12 malabsorption jeopardizes sufficient vitamin B12 status. Our study identified the male elderly population as especially at risk for low vitamin B12 status. Particularly in the elderly, monitoring for vitamin B12 is very relevant, and supplementation could be considered.

Supplementary Materials: The following are available online at https://www.mdpi.com/article/10.3390/nu14020334/s1. Supplementary Table S1. Mean (SD) and model-adjusted mean (95% CI) levels of usual daily intake of vitamin B12 by sex, residential area, education, financial status, employment, vegetarian/vegan diet, smoking status, dietary restrictions, BMI, and IPAQ score for different age cohorts. Supplementary Table S2. The relative contributions of selected food categories to usual daily dietary vitamin B12 intake among different age groups (% of total dietary vitamin B12 intake). Supplementary Table S3. Mean (SD) and model-adjusted mean (95% CI) serum homocysteine concentration (µmol/L) by age, sex, residential area, education, financial status, smoking status, BMI, IPAQ, diet, use of multivitamin supplements, serum folate concentration (nmol/L), and serum vitamin B12 concentration (pmol/L).

Author Contributions: I.P. was responsible for assuring the Nutrihealth study set-up of the analyses and participated in the data analyses. J.O. was responsible for the data collection in the Nutrihealth study. J.O., K.Z. and A.O. were responsible for the sample collection and laboratory analyses in the Nutrihealth study. Ž.L. wrote the first draft of the manuscript and prepared the submission. H.H. prepared data and performed data analyses. N.G. participated in data interpretation and discussion. A.K. and K.Ž. participated in study conduction and data interpretation. M.H. conducted food-matching to estimate nutritional composition of foods. M.G. and U.B. were responsible for SI.Menu study design and food consumption data. B.K.S. was responsible for information technology. All authors have read and agreed to the published version of the manuscript.

Funding: The SI.Menu 2017/18 national food consumption survey was funded by the European Food Safety Authority (EFSA contract No. OC/EFSA/DATA/2014/02-LOT2-CT03), Ministry of Health of the Republic of Slovenia, and Slovenian Research Agency. The Nutrihealth study was part of the research project L3-8213 "Children's and adults' nutrition as a protective or health-risk factor", funded by the Slovenian Research Agency and Ministry of Health of the Republic of Slovenia. The study was conducted within research programs P3-0395 "Nutrition and Public Health" and P2-0098 "Computer Systems and Structures", and "Infrastructure program for monitoring of the composition and labelling of foods" (IO-0054), funded by the Slovenian Research Agency. We also acknowledge that the contribution of Ž.L. has been supported by the Slovenian Research Agency, which has been funding the post-doctoral research project Z3-3213 "Dietary challenges in the population of nursing homes' residents".

Institutional Review Board Statement: The study was conducted according to the guidelines of the Declaration of Helsinki. SI.Menu study was approved by the National Medical Ethics Committee, Ljubljana, Slovenia (KME 53/07/16; approval No. 0120-337/2016 issued on 19 July 2016). Nutrihealth study was conducted as an extension of the SI.Menu study. Study protocol was also approved by the Slovenian National Medical Ethics Committee, Ljubljana, Slovenia (KME 72/07/16; approval No. 0120-337/2016-4 issued on 7 July 2017). Study was registered at ClinicalTrials.gov (ID: NCT03284840).

Informed Consent Statement: Informed consent was obtained from all subjects involved in the study.

Data Availability Statement: The data presented in this study are available on request from the corresponding author.

Acknowledgments: We acknowledge the support of all SI.Menu project partners and their researchers, particularly Nataša Delfar, Metka Zaletel, Darja Lavtar, Ada Hočevar Grom, and Ivan Eržen (National Institute of Public Health, Ljubljana, Slovenia); Petra Golja and Katja Zdešar Kotnik (Group of Anthropology at Biotechnical Faculty of University of Ljubljana, Slovenia); Stojan Kostanjevec (University of Ljubljana, Faculty of Education, Ljubljana, Slovenia); Majda Pajnkihar (University of Maribor, Faculty of Health Sciences, Maribor, Slovenia); Tamara Poklar Vatovec (University of Primorska, Faculty of Health Sciences, Koper, Slovenia); and Evgen Benedik (University Medical Centre Ljubljana, University Children' s Hospital, Ljubljana, Slovenia). We acknowledge the support of Veronika Belec and her team and interviewers at GfK Research Institute (Slovenia) for conducting the survey field work. We also acknowledge Irena Rogelj (Biotechnical Faculty, University of Ljubljana) for support in project management and funding acquisition, and the regional community healthcare centers and hospitals for the collection of biological samples in the Nutrihealth study. We are grateful to the staff of UKC Ljubljana (Troha Poljančič Vera, Gaberšček Simona) and all medical laboratory teams across Slovenia led by Spomenka Lajtner (Community Health Center Ljubljana); Jana Žnidaršič (Community Health Center Cerknica), Katarina Furar (Community Health Center Ivančna Gorica); Tomaž Vargazon (Community Health Center Kamnik); Ljubica Štefanič (Community Health Center Kočevje); Irena Sluga (Community Health Center Litija); Jolanda Prah (Community Health Center Maribor); Jelica Štanta (Community Health Center Gornja Radgona); Ana Gruškovnjak (Community Health Center Ljutomer); Edita Ružič (Community Health Center Murska Sobota); Mojca Završnik (Hospital Ptuj); Tanja Lađić (General Hospital Slovenj Gradec); Tina Kochler (Community Health Center Slovenska Bistrica); Mojca Kozmelj (Community Health Center Celje); Maja Škerjanc (Community Health Center Velenje); Branka Svetic (Community Health Center Kranj); Sabina Nabernik (Community Health Center Jesenice); Marko Rudež (Community Health Center Škofja Loka); Vanja Pahor (General Hospital Izola); Damjana Bogataj (Community Health Center Ajdovščina); Andreja Seljak (Community Health Center Idrija); Kristina Gruden (Community Health Center Nova Gorica); Valentina Buda (Community Health Center Postojna); Polonca Stopar (Community Health Center Sežana); Zalka Murovec (Community Health Center Tolmin); Danijela Furlan (General Hospital Novo mesto); and Andreja Kerin (General hospital Brežice). We also acknowledge Sanja Krušič (Nutrition Institute, Ljubljana, Slovenia) and Naska Kvarantan (Wageningen University and Research, Netherlands; Nutrition Institute, Ljubljana, Slovenia) for support in food-matching to estimate the nutritional compositions of foods and Nina Zupanič (Nutrition Institute, Ljubljana, Slovenia) for support in the literature review.

Conflicts of Interest: The authors declare no conflict of interest. The funders had no role in the design of the study; in the collection, analyses, or interpretation of data; in the writing of the manuscript; or in the decision to publish the results. I.P. has led and participated in various other research projects in the area of nutrition, public health, and food technology, which were (co)funded by the Slovenian Research Agency, Ministry of Health of the Republic of Slovenia, the Ministry of Agriculture, Forestry and Food of the Republic of Slovenia, and in cases of specific applied research projects, also by food businesses. I.P., M.G., and U.B. are members of a workgroup for food reformulation at the Ministry of Health of the Republic of Slovenia.

References

1. Green, R.; Allen, L.H.; Bjørke-Monsen, A.L.; Brito, A.; Guéant, J.L.; Miller, J.W.; Molloy, A.M.; Nexo, E.; Stabler, S.; Toh, B.H.; et al. Vitamin B(12) deficiency. *Nat. Rev. Dis. Primers* **2017**, *3*, 17040. [CrossRef]
2. Castellanos-Sinco, H.B.; Ramos-Peñafiel, C.O.; Santoyo-Sánchez, A.; Collazo-Jaloma, J.; Martínez-Murillo, C.; Montaño-Figueroa, E.; Sinco-Ángeles, A. Megaloblastic anaemia: Folic acid and vitamin B12 metabolism. *Rev. Med. Hosp. Gen. México* **2015**, *78*, 135–143. [CrossRef]
3. Lachner, C.; Steinle, N.I.; Regenold, W.T. The Neuropsychiatry of Vitamin B12 Deficiency in Elderly Patients. *J. Neuropsychiatry Clin. Neurosci.* **2012**, *24*, 5–15. [CrossRef]
4. Carmel, R. Current concepts in cobalamin deficiency. *Annu. Rev. Med.* **2000**, *51*, 357–375. [CrossRef] [PubMed]
5. Allen, L.H. Causes of vitamin B12 and folate deficiency. *Food Nutr. Bull.* **2008**, *29*, S20–S34. [CrossRef] [PubMed]
6. Shane, B.; Stokstad, E.L. Vitamin B12-folate interrelationships. *Annu. Rev. Nutr.* **1985**, *5*, 115–141. [CrossRef]
7. Refsum, H.; Ueland, P.; Nygård, O.; Vollset, S. Homocysteine and cardiovascular disease. *Annu. Rev. Med.* **1998**, *49*, 31–62. [CrossRef] [PubMed]
8. Woo, K.S.; Chook, P.; Lolin, Y.I.; Cheung, A.S.; Chan, L.T.; Sun, Y.Y.; Sanderson, J.E.; Metreweli, C.; Celermajer, D.S. Hyperhomocyst(e)inemia is a risk factor for arterial endothelial dysfunction in humans. *Circulation* **1997**, *96*, 2542–2544. [CrossRef] [PubMed]
9. Welch, G.N.; Loscalzo, J. Homocysteine and atherothrombosis. *NEJM* **1998**, *338*, 1042–1050. [CrossRef]

10. Clarke, R.; Daly, L.; Robinson, K.; Naughten, E.; Cahalane, S.; Fowler, B.; Graham, I. Hyperhomocysteinemia: An independent risk factor for vascular disease. *NEJM* **1991**, *324*, 1149–1155. [CrossRef]
11. Martens, J.-H.; Barg, H.; Warren, M.A.; Jahn, D. Microbial production of vitamin B 12. *Appl. Microbiol. Biotechnol.* **2002**, *58*, 275–285. [CrossRef]
12. Watanabe, F. Vitamin B12 sources and bioavailability. *Exp. Biol. Med.* **2007**, *232*, 1266–1274. [CrossRef] [PubMed]
13. Melini, F.; Melini, V.; Luziatelli, F.; Ficca, A.G.; Ruzzi, M. Health-Promoting Components in Fermented Foods: An Up-to-Date Systematic Review. *Nutrients* **2019**, *11*, 1189. [CrossRef]
14. EC. Regulation (EC) No 1925/2006 of the European Parliament and of the Council of 20 December 2006 on the addition of vitamins and minerals and of certain other substances to foods. *OJEU* **2006**, *50*, 26–38.
15. Scott, J. Bioavailability of vitamin B12. *Eur. J. Clin. Nutr.* **1997**, *51*, S49–S53.
16. Squires, M.; Naber, E. Vitamin profiles of eggs as indicators of nutritional status in the laying hen: Vitamin B12 study. *Poult. Sci.* **1992**, *71*, 2075–2082. [CrossRef]
17. Doets, E.L.; Veld, P.H.I.; Szczecińska, A.; Dhonukshe-Rutten, R.A.; Cavelaars, A.E.; vam't Veer, P.; Brzozowska, A.; de Groot, L.C. Systematic review on daily vitamin B12 losses and bioavailability for deriving recommendations on vitamin B12 intake with the factorial approach. *Ann. Nutr. Metab.* **2013**, *62*, 311–322. [CrossRef]
18. Heyssel, R.M.; Bozian, R.C.; Darby, W.J.; Bell, M.C. Vitamin B12 turnover in man. The assimilation of vitamin B12 from natural foodstuff by man and estimates of minimal daily dietary requirements. *Am. J. Clin. Nutr.* **1966**, *18*, 176–184. [CrossRef]
19. Gille, D.; Schmid, A. Vitamin B12 in meat and dairy products. *Nutr. Rev.* **2015**, *73*, 106–115. [CrossRef]
20. Lešková, E.; Kubíková, J.; Kováčiková, E.; Košická, M.; Porubská, J.; Holčíková, K. Vitamin losses: Retention during heat treatment and continual changes expressed by mathematical models. *J. Food Compos. Anal.* **2006**, *19*, 252–276. [CrossRef]
21. Bakaloudi, D.R.; Halloran, A.; Rippin, H.L.; Oikonomidou, A.C.; Dardavesis, T.I.; Williams, J.; Wickramasinghe, K.; Breda, J.; Chourdakis, M. Intake and adequacy of the vegan diet. A systematic review of the evidence. *Clin. Nutr.* **2021**, *40*, 3503–3521. [CrossRef]
22. Pawlak, R.; Lester, S.E.; Babatunde, T. The prevalence of cobalamin deficiency among vegetarians assessed by serum vitamin B12: A review of literature. *Eur. J. Clin. Nutr.* **2014**, *68*, 541–548. [CrossRef]
23. Andrès, E.; Loukili, N.H.; Noel, E.; Kaltenbach, G.; Abdelgheni, M.B.; Perrin, A.E.; Noblet-Dick, M.; Maloisel, F.; Schlienger, J.-L.; Blicklé, J.-F. Vitamin B12 (cobalamin) deficiency in elderly patients. *CMAJ* **2004**, *171*, 251–259. [CrossRef]
24. Conzade, R.; Koenig, W.; Heier, M.; Schneider, A.; Grill, E.; Peters, A.; Thorand, B. Prevalence and predictors of subclinical micronutrient deficiency in german older adults: Results from the population-based KORA-age study. *Nutrients* **2017**, *9*, 1276. [CrossRef]
25. Stover, P.J. Vitamin B12 and older adults. *Curr. Opin. Clin. Nutr.* **2010**, *13*, 24. [CrossRef] [PubMed]
26. Wolters, M.; Ströhle, A.; Hahn, A. Cobalamin: A critical vitamin in the elderly. *Prev. Med.* **2004**, *39*, 1256–1266. [CrossRef]
27. Toh, B.-H.; van Driel, I.R.; Gleeson, P.A. Pernicious anemia. *NEJM* **1997**, *337*, 1441–1448. [CrossRef] [PubMed]
28. Chapman, L.E.; Darling, A.L.; Brown, J.E. Association between metformin and vitamin B12 deficiency in patients with type 2 diabetes: A systematic review and meta-analysis. *Diabetes Metab. J.* **2016**, *42*, 316–327. [CrossRef]
29. Wiersinga, W.J.; de Rooij, S.E.; Huijmans, J.G.; Fischer, C.; Hoekstra, J.B. Diagnosis of vitamin B12 deficiency revised. *Ned. Tijdschr. Voor Geneeskd.* **2005**, *149*, 2789–2794.
30. Carmel, R.; Green, R.; Jacobsen, D.W.; Rasmussen, K.; Florea, M.; Azen, C. Serum cobalamin, homocysteine, and methylmalonic acid concentrations in a multiethnic elderly population: Ethnic and sex differences in cobalamin and metabolite abnormalities. *Am. J. Clin. Nutr.* **1999**, *70*, 904–910. [CrossRef] [PubMed]
31. Stabler, S.P. Vitamin B12 deficiency. *NEJM* **2013**, *368*, 149–160. [CrossRef]
32. Obeid, R.; Heil, S.G.; Verhoeven, M.M.A.; van den Heuvel, E.G.H.M.; de Groot, L.C.P.G.M.; Eussen, S.J.P.M. Vitamin B12 Intake From Animal Foods, Biomarkers, and Health Aspects. *Front. Nutr.* **2019**, *6*, 93. [CrossRef] [PubMed]
33. Hannibal, L.; Lysne, V.; Bjørke-Monsen, A.-L.; Behringer, S.; Grünert, S.C.; Spiekerkoetter, U.; Jacobsen, D.W.; Blom, H.J. Biomarkers and Algorithms for the Diagnosis of Vitamin B12 Deficiency. *Front. Mol. Biosci* **2016**, *3*, 27. [CrossRef]
34. de Benoist, B. Conclusions of a WHO Technical Consultation on folate and vitamin B12 deficiencies. *Food Nutr. Bull.* **2008**, *29*, S238–S244. [CrossRef]
35. Green, R. Indicators for assessing folate and vitamin B-12 status and for monitoring the efficacy of intervention strategies. *Am. J. Clin. Nutr.* **2011**, *94*, 666S–672S. [CrossRef] [PubMed]
36. Aparicio-Ugarriza, R.; Palacios, G.; Alder, M.; González-Gross, M. A review of the cut-off points for the diagnosis of vitamin B12 deficiency in the general population. *CCLM* **2015**, *53*, 1149–1159. [CrossRef] [PubMed]
37. Ueland, P.M.; Refsum, H.; Stabler, S.P.; Malinow, M.R.; Andersson, A.; Allen, R.H. Total homocysteine in plasma or serum: Methods and clinical applications. *Clin. Chem.* **1993**, *39*, 1764–1779. [CrossRef] [PubMed]
38. WHO. *Vitamin and Mineral Requirements in Human Nutrition*, 2nd ed.; World Health Organization: Geneva, Switzerland, 2005.
39. Nordic Co-Operation. Nordic Nutrition Recommendations 2012: Integrating nutrition and physical activity. *Nord. Counc. Minist. Cph. Den.* **2014**, 627.
40. NIJZ. Referenčne vrednosti za energijski vnos ter vnos hranil. Available online: https://www.nijz.si/sites/www.nijz.si/files/uploaded/referencne_vrednosti_2020_3_2.pdf (accessed on 3 February 2021).

41. D-A-CH. Reference Values DACH. Available online: https://www.sge-ssn.ch/grundlagen/lebensmittel-und-naehrstoffe/naehrstoffempfehlungen/dachreferenzwerte/ (accessed on 14 July 2021).
42. Ströhle, A.; Richter, M.; González-Gross, M.; Neuhäuser-Berthold, M.; Wagner, K.H.; Leschik-Bonnet, E.; Egert, S. The Revised D-A-CH. Reference Values for the Intake of Vitamin B(12): Prevention of Deficiency and Beyond. *Mol. Nutr. Food Res.* **2019**, *63*, e1801178. [CrossRef]
43. EFSA. Scientific opinion on dietary reference values for cobalamin (vitamin B12). *EFSA J.* **2015**, *13*, 4150.
44. McLean, E.; de Benoist, B.; Allen, L.H. Review of the Magnitude of Folate and Vitamin B12 Deficiencies Worldwide. *Food. Nutr. Bull.* **2008**, *29*, S38–S51. [CrossRef] [PubMed]
45. Allen, L.H. Folate and Vitamin B 12 Status in the Americas. *Nutr. Rev.* **2004**, *62*, S29–S33. [CrossRef]
46. Arazo-Rusindo, M.C.; Zúñiga, R.N.; Cortés-Segovia, P.; Benavides-Valenzuela, S.; Pérez-Bravo, F.; Castillo-Valenzuela, O.; Mariotti-Celis, M.S. Nutritional Status and Serum Levels of Micronutrients in an Elderly Group Who Participate in the Program for Complementary Food in Older People (PACAM) from the Metropolitan Region, Santiago de Chile. *Nutrients* **2022**, *14*, 3. [CrossRef]
47. Hunt, A.; Harrington, D.; Robinson, S. Vitamin B12 deficiency. *BMJ* **2014**, *349*, g5226. [CrossRef]
48. Dhonukshe-Rutten, R.; De Vries, J.; De Bree, A.; Van Der Put, N.; Van Staveren, W.; De Groot, L. Dietary intake and status of folate and vitamin B12 and their association with homocysteine and cardiovascular disease in European populations. *Eur. J. Clin. Nutr.* **2009**, *63*, 18–30. [CrossRef]
49. Mijatov, M.A.; Mičetić-Turk, D. Dietary Intake In Adult Female Coeliac Disease Patients In Slovenia: Prehranski Vnos Odraslih Bolnic S Celiakijo V Sloveniji. *Zdr Varst* **2016**, *55*, 86–93. [CrossRef]
50. Gregorič, M.; Zdešar Kotnik, K.; Pigac, I.; Gabrijelčič Blenkuš, M. A Web-Based 24-H Dietary Recall Could Be a Valid Tool for the Indicative Assessment of Dietary Intake in Older Adults Living in Slovenia. *Nutrients* **2019**, *11*, 2234. [CrossRef] [PubMed]
51. Jakše, B.; Jakše, B.; Godnov, U.; Pinter, S. Nutritional, Cardiovascular Health and Lifestyle Status of 'Health Conscious' Adult Vegans and Non-Vegans from Slovenia: A Cross-Sectional Self-Reported Survey. *Int. J. Environ.* **2021**, *18*, 5968. [CrossRef]
52. Širca Čampa, A.; Hren, I.; Sedmak, M.; Brecelj, J.; Fidler Mis, N.; Krišnik, C.; Koletzko, B. Nutrition of mothers with hypogalactia in Slovenia. *J. Pediatr. Gastroenterol. Nutr.* **2006**, *42*, E104.
53. EFSA. Guidance on the EU Menu methodology. *EFSA J.* **2014**, *12*, 3944.
54. Gregorič, M.; Blaznik, U.; Delfar, N.; Zaletel, M.; Lavtar, D.; Seljak, B.K.; Golja, P.; Kotnik, K.Z.; Pravst, I.; Mis, N.F. Slovenian national food consumption survey in adolescents, adults and elderly. *EFSA Supporting Publ.* **2019**, *16*, 1729E.
55. Blenkuš, M.G.; Gregorič, M.; Tivadar, B.; Koch, V.; Kostanjevec, S.; Turk, V.F.; Žalar, A.; Lavtar, D.; Kuhar, D.; Rozman, U. *Prehrambene navade odraslih prebivalcev Slovenije z vidika varovanja zdravja;* University of Ljubljana, Faculty of Education: Ljubljana, Slovenija, 2009; p. 183.
56. Koch, V.; Pokorn, D.; Salobir, K. Prehrambene navade odraslih prebivalcev Slovenije z vidika varovanja zdravja. Ph.D. Thesis, University of Ljubljana, Biotechnical Faculty, Ljubljana, Slovenija, 1997.
57. Jacques, P.; Sulsky, S.; Sadowski, J.; Phillips, J.; Rush, D.; Willett, W. Comparison of micronutrient intake measured by a dietary questionnaire and biochemical indicators of micronutrient status. *Am. J. Clin. Nutr.* **1993**, *57*, 182–189. [CrossRef] [PubMed]
58. Koehler, K.M.; Romero, L.; Stauber, P.; Pareo-Tubbeh, S.L.; Liang, H.C.; Baumgartner, R.N.; Garry, P.J.; Allen, R.H.; Stabler, S.P. Vitamin supplementation and other variables affecting serum homocysteine and methylmalonic acid concentrations in elderly men and women. *J. Am. Coll. Nutr.* **1996**, *15*, 364–376. [CrossRef] [PubMed]
59. Hribar, M.; Hristov, H.; Gregorič, M.; Blaznik, U.; Zaletel, K.; Oblak, A.; Osredkar, J.; Kušar, A.; Žmitek, K.; Rogelj, I.; et al. Nutrihealth Study: Seasonal Variation in Vitamin D Status Among the Slovenian Adult and Elderly Population. *Nutrients* **2020**, *12*, 1838. [CrossRef]
60. Lee, P.H.; Macfarlane, D.J.; Lam, T.H.; Stewart, S.M. Validity of the international physical activity questionnaire short form (IPAQ-SF): A systematic review. *Int. J. Behav. Nutr.* **2011**, *8*, 115. [CrossRef]
61. De Onis, M.; Onyango, A.; Borghi, E.; Siyam, A.; Blössner, M.; Lutter, C. Worldwide implementation of the WHO child growth standards. *Public Health Nutr.* **2012**, *15*, 1603–1610. [CrossRef] [PubMed]
62. De Onis, M.; Onyango, A.W.; Borghi, E.; Siyam, A.; Nishida, C.; Siekmann, J. Development of a WHO growth reference for school-aged children and adolescents. *Bull. World Health Organ.* **2007**, *85*, 660–667. [CrossRef]
63. Vede, T. *Izdelava in validacija slikovnega gradiva za določanje vnosa živil;* Design and Validation of Food Picture Book; University of Ljubljana, Biotechnical Faculty: Ljubljana, Slovenia, 2016.
64. Korošec, M.; Golob, T.; Bertoncelj, J.; Stibilj, V.; Seljak, B.K. The Slovenian food composition database. *Food Chem.* **2013**, *140*, 495–499. [CrossRef]
65. Fineli. *Finnish food Composition Database;* Finnish Public Health Institute, Nutrition Unit Helsinki: Helsinki, Finland, 2004.
66. Bodner-Montville, J.; Ahuja, J.K.; Ingwersen, L.A.; Haggerty, E.S.; Enns, C.W.; Perloff, B.P. USDA food and nutrient database for dietary studies: Released on the web. *J. Food Compos. Anal.* **2006**, *19*, S100–S107. [CrossRef]
67. Haubrock, J.; Nöthlings, U.; Volatier, J.L.; Dekkers, A.; Ocké, M.; Harttig, U.; Illner, A.K.; Knüppel, S.; Andersen, L.F.; Boeing, H. Estimating usual food intake distributions by using the multiple source method in the EPIC-Potsdam Calibration Study. *J. Nutr.* **2011**, *141*, 914–920. [CrossRef]

68. Pravst, I.; Lavriša, Ž.; Hribar, M.; Hristov, H.; Kvarantan, N.; Seljak, B.K.; Gregorič, M.; Blaznik, U.; Gregorič, N.; Zaletel, K.; et al. Dietary Intake of Folate and Assessment of the Folate Deficiency Prevalence in Slovenia Using Serum Biomarkers. *Nutrients* **2021**, *13*, 3860. [CrossRef] [PubMed]
69. Bailey, L.B.; Stover, P.J.; McNulty, H.; Fenech, M.F.; Gregory, J.F., III; Mills, J.L.; Pfeiffer, C.M.; Fazili, Z.; Zhang, M.; Ueland, P.M.; et al. Biomarkers of Nutrition for Development—Folate Review. *J. Nutr.* **2015**, *145*, 1636S–1680S. [CrossRef] [PubMed]
70. Black, A.E. Critical evaluation of energy intake using the Goldberg cut-off for energy intake: Basal metabolic rate. A practical guide to its calculation, use and limitations. *Int. J. Obes.* **2000**, *24*, 1119–1130. [CrossRef]
71. Harris, J.A.; Benedict, F.G. A biometric study of human basal metabolism. *Proc. Natl. Acad. Sci. USA* **1918**, *4*, 370. [CrossRef]
72. Roza, A.M.; Shizgal, H.M. The Harris Benedict equation reevaluated: Resting energy requirements and the body cell mass. *Am. J. Clin. Nutr.* **1984**, *40*, 168–182. [CrossRef]
73. Zupanič, N.; Hristov, H.; Gregorič, M.; Blaznik, U.; Delfar, N.; Koroušić Seljak, B.; Ding, E.L.; Fidler Mis, N.; Pravst, I. Total and Free Sugars Consumption in a Slovenian Population Representative Sample. *Nutrients* **2020**, *12*, 1729. [CrossRef]
74. Harttig, U.; Haubrock, J.; Knüppel, S.; Boeing, H. The MSM program: Web-based statistics package for estimating usual dietary intake using the Multiple Source Method. *Eur. J. Clin. Nutr.* **2011**, *65* (Suppl. S1), S87–S91. [CrossRef]
75. Kolenikov, S. Calibrating Survey Data using Iterative Proportional Fitting (Raking). *Stata J.* **2014**, *14*, 22–59. [CrossRef]
76. Dunford, E.; Webster, J.; Metzler, A.B.; Czernichow, S.; Ni Mhurchu, C.; Wolmarans, P.; Snowdon, W.; L'Abbe, M.; Li, N.; Maulik, P.K.; et al. International collaborative project to compare and monitor the nutritional composition of processed foods. *Eur. J. Prev. Cardiol.* **2012**, *19*, 1326–1332. [CrossRef]
77. Rippin, H.L.; Hutchinson, J.; Jewell, J.; Breda, J.J.; Cade, J.E. Adult nutrient intakes from current national dietary surveys of European populations. *Nutrients* **2017**, *9*, 1288. [CrossRef]
78. Vrentzos, G.E.; Papadakis, J.A.; Malliaraki, N.; Zacharis, E.A.; Mazokopakis, E.; Margioris, A.; Ganotakis, E.S.; Kafatos, A. Diet, serum homocysteine levels and ischaemic heart disease in a Mediterranean population. *Br. J. Nutr.* **2004**, *91*, 1013–1019. [CrossRef]
79. Gregoric, M.; Blaznik, U.; Turk, V. *Razlicni Vidiki Prehranjevanja Prebivalcev Slovenije (V Starosti Od 3 Mesecev Do 74 Let)*; NIJZ: Ljubljana, Slovenija, 2020.
80. Hafner, E.; Pravst, I. Evaluation of the Ability of Nutri-Score to Discriminate the Nutritional Quality of Prepacked Foods Using a Sale-Weighting Approach. *Foods* **2021**, *10*, 1689. [CrossRef] [PubMed]
81. Schwartz, M.B.; Vartanian, L.R.; Wharton, C.M.; Brownell, K.D. Examining the Nutritional Quality of Breakfast Cereals Marketed to Children. *J. Am. Diet. Assoc.* **2008**, *108*, 702–705. [CrossRef]
82. Lavriša, Ž.; Pravst, I. Marketing of Foods to Children through Food Packaging Is Almost Exclusively Linked to Unhealthy Foods. *Nutrients* **2019**, *11*, 1128. [CrossRef] [PubMed]
83. Sans, P.; Combris, P. World meat consumption patterns: An overview of the last fifty years (1961–2011). *Meat Sci.* **2015**, *109*, 106–111. [CrossRef] [PubMed]
84. Darmon, N.; Ferguson, E.L.; Briend, A. Impact of a Cost Constraint on Nutritionally Adequate Food Choices for French Women: An Analysis by Linear Programming. *J. Nutr. Educ. Behav.* **2006**, *38*, 82–90. [CrossRef]
85. Villamor, E.; Mora-Plazas, M.; Forero, Y.; Lopez-Arana, S.; Baylin, A. Vitamin B-12 Status Is Associated with Socioeconomic Level and Adherence to an Animal Food Dietary Pattern in Colombian School Children. *J. Nutr.* **2008**, *138*, 1391–1398. [CrossRef] [PubMed]
86. Joubert, L.M.; Manore, M.M. The role of physical activity level and B-vitamin status on blood homocysteine levels. *Med. Sci. Sports Exerc.* **2008**, *40*, 1923–1931. [CrossRef]
87. Zhao, J.; Li, Z.; Hou, C.; Sun, F.; Dong, J.; Chu, X.; Guo, Y. Gender differences in risk factors for high plasma homocysteine levels based on a retrospective checkup cohort using a generalized estimating equation analysis. *Lipids Health Dis.* **2021**, *20*, 31. [CrossRef]
88. Clarke, R.; Refsum, H.; Birks, J.; Evans, J.G.; Johnston, C.; Sherliker, P.; Ueland, P.M.; Schneede, J.; McPartlin, J.; Nexo, E. Screening for vitamin B-12 and folate deficiency in older persons. *Am. J. Clin. Nutr.* **2003**, *77*, 1241–1247. [CrossRef]
89. Bao, F.; Cui, M.; Shi, X.; Ju, S.; Cong, H. Distribution characteristics and influencing factors of homocyteine in an apparently healthy examined population. *BMC Cardiovasc. Disord.* **2021**, *21*, 429. [CrossRef] [PubMed]
90. Allen, L.H. How common is vitamin B-12 deficiency? *Am. J. Clin. Nutr.* **2009**, *89*, 693S–696S. [CrossRef]
91. Wong, C. Vitamin B12 deficiency in the elderly: Is it worth screening. *Hong Kong Med. J.* **2015**, *21*, 155–164. [CrossRef]
92. Adams, N.E.; Bowie, A.J.; Simmance, N.; Murray, M.; Crowe, T.C. Recognition by medical and nursing professionals of malnutrition and risk of malnutrition in elderly hospitalised patients. *Nutr. Diet.* **2008**, *65*, 144–150. [CrossRef]
93. Bruins, M.J.; Van Dael, P.; Eggersdorfer, M. The Role of Nutrients in Reducing the Risk for Noncommunicable Diseases during Aging. *Nutrients* **2019**, *11*, 85. [CrossRef]
94. De Groot, C.; Van Den Broek, T.; Van Staveren, W. Energy intake and micronutrient intake in elderly Europeans: Seeking the minimum requirement in the SENECA study. *Age Ageing* **1999**, *28*, 469–474. [CrossRef] [PubMed]
95. Donini, L.M.; Scardella, P.; Piombo, L.; Neri, B.; Asprino, R.; Proietti, A.; Carcaterra, S.; Cava, E.; Cataldi, S.; Cucinotta, D. Malnutrition in elderly: Social and economic determinants. *J. Nutr. Health Aging* **2013**, *17*, 9–15. [CrossRef] [PubMed]

96. Dullemeijer, C.; Souverein, O.W.; Doets, E.L.; van der Voet, H.; van Wijngaarden, J.P.; de Boer, W.J.; Plada, M.; Dhonukshe-Rutten, R.A.; In't Veld, P.H.; Cavelaars, A.E.; et al. Systematic review with dose-response meta-analyses between vitamin B-12 intake and European Micronutrient Recommendations Aligned's prioritized biomarkers of vitamin B-12 including randomized controlled trials and observational studies in adults and elderly persons. *Am. J. Clin. Nutr.* **2012**, *97*, 390–402. [CrossRef]
97. Hinds, H.E.; Johnson, A.A.; Webb, M.C.; Graham, A.P. Iron, Folate, and Vitamin B12 Status in the Elderly by Gender and Ethnicity. *J. Natl. Med. Assoc.* **2011**, *103*, 870–878. [CrossRef]
98. Kulnik, D.; Elmadfa, I. Assessment of the nutritional situation of elderly nursing home residents in Vienna. *Ann. Nutr. Metab.* **2008**, *52* (Suppl. S1), 51–53. [CrossRef]
99. Novaković, R.; Cavelaars, A.; Geelen, A.; Nikolić, M.; Altaba, I.I.; Viñas, B.R.; Ngo, J.; Golsorkhi, M.; Medina, M.W.; Brzozowska, A.; et al. Review Article Socio-economic determinants of micronutrient intake and status in Europe: A systematic review. *Public Health Nutr.* **2013**, *17*, 1031–1045. [CrossRef] [PubMed]
100. Björkegren, K.; Svärdsudd, K. Elevated serum levels of methylmalonic acid and homocysteine in elderly people. A population-based intervention study. *J. Intern. Med.* **1999**, *246*, 317–324. [CrossRef] [PubMed]
101. Wolters, M.; Hermann, S.; Hahn, A. B vitamin status and concentrations of homocysteine and methylmalonic acid in elderly German women. *Am. J. Clin. Nutr.* **2003**, *78*, 765–772. [CrossRef]
102. Holt, P.R. Intestinal Malabsorption in the Elderly. *J. Dig. Dis.* **2007**, *25*, 144–150. [CrossRef]
103. Lewerin, C.; Jacobsson, S.; Lindstedt, G.; Nilsson-Ehle, H. Serum biomarkers for atrophic gastritis and antibodies against Helicobacter pylori in the elderly: Implications for vitamin B12, folic acid and iron status and response to oral vitamin therapy. *Scand. J. Gastroenterol.* **2008**, *43*, 1050–1056. [CrossRef]
104. Selhub, J.; Bagley, L.C.; Miller, J.; Rosenberg, I.H. B vitamins, homocysteine, and neurocognitive function in the elderly. *Am. J. Clin. Nutr.* **2000**, *71*, 614S–620S. [CrossRef]
105. Joo, Y.-E.; Park, H.-K.; Myung, D.-S.; Baik, G.-H.; Shin, J.-E.; Seo, G.-S.; Kim, G.H.; Kim, H.U.; Kim, H.Y.; Cho, S.-I.; et al. Prevalence and risk factors of atrophic gastritis and intestinal metaplasia: A nationwide multicenter prospective study in Korea. *Gut Liver* **2013**, *7*, 303–310. [CrossRef] [PubMed]
106. Raza, M.; Bhatt, H. *Atrophic Gastritis*; StatPearls Publishing: Treasure Island, FL, USA, 2020.
107. Weck, M.N.; Brenner, H. Prevalence of chronic atrophic gastritis in different parts of the world. *Cancer Epidemiol. Biomark. Prev.* **2006**, *15*, 1083–1094. [CrossRef]
108. Russell, R.M.; Rasmussen, H.; Lichtenstein, A.H. Modified Food Guide Pyramid for People over Seventy Years of Age. *J Nutr* **1999**, *129*, 751–753. [CrossRef]
109. Selhub, J.; Paul, L. Folic acid fortification: Why not vitamin B12 also? *BioFactors* **2011**, *37*, 269–271. [CrossRef]
110. Allen, L.H.; Rosenberg, I.H.; Oakley, G.P.; Omenn, G.S. Considering the case for vitamin B12 fortification of flour. *Food Nutr. Bull.* **2010**, *31*, S36–S46. [CrossRef] [PubMed]
111. Dharmarajan, T.S.; Norkus, E.P. Approaches to vitamin B12 deficiency. *Postgrad. Med.* **2001**, *110*, 99–105. [CrossRef] [PubMed]
112. Green, R. Vitamin B(12) deficiency from the perspective of a practicing hematologist. *Blood* **2017**, *129*, 2603–2611. [CrossRef] [PubMed]

nutrients

MDPI

Article

The Intake of Phosphorus and Nitrites through Meat Products: A Health Risk Assessment of Children Aged 1 to 9 Years Old in Serbia

Jelena Milešević [1], Danijela Vranić [2], Mirjana Gurinović [1], Vladimir Koričanac [2], Branka Borović [2], Milica Zeković [1], Ivana Šarac [1], Dragan R. Milićević [2,*] and Maria Glibetić [1]

1 Centre of Research Excellence in Nutrition and Metabolism, Institute for Medical Research, National Institute of Republic of Serbia, University of Belgrade, Tadeusa Koscuska, 111000 Belgrade, Serbia; jelena.milesevic@gmail.com (J.M.); mirjana.gurinovic@gmail.com (M.G.); zekovicmilica@gmail.com (M.Z.); ivanasarac@yahoo.com (I.Š.); mglibetic@gmail.com (M.G.)

2 Institute of Meat Hygiene and Technology, Kaćanskog 13, 11040 Belgrade, Serbia; danijela.vranic@inmes.rs (D.V.); vladimir.koricanac@inmes.rs (V.K.); branka.borovic@inmes.rs (B.B.)

* Correspondence: dragan.milicevic@inmes.rs

Abstract: This study provides the data on dietary exposure of Serbian children to nitrites and phosphorus from meat products by combining individual consumption data with available analytical data of meat products. A total of 2603 and 1900 commercially available meat products were categorized into seven groups and analysed for nitrite and phosphorous content. The highest mean levels of nitrite content, expressed as $NaNO_2$, were found in finely minced cooked sausages (40.25 ± 20.37 mg/kg), followed by canned meat (34.95 ± 22.12 mg/kg) and coarsely minced cooked sausages (32.85 ± 23.25 mg/kg). The EDI (estimated daily intake) of nitrites from meat products, calculated from a National Food Consumption Survey in 576 children aged 1–9 years, indicated that the Serbian children population exceeded the nitrite ADI (acceptable daily intake) proposed by EFSA (European Food Safety Authority) in 6.4% of children, with a higher proportion in 1–3-year-old participants. The mean phosphorus concentration varied from 2.71 ± 1.05 g/kg to 6.12 ± 1.33 g/kg in liver sausage and pate and smoked meat products, respectively. The EDI of phosphorus from meat products was far below the ADI proposed by EFSA, indicating that the use of phosphorus additives in Serbian meat products is generally in line with legislation.

Keywords: nitrite; phosphorus; food additives; dietary intake; children; Serbia

Citation: Milešević, J.; Vranić, D.; Gurinović, M.; Koričanac, V.; Borović, B.; Zeković, M.; Šarac, I.; Milićević, D.R.; Glibetić, M. The Intake of Phosphorus and Nitrites through Meat Products: A Health Risk Assessment of Children Aged 1 to 9 Years Old in Serbia. *Nutrients* **2022**, *14*, 242. https://doi.org/10.3390/nu14020242

Academic Editor: Roberto Iacone

Received: 14 December 2021
Accepted: 30 December 2021
Published: 6 January 2022

1. Introduction

Results of the latest Global Burden of Diseases, Injuries, and Risk Factors Study (GBD) from 2019 have shown that the diet is still one of the most important risk factors for attributable mortality and disability-adjusted life years (DALYs) [1]. Meat and processed meat products (ham, sausages, bacon, frankfurters, salami, etc.) are often perceived as unhealthy by consumers due to their intakes having been positively associated with the risk of several major chronic diseases. According to several reports of the World Cancer Research Fund (WCRF) [2], there is convincing evidence that the consumption of red and particularly processed meat is associated with cancer risk. Additionally, the International Agency on Research on Cancer (IARC) has classified processed meat as carcinogenic to humans (Group 1 carcinogen), based on sufficient evidence in humans that the consumption of processed meat causes colorectal cancer (CRC) and that red meat consumption is probably carcinogenic to humans (Group 2A) [3]. Red meat and processed meats contain multiple substances that are potentially carcinogenic, including nitrates, nitrites, polycyclic aromatic hydrocarbons (PAHs), and heterocyclic amines (HCAs), resulting from cooking or processing.

Food additives are intentionally added to food products for a technological purpose in the manufacture, processing, preparation, treatment, packaging, transport, or storage of food. Nitrates and nitrites, in the form of sodium and potassium salts, are widely used as preservatives in meat production (E249–E252). From the technological point of view, the main reason for adding nitrites and/or nitrates in the processing of meat products is to improve the quality (stabilize red meat colour and texture; may also contribute to the product flavour characterization) [4–6] and durability, due to retardation of the oxidative rancidity [7]; the additions and may also contribute to the safety of products by inhibiting the growth and reproduction of bacteria *Clostridium botulinum, Clostridium perfringens, Staphylococcus aureus*, and *Bacillus cereus* [7,8]. They are also used in other processed foods, such as cheese and fish, for preservation purposes [9]. Besides this, nitrites and nitrates are naturally present in vegetables [10], and they are in water as residues of contamination of ground water and surface water as a result of manuring and fertilisation practices.

Although nitrites and nitrates are not carcinogenic, due to possible formation of carcinogenic nitrosamines under certain conditions, a diet high in nitrates and nitrites is associated with increased risk of colorectal cancers (CRC) [2,3,11]. Ingested nitrate is reduced to nitrite by the bacterial flora in the mouth and digestive tract. Further, nitrite reacts with amines, amides, or amino acids and other nitrosation precursors in the gastrointestinal tract to form carcinogenic *N*-nitroso compounds (NOCs) [12,13]. Carcinogenic *N*-nitroso compounds can be found in many processed types of meat and can be endogenously formed after ingesting red meat in the human intestines by the bacterial flora. Because endogenous nitrosation is estimated to account for 45–75% of total NOC exposure [14], dietary intake of nitrate and nitrite, precursors for endogenous nitrosation, may be important colorectal cancer risk factors. There is strong evidence of an association between red and processed meat consumption and the risk of colorectal cancer [15,16]. In addition, processed meat products, and especially cured meat, already contain preformed NOCs [17]. Therefore, they are the major source of nitrites and N-nitrosamines in human dietary intake. Another aspect of CRC is the fact that red meat is also the source of heme iron, which participates in the abovementioned process of endogenous N-nitrosation in the intestine. Moreover, that nitrite could oxidize haemoglobin to methaemoglobin, which cannot bind and transport oxygen to tissues, leading to acute or chronic toxicity such as methemoglobinemia [18,19]. Nitrate also competitively inhibits iodide uptake by the thyroid [20,21], possibly affecting thyroid hormone production and potentially resulting in thyroid tumour promotion. Thus, the World Cancer Research Fund and the American Institute for Cancer Research (WCRF/AICR) recommend consuming <500 g/week of red meat and <50 g/day of processed meat [16].

The current Serbian legislation has restricted the maximum amount of nitrate or nitrite that may be added in processed meat expressed as $NaNO_2$ or $NaNO_3$ to 100 and 150 mg/kg, depending on the type of product [22,23], whereas regulations in Europe state that the maximum amount of nitrite and nitrates that may be added to the processed meat during manufacturing should be from 50 to 180 mg/kg, with a number of exemptions [19]. The Joint Expert Committee on Food Additives (JECFA) established an acceptable daily intake (ADI) for nitrite of 0–0.07 mg per kg body weight per day (mg/kg bw/day), expressed as nitrite ion based on a no-effect level (NOEL) of 6.7 mg/kg bw/day for effects on the heart and lungs in a 2-year study in rats, and a safety factor of 100 [19,24,25].

Among the food additives, food-grade phosphates are widely used as an additive compound in various products. Phosphorus additives (E338–341, E343, E450–452) are increasingly being used in processed and fast foods, especially in the meat industry, cheeses, baked products, and beverages for several technological purposes. They increase water holding capacity (WHC), preserve moisture or colour, emulsify ingredients, and enhance flavour, as well as stabilize foods. Despite its technological benefits, it has been estimated that 50% of daily phosphorus (P) intake in the Western world is from "hidden phosphorus" as a food additive [26]. Furthermore, phosphorus in food additives is rapidly and almost completely absorbed, whereas a natural constituent of protein-bound phosphorus is more slowly and less efficiently absorbed (60%) [27]. An association between excessive phosphate intake, high serum phosphate levels (hyperphosphatemia), and cardiovascular morbidity and mortality in patients with chronic kidney disease and bone health complications has long been known [28]. Due to the increased consumption of processed foods, high phosphorus intake from additives should be taken into account as a potential public health concern. Thus, phosphates are included in the list of food additives that must be reduced in meat preparations. These additives were recently evaluated by the Scientific Committee for Food [28], which derived a group acceptable daily intake (ADI) for phosphates expressed as phosphorus of 40 mg/kg bw/day. The Panel concluded that this ADI is safe for healthy adults because it is below the doses at which clinically relevant adverse effects were reported in short-term and long-term studies in humans. In addition, European Directives on food additives [29] require that Member States monitor intakes to ensure that consumers do not have an excessive intake of a given food additive, which could lead to health hazards. The Serbian standard maximum limit for total phosphates expressed as P_2O_5 in meat products is <8 g/kg [22] or ≤5 g/kg of added phosphorus [23].

Chronic noncommunicable diseases such as cardiovascular diseases and cancer are a national public health problem and are the leading causes of death in Serbia (47.3% and 17.8%, respectively). In Serbia, they constitute the major contributor to the burden of disease in terms of DALYs or mortality [30]. Meat products such as ham, sausages, bacon, frankfurters, salami, etc., are widely consumed by all groups of the population in Serbia, at home or in fast foods restaurant. Although comprehensive national surveys of assessment of total dietary phosphorus intake and food additives such as nitrates and nitrites in Serbia are scarce, studying the nutritional status as well as a lifestyle of the population, particularly children, is fundamental to design national guidelines and public health policies.

Therefore, the objective of this study was to determine nitrites and phosphorus content in processed meat products. Based on the analysis results, dietary exposure of the Serbian children population to nitrites and phosphorus was then estimated and discussed.

2. Materials and Methods

2.1. Reagents

All chemicals and reagents used were of analytical reagent grade from Merck Co. (Darmstadt, Germany) unless otherwise stated.

2.2. Meat Products and Sample Preparation

In the present study, a total of 2603 meat product samples, produced by the Serbian meat industry or imported (241 bacon, 362 canned meat, 353 coarsely minced cooked sausages, 683 dry fermented sausages, 580 finely minced cooked sausages, 46 liver sausage and pate, and 338 smoked meat products), were obtained from different regions from the Serbian retail market during 2018–2021 and were analysed for nitrite content. In most of the meat products, all parameters of quality defined by the legislation were examined, and in a smaller number, analyses were carried out as per the client's request.

In the same period of investigation, a total of 1900 meat product samples, categorized into six groups including bacon, canned meat, coarsely minced cooked sausages, finely minced cooked sausages, liver sausage and pate, and smoked meat products were analysed for phosphorous content.

All meat products were kept at refrigeration temperature and analysed within 48 h. If the analyses were not conducted on the same day, the samples were stored in a refrigerator at 4 °C until required for testing.

The analysed samples were thawed and blended in a commercial kitchen blender unit (Homogenizator Blixer 2, Robot Coupe, Vincennes, France (2.9 L) 700 w, 3000 rpm). For each sample, two composite samples were prepared. All samples were then analysed in duplicate.

2.3. Determination of Nitrite Content

The content of nitrite in examined meat products was determined according to the standard ISO procedure [31]. A representative sample amount (~10 g) was then measured into 300 mL flask using an analytical balance (Mettler, AE 200, USA), followed by the addition of a solution of hydrous sodium borate, $Na_2B_4O_7.10H_2O$ (50 g/L) and 100 mL deionised water at 70.0 ± 0.2 °C. Residual nitrite extraction was achieved by keeping the samples in a hot water bath, at the temperature of boiling, for 15 min, and every 5 min, flasks were shaken vigorously. After cooling, 2 mL of each Carrez solution (Carezz reagent I and Carezz reagent II) were added and mixed thoroughly. Samples were then diluted to 200 mL with deionized water. Samples were filtered through quantitative cellulose filters (pore size <5 µm). Colour generation was achieved by transferring an aliquot of the filtrate (25 mL) to a 100 mL volumetric flask and adding 10 mL of the sulphanilamide solution and then 6 mL conc. HCl. Flasks were stored in the dark for 5 min. Subsequently, 2 mL solution of N-naftil-1-ethylenediamine-chloride (0.25 g/250 mL) was added to each flask and moved to the dark for 3 min. Thereafter, samples were diluted to 100 mL. Absorbance was measured at 538 nm using a spectrophotometer (UV/VIS Spectrophotometer, Jenway 6405). A procedural blank was run with every batch of samples.

Calibration curves were generated using the concentration levels ranging from 2.5 to 10 $NaNO_3$ µg mL^{-1}, Y = 0.0669X + 0.024: R^2 = 0.999. A recovery study of the analytical procedure was carried out by spiking several already analysed samples with standard solutions, and recovery rates were found to be between 87% and 94%. The nitrite content is expressed as $NaNO_2$ (mg·kg^{-1}), following: c × 2000/ m × V, where c is the concentration of $NaNO_2$ (µg/mL) from the calibration curve, m is the mass of sample (g) for analysing, and V is a volume of an aliquot of the filtrate used for spectrometric determination.

2.4. Phosphorus Measurement

The total phosphorus content, expressed as P_2O_5 (g/kg), in examined meat products was determined according to the standard ISO procedure [32]. In brief, ~5 g portion of samples (measured using an analytical balance (Mettler, AE 200, USA) was ashed at the maximum temperature of 500 °C in a muffle furnace (LE 14/11/R7, Nabertherm). On completion of the digestion, the white ash was dissolved by heating with dilute nitric acid (1 + 1, v/v) and quantitatively transferred in 100 mL flask. Then, made up by the addition of deionized water, and after mixing, the solution was then filtered, and the first 5 to 10 mL were discarded.

Aliquots (20 mL) of treated solution were pipetted into 100 mL volumetric flasks and mixed thoroughly with 30 mL ammonium heptamolybdate solution 50g L^{-1}. The resulting solution was then diluted to the volume with deionized water. After 15 min at room temperature, the absorbance was read against a reagent blank at 430 ± 2 nm using a UV-visible spectrophotometer (UV/VIS Spectrophotometer, Jenway 6405). The standard curve was determined under the same conditions as those for the samples using potassium dihydrogen phosphate as a standard (10–60 P_2O_5 µg/mL; Y = 0.0187X − 0.0096: R^2 = 0.9999). A recovery study of the analytical procedure was carried out by spiking several already analysed samples with standard solutions, and recovery rates were found to be between 89% and 95%. The total phosphorus content, expressed as P_2O_5 (g/kg), following: c/20 m, where c is the concentration of P_2O_5 (µg/mL) from the calibration curve, and m is the mass of the sample (g) for analysis.

2.5. National Food Consumption Survey on Toddlers and Children

The National Food Consumption Survey on toddlers and children, in compliance with the EFSA EU Menu methodology [33,34], was conducted between 2017 and 2021 and included a total of 576 participants, comprising 290 toddlers aged from 1 to 2.9 years old and 286 children aged from 3 to 9 years old. Data collected included: general questionnaire, body weight and height measurements, age-appropriate Food Propensity Questionnaire (FPQ), and twice repeated 24 h food record. The consumed portion sizes were estimated based on natural units, household measures, packaging information, and a validated national Food Atlas for Portion Size Estimation [35]. Frequency of consumption of processed meat products was explored by FPQ, which categorized consumption in seven frequency groups (never, less than once a month, 1–3/month, once per week, 2–3/week, 4–5/week, 6–7/week). Data were analysed using a nutritional software tool DIET ASSESS & PLAN (DAP) [36] and the Serbian Food Composition Database, developed in compliance with EuroFIR standards [37]. Weight measurements were performed without shoes and jackets using a digital balance where data were recorded to the nearest 0.1 kg or from the most recent paediatric report not older than 3 months. Minimum sample size (n = 130) was determined by EFSA EU Menu methodology [38]. The study sample distribution according to age and gender and body weight expressed as a mean value with standard deviations is shown in Table 1.

Table 1. Characteristics of the study sample.

Age Group	*N*		Age (Year) Median (ICR)		Body Weight (kg) Mean ± SD	
	Male	Female	Male	Female	Male	Female
Toddlers, 1–3 years	149	141	2.0 (1.5–2.6)	2.0 (1.4–2.5)	13.92 ± 2.81	13.92 ± 2.76
Children, 3–9 years	147	139	6.1 (4.4–7.9)	6.2 (5.0–7.8)	23.99 ± 6.98	23.89 ± 6.63
Total sample	576					

N—number of participants; ICR—interquartile range

2.6. Exposure Assessment and Risk Characterization

Exposure was calculated for all the processed meat categories and both study groups according to their gender and age by the deterministic approach involving the average probable daily intake (APDI) method [39]. Nonbrand loyal scenario as the most relevant exposure scenario for the safety evaluation of phosphates was used [28]. $NaNO_2$ concentrations were first converted to nitrite (66.65% of $NaNO_2$), then multiplied by the processed meat consumption rate (kg/day) and divided by the average body weight. Results are shown in Section 3.

The same pattern was used for phosphorus content where P content was 43.64% of P_2O_5. Results are also shown in Section 3. Within the general framework of chemical risk assessment, a difficult step in dietary exposure evaluation is handling concentration data reported to be below the limit of detection (LOD). These data are known as nondetects, and the resulting occurrence distribution is left-censored. The left-censored data (data below LOD and LOQ) were processed by applying the substitution method of EFSA [40]. According to this guidance, for dietary exposure assessments, two exposure scenarios were considered.

The mean values and 95th (P95) percentiles of phosphorus and nitrite content of the total dataset and processed meat categories were used to estimate daily intakes. The intake was calculated as the average daily intake for each studied group and expressed in mg/kg bw/day. The 95th (P95) percentiles of phosphorus and nitrite content were used to calculate the highest exposure scenario [41].

To evaluate the adequacy of intakes, the calculated estimated daily intake (EDI) of nitrites and total phosphorus were compared with the ADIs proposed by the EFSA [19,28].

2.7. Statistical Analysis

For statistical evaluation on data, Minitab 17 Ink statistical software was used (Minitab Ink., Coventry, UK). The results are presented as mean and standard deviation of the mean (SD).

3. Results and Discussion

3.1. Processed Meat Products Consumption among Children 1–9 Years Old

In Table 2 is given the average daily consumption of specific meat products according to 24 h food record data from the national survey.

Table 2. The average intake of processed meat products by children population (g/day).

Age Group		Bacon	Canned Meat	CMCS	DFS	FMCS	LSP	SMP	Total
					Mean ± SD (g/day)				
Toddlers,	M	2.17 ± 5.55	2.37 ± 6.40	0.92 ± 6.79	1.06 ± 3.90	6.49 ± 16.41	0.86 ± 3.42	0.40 ± 2.38	14.28 ± 44.86
1–3 years	F	2.87 ± 12.71	2.17 ± 6.08	0.50 ± 3.08	1.13 ± 4.25	5.68 ± 14.41	1.41 ± 4.20	0.75 ± 3.03	14.51 ± 47.76
Children,	M	1.24 ± 5.37	8.01 ± 13.74	0.05 ± 0.64	1.75 ± 6.18	11.11 ± 19.71	3.22 ± 7.51	0.95 ± 3.89	26.33 ± 57.06
3–9 years	F	2.93 ± 10.00	8.32 ± 11.51	-	2.65 ± 11.53	7.66 ± 18.74	3.21 ± 7.78	1.54 ± 7.34	26.30 ± 66.89
Average		2.29 ± 8.89	5.20 ± 10.39	0.37 ± 3.80	1.64 ± 7.10	7.76 ± 17.53	2.16 ± 6.11	0.90 ± 4.54	20.31 ± 58.35

M—male, F—female, CMCS—coarsely minced cooked sausages, DFS—dry fermented sausages, FMCS—finely minced cooked sausages, LSP—liver sausage and pate, SMP—smoked meat products.

The average cumulative daily consumption of processed meat products per child was 20.31 ± 58.35 g/day, but it varied between the age groups: in younger children it was 14.39 ± 46.31 g/day, while in the older group it was 26.32 ± 61.97 g/day, almost doubled, with negligible difference between males and females (Table 2).

According to data from 24 h food records, 386 (i.e., 67.01%) participants consumed meat products in this survey. The percentage of consumers was higher in the older children group (3–9 years): 76.87% for males and 79.86% for females, compared with the younger children group: 54.36% for males and 57.45% for females.

Additional analysis determined frequency of consumption of meat products within the children's groups. For this analysis, the FPQ records were used, which documented a frequent consumption of processed meat products (i.e., 2–3/week and more) in 62.67% of children, in total (Figure 1). However, the frequency was higher in the older group, where 73.47% and 73.38% of males and females, respectively, were frequent consumers, while in the younger group, 51.01% and 53.19% of males and females, respectively, were frequent consumers. These data are in accordance with the results obtained from 24 h food records.

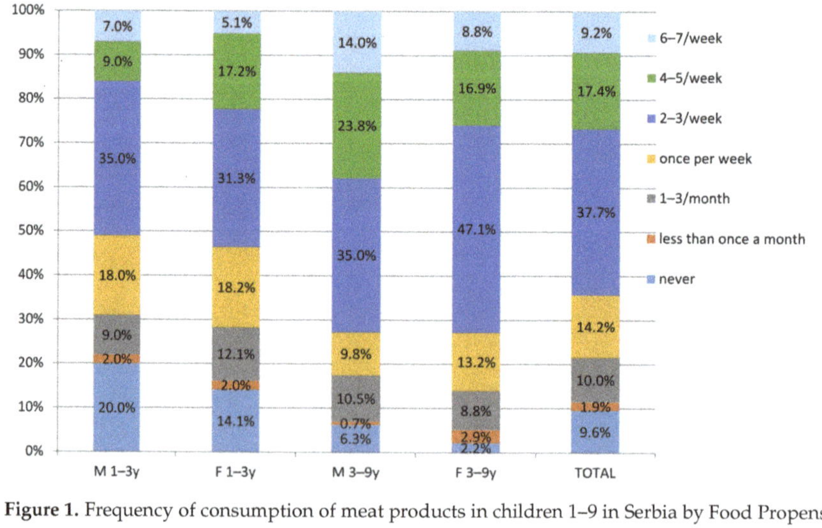

Figure 1. Frequency of consumption of meat products in children 1–9 in Serbia by Food Propensity Questionnaire (FPQ).

In all child groups, the highest consumption was noted for finely minced cooked sausages and canned meat, followed by bacon and liver sausages and pate (Table 2).

3.2. Nitrite Content in Processed Meat Products

The nitrite content for the 2603 meat products used in the exposure assessment is presented in Table 3, while the frequency distribution categorized by product type is given in Figure 2. The overall nitrite detection rate was 85.97%, with nitrite detected in 2238 samples of the seven categories of meat products. Among the seven food categories, finely minced cooked sausages had the highest detection rate (99.6%) and the highest overall mean concentration of nitrite at 40.25 ± 20.37 mg/kg, followed by canned meat, coarsely minced cooked sausages, smoked meat products, bacon, liver sausage and pate, and dry fermented sausages, with overall mean concentrations of 34.95, 32.85, 23.09, 15.26, 12.62, and 1.50 mg/kg, respectively. In terms of nitrite content of these products, heterogeneous results were obtained (Figure 2 and Table 3). A high standard deviation, almost the same as the mean value (23.31 ± 23.75 mg/kg), was caused by a high nitrite content distribution, which ranged from 0.05–180.25 mg/kg. Smoked meat product samples had the highest residual nitrite content (180.25 mg/kg) within the studied meat samples, followed by coarsely minced cooked sausages (113.51 mg/kg), finely minced cooked sausages (101.19 mg/kg), bacon (100.38 mg/kg), and canned meat 93.45 (mg/kg). The lowest residual nitrite content was found in dry fermented sausages (24.88 mg/kg). Following the Serbian and EU regulations [29], the compliance levels were estimated at 99.9%. One sample exceeded maximum limits for the addition of nitrite (in the category of smoked meat products, 180 mg/kg). The results obtained in the present study are very similar to the information available in the literature [42,43], where it was reported a 91% detection rate for nitrite in processed meats. Concerning overall mean residual nitrite content (23.31 ± 23.75 mg/kg), this level is comparable to the median level of 27 mg/kg nitrite in meat products reported in a German food monitoring study [44].

Table 3. Mean levels and ranges of nitrite content expressed as $NaNO_2$ in examined processed meat products (mg/kg).

Meat Product	*N*	*n* (%)	Mean ± SD (mg/kg)	Q1 (mg/kg)	Q2 (mg/kg)	Q3 (mg/kg)	P95 (mg/kg)	Range (mg/kg)	MPL (mg/kg) [23]
Bacon	241	228 (94.6)	15.26 ± 18.74	2.74	6.78	21.00	53.66	0.33–100.38	
Canned meat	362	348 (96)	34.95 ± 22.12	18.00	33.77	52.28	71.1	0.08–93.45	
Coarsely minced cooked sausages	353	345 (97.7)	32.85 ± 23.25	11.31	31.31	51.01	71.14	0.05–113.51	
Dry fermented sausages	683	374 (54.7)	1.50 ± 2.48	0.00	0.41	1.95	6.15	0.05–24.88	150
Finely minced cooked sausages	580	578 (99.6)	40.25 ± 20.37	26.75	40.72	54.67	71.75	0.09–101.19	
Liver sausage and pate	46	42 (91.3)	12.62 ± 9.01	5.10	11.07	19.95	26.12	0.84–30.39	
Smoked meat products	338	322 (95.2)	23.09 ± 22.80	4.67	18.51	35.10	61.70	0.10–180.25	
Average	2603	2238 (85.97)	23.31 ± 23.75	1.72	15.97	41.08	67.10	0.05–180.25	

N—total number of analysed samples; *n*—number of samples that contained nitrite (%); 1st quartile (Q1), 25% of the data are less than or equal to this value; 2nd quartile (Q2), the median 50% of the data are less than or equal to this value; 3rd quartile (Q3), 75% of the data are less than or equal to this value; P95-95th percentile. MPL—maximum amount of nitrites that may be added during manufacturing [23]; LOQ—limit of quantification = 0.03 mg/kg.

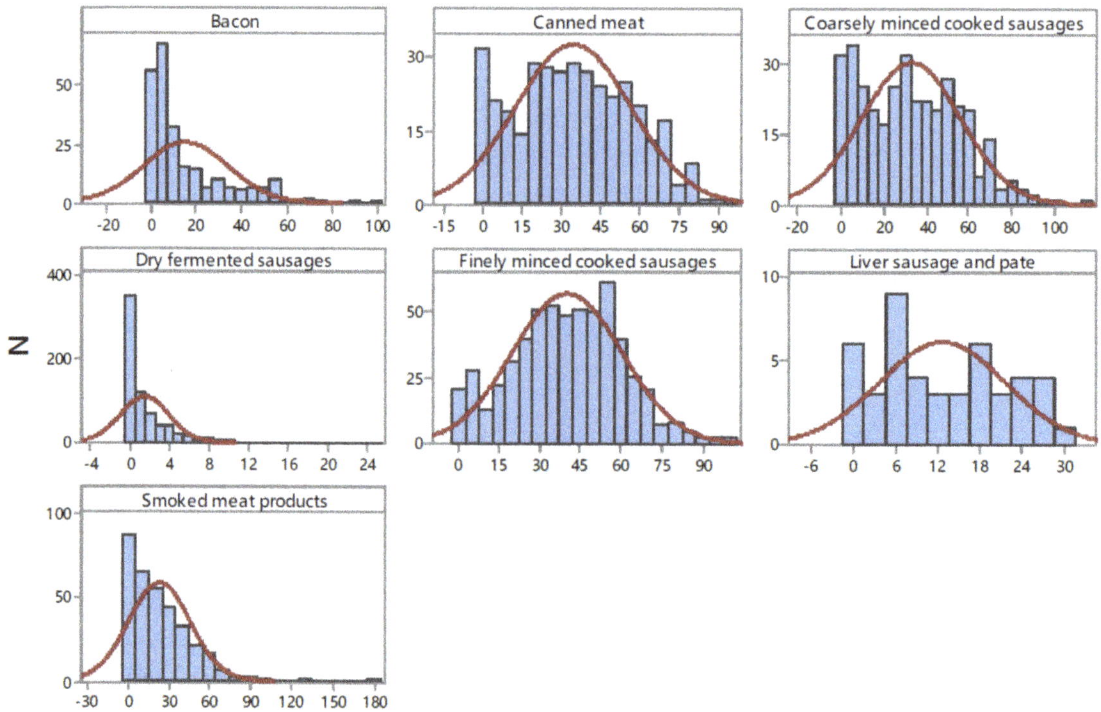

Figure 2. Distribution of $NaNO_2$ content in examined processed meat products (mg kg^{-1}). N—number of samples.

The levels of residual nitrite and nitrate in processed meat products are variable depending on the time and temperature used during processing and storing, the initial addition of nitrite and nitrate, the composition of the meat, pH, addition of antioxidant components such as ascorbate, and the presence of microorganisms [4,45]. Honikel [4] estimated that the decline in nitrite levels due to heating during manufacturing is about 35% of the added level, and thereafter, there is a continuing decrease in nitrite levels during storage. The results obtained show distributions of nitrites differed in the four product types (Figure 2), which is related to the stability of the nitrite content in the different meat products, i.e., is the consequence of their differing preparation and processing methods [46]. Furthermore, different cured meat products may require a different ratio of nitrite and nitrate as preservatives. In fermented sausages, the content of nitrite decreased during the ripening of sausages as a result of the process that takes place in the sausage, i.e., reduction of nitrite content is significant where the main process is nitrite conversion into nitrates in the weak acid environment. In fermented sausages, the presence of nitrite becomes latent because the process is reversible and nitrates, under certain conditions, can revert into nitrites [47].

3.3. Estimated Dietary Intake of Nitrites from Processed Meat Products

Dietary intake of nitrites from processed meat product in the Serbian children population is estimated according to average daily intake of meat products in this population and the content of nitrites salts, i.e., nitrite ion (NO_2), in these products. Daily intakes of nitrites were expressed as EDI in mg/kg bw/day and are presented in Table 4.

The total cumulative EDI for nitrite ion from all processed meat products for the whole population was 0.021 mg/kg bw/day, which contributed 29.52% to the ADI (Table 4). When the 95th percentile (P95th) of nitrite ion content in meat products was used to calculate

EDI (representing the highest risk exposure scenario [41]), the total EDI for the whole population was 0.042 mg/kg bw/day, contributing to almost two-thirds (59.96%) of ADI. The total cumulative EDI for different child groups was similar, ranging from 0.019 mg/kg bw/day (in females 1–3 years old) to 0.023 mg/kg bw/day (in males 3–9 years old). In terms of contribution to the ADI, the contribution of nitrites from meat products varied from 27.46% of ADI (in females 1–3 years old) to 32.27% of ADI (in males 3–9 years old), and when the highest exposure scenario was considered (EDI P95th), it varied from 57.54% to 63.12% of ADI (Table 4). In general, there was no substantial difference in total EDI between different age groups, in spite of much lower consumption of processed meat products in the younger group, which can be explained by the lower body weight in the younger group and expression of EDI per kilogram body weight.

Table 5 presents the analysis of EDI in the consumer group (i.e., only consumers were included). As we already mentioned in Section 3.1, the percentage of consumers was higher in the older children group than in the younger children group (e.g., 76.87% vs. 55.90%). Moreover, the daily intake of processed meat products was almost doubled in the older children group (Table 2). Nevertheless, the total EDI for nitrite from processed meat products per consumer was higher (almost doubled) in the younger children group: 0.037 mg/kg bw/day for both males and females, compared with the older children group: 0.021 mg/kg bw/day for both males and females (Table 5). Similarly, contribution of nitrites from processed meat products to ADI was higher in the younger group: 53.13%, when mean EDI was considered (and 107.60%, when highest exposure scenario was considered), compared with the older group: about 30.48%, when mean EDI was considered (and 62.93%, when highest exposure scenario was considered). We did not observe significant differences between males and females in each age group (Tables 4 and 5).

On average, in the whole child population (with both consumers and nonconsumers included), the percentage of those who exceeded ADI was 6.25% (and 17.01%, when the highest exposure scenario was considered). When only consumers were regarded, the percentage of those who exceeded ADI was 9.33% (i.e., 25.39%, when the highest exposure scenario was considered). Again, the proportion of children who exceeded ADI was higher in the younger group: 12.35% and 38.79%, for mean and highest exposure scenario, respectively, compared with 7.14% and 15.63%, in the older group. The reason for this discrepancy between the processed meat products intake and EDI per consumer in different age groups is that EDI is expressed per kilogram body weight (and body weight was much lower in the younger group) and that the proportion of consumers in younger group was lower. In all child groups, the highest contribution to total EDI derived from finely minced cooked sausages and canned meat (Table 4 and Figure 3), which is related to both the highest content of nitrites in those products (Table 3) and the highest consumption of those products by the children (Table 2).

We compared our results with the findings from other countries and surveys. Similar values of EDI in child populations were observed in many other studies. In the latest EFSA report [19] in which were given results from 15 EU countries (including Austria, Belgium, Bulgaria, Czech Republic, Denmark, Finland, France, Germany, Greece, Italy, Latvia, Netherlands, Spain, Sweden, and UK) involving 13,637 children, the mean EDI was 0.015 mg/kg bw/day in the 1–3 years old group, and 0.016 mg/kg bw/day in the 3–9 age group. The highest exposure for the 1–3-year-old group was noted in Denmark (0.027 mg/kg bw/day), and for the 3–9-year-old group was in Czech Republic (0.023 mg/kg bw/day). (For details on all included countries and surveys, please refer to the EFSA report) [19]. In a very recent Estonian study on children aged 1-10 years, mean nitrite intakes were 0.015 and 0.016 mg/kg bw/day and reached 21.9% and 22.9% of the ADI among children aged 1–3 years and 3–10 years, respectively. ADI was exceeded in 3.1% of children, predominantly in the younger age group [48]. In a study from Sudan [49], much higher intakes were noted (ranging from 0.026–0.128 mg/kg bw/day), despite much lower legal limits (100 mg/kg). In contrast, in another recent study in Korea, much lower nitrite intake was noted, which reached only 0.8% of the ADI for all subjects and 2.8% for the consumer group, which can be attributed to different dietary habits and significantly lower legal limits for nitrite content in processed meat products (70 mg/kg) [50].

In our study, 6.25% of all participants and 9.33% of consumers exceeded ADI. In the population of toddlers (1–3 years old), these proportions were even higher—12.35% (and reaching 40% at the highest exposure scenario). These values are alarmingly high, considering that nitrites in processed meat products are just one of numerous other natural and artificial sources of nitrites in the diet, including both food and water supplies. In the mentioned EFSA report [19], only 17% of total daily nitrite intake (i.e., EDI) is derived from food additives in processed meat products in the general population. Considering children, in the population of 1–3 years old, 8.9–27.0% of daily nitrite intake is derived from processed meat products, while in the population of 3–9 years old, 3.2–25.8% of daily nitrite intake is derived from these products [19]. The highest contribution of processed meat products for toddlers was in Denmark and, for other children, in Germany and Austria. A more recent Dutch study suggests that nitrates and nitrites in processed meat product account only for 8–9% of total daily intake of these compounds in the general population [10]. The majority of nitrates and nitrites in the diet derive from composite food, fruit and fruit products [19], vegetables and vegetable products, particularly leafy vegetables such as spinach and rucola [10], poultry, livestock meat, and cheese (and, in toddlers, additionally from food for infants and toddlers). Therefore, the EFSA Panel noted that the ADI would be exceeded for EU toddlers and children, both at the mean and at the highest exposure level, if all sources of dietary nitrite exposure (food additives, natural presence, and contamination) were considered [19]. Similarly, the actual EDI for nitrites in the Serbian children population might be much higher than proposed ADI when the whole diet is considered and should be the objective of future research.

As mentioned above, the finely minced meat products and canned meat contributed the most to total daily nitrite intake from processed meat products because of both the highest content of nitrites and the highest consumption among all analysed meat products categories (Table 4). These results are in accordance with the EFSA [19], where the most important contributors to the total cumulative exposure in all age groups were sausages and preserved meat, while pastes, pates, and terrines and meat specialities contributed less. In a recent US study [41], cured, cooked sausages, and whole-muscle brine-cured products were also the most important contributors to processed meat nitrite intake among children. In addition, in a recent Italian study, cooked ham and wurstel contributed the most to total EDI from processed meat products [19]. In the study from Korea, ham, sausage, and bacon were found to contribute the most to total daily nitrite intake from processed meat products [50].

Our data indicate need for changes in the consumption habits and legislation to decrease the nitrite exposure from processed meat products, particularly in the population of younger children. For example, the recent Estonian study showed that the exposure of children to nitrites declined over last 10 years as a result of changes in food preferences and decreased usage of nitrite in cured meat products by meat industries [48]. In line with that, the Denmark government in 2010 reduced maximal permitted levels of nitrites in processed meat food products to 60 mg/kg, leading to a much lower exposure (almost halved) than in other EU countries [19].

High exposure to nitrites in the long-term can be connected with several health risks. Particularly young children population may be on significant risk for long term over-consumption. Some harmful effects of nitrites on humans, such as methemoglobinemia (leading to cyanosis or anaemic hypoxia, which can be potentially lethal, especially in infants and children, who have lower nicotinamide adenine dinucleotide (NADH) cyb5r reductase activity, which converts methaemoglobin to haemoglobin), and cancerogenic nitrosamines formation, are well documented [19,51]. Related to cancerogenic nitrosamines, in 2010 the IARC, declared, "There is sufficient evidence in experimental animals for the carcinogenicity of nitrite in combination with amines or amide. There is limited evidence in experimental animals for the carcinogenicity of nitrite per se. Ingested nitrate and nitrite under conditions that result in endogenous nitrosation is probably carcinogenic to humans (Group 2A)" [52], while in 2018, the IARC re-evaluated their statements and declared: "There is sufficient evidence in humans for the carcinogenicity of consumption of processed meat. Consumption of processed meat causes cancer of the colorectum. Positive associations have been observed between consumption of processed meat and cancer of the stomach. There is inadequate evidence in experimental animals for the carcinogenicity of consumption of processed meat. Consumption of processed meat is carcinogenic to humans (Group 1)" [53]. Even though there is sufficient evidence for colorectal carcinoma, there is still insufficient evidence for other cancers: oesophageal, gastric, lung, non-Hodgkin's lymphoma, thyroid, pancreatic, liver, ovarian, prostate, bladder, renal cancer, and brain tumours and glioma [19]. Furthermore, there is no sufficient evidence for risk of diabetes type 1 [54]. Nevertheless, some clinical studies have also demonstrated beneficial effects of dietary nitrates and nitrites, through their conversion to nitric oxide, especially in cardiovascular, immune, metabolic, and neural and reproductive health, and consumption of nitrites above the legislative limits are questioned as being less harmful [51,55].

Table 4. EDI and risk characterization of nitrite ion (NO_2^-) intake through processed meat products.

Meat Product	Age Group	Gender	ADC (g/day)	Nitrite Ion (NO_2^-) Content		EDI (mg/kg bw/day)		Contribution to ADI (%)		ADI (mg/kg bw/day)
				Mean ± SD (mg/kg)	P95th (mg/kg)	Mean	P95th	Mean	P95th	
Bacon	Toddlers, 1–3 years	M	2.17	10.17 ± 12.50	35.77	0.002	0.006	2.26	7.97	0.07
		F	2.87			0.002	0.007	3.00	10.55	
	Children, 3–9 years	M	1.24			0.001	0.002	0.75	2.64	
		F	2.93			0.001	0.004	1.78	6.27	
Canned meat	Toddlers, 1–3 years	M	2.37	23.30 ± 14.75	47.4	0.004	0.008	5.67	11.54	
		F	2.17			0.004	0.007	5.18	10.53	
	Children, 3–9 years	M	8.01			0.008	0.016	11.12	22.62	
		F	8.32			0.008	0.017	11.59	23.58	
Coarsely minced cooked sausages	Toddlers, 1–3 years	M	0.92	21.90 ± 15.50	47.43	0.001	0.003	2.07	4.49	
		F	0.50			0.001	0.002	1.12	2.42	
	Children, 3–9 years	M	0.05			0.000	0.000	0.07	0.15	
		F	0.00			0.000	0.000	0.00	0.00	
Dry fermented sausages	Toddlers, 1–3 years	M	1.06	1.00 ± 1.65	4.11	0.000	0.000	0.11	0.45	
		F	1.13			0.000	0.000	0.12	0.48	
	Children, 3–9 years	M	1.75			0.000	0.000	0.10	0.43	
		F	2.65			0.000	0.000	0.16	0.65	
Finely minced cooked sausages	Toddlers, 1–3 years	M	6.49	26.83 ± 13.58	47.82	0.013	0.022	17.88	31.87	
		F	5.68			0.011	0.020	15.65	27.90	
	Children, 3–9 years	M	11.11			0.012	0.022	17.75	31.64	
		F	7.66			0.009	0.015	12.29	21.90	
Liver sausage and pate	Toddlers, 1–3 years	M	0.86	8.41 ± 6.0	17.41	0.001	0.001	0.74	1.54	
		F	1.41			0.001	0.002	1.22	2.52	
	Children, 3–9 years	M	3.22			0.001	0.002	1.61	3.34	
		F	3.21			0.001	0.002	1.61	3.34	
Smoked meat products	Toddlers, 1–3 years	M	0.40	15.39 ± 15.19	41.13	0.000	0.001	0.62	1.67	
		F	0.75			0.001	0.002	1.18	3.15	
	Children, 3–9 years	M	0.95			0.001	0.002	0.87	2.32	
		F	1.54			0.001	0.003	1.42	3.78	
Average	Toddlers, 1–3 years	M	2.04	15.54 ± 15.84	44.86	0.003	0.006	4.20	8.50	
		F	2.07			0.003	0.006	3.92	8.22	
	Children, 3–9 years	M	3.76			0.003	0.006	4.61	9.02	
		F	3.76			0.003	0.006	4.12	8.50	
Total	Toddlers, 1–3 years	M	14.28	15.54 ± 15.84	44.86	0.021	0.042	29.37	59.52	
		F	14.51			0.019	0.040	27.46	57.54	
	Children, 3–9 years	M	26.33			0.023	0.044	32.27	63.12	
		F	26.30			0.020	0.042	28.85	59.52	

Nitrate is expressed as nitrate ion (66.65% of $NaNO_2$); ADC–average daily consumption of meat products (Table 1); EDI–estimated daily intake; ADI–acceptable daily intake [19,24].

Table 5. EDI and risk characterization of nitrite ion (NO_2^-) intake in consumers of processed meat products.

Age Group	Gender	N	% of Total Subjects	EDI (mg/kg bw/day)		Contribution to ADI (%)		N (%) Above ADI	
				Mean	P95th	Mean	P95th	Mean	P95th
Toddlers, 1–3 years	M	81	54.36	0.038	0.076	53.89	109.00	10 (12.35)	32 (39.51)
	F	81	57.45	0.037	0.074	52.37	106.18	10 (12.35)	31 (38.27)
Children, 3–9 years	M	113	76.87	0.021	0.044	30.33	62.57	8 (7.08)	17 (15.04)
	F	111	79.86	0.021	0.044	30.64	63.29	8 (7.21)	18 (16.22)
Average/Total		386	67.01	0.028	0.057	39.99	81.67	36 (9.33)	98 (25.39)

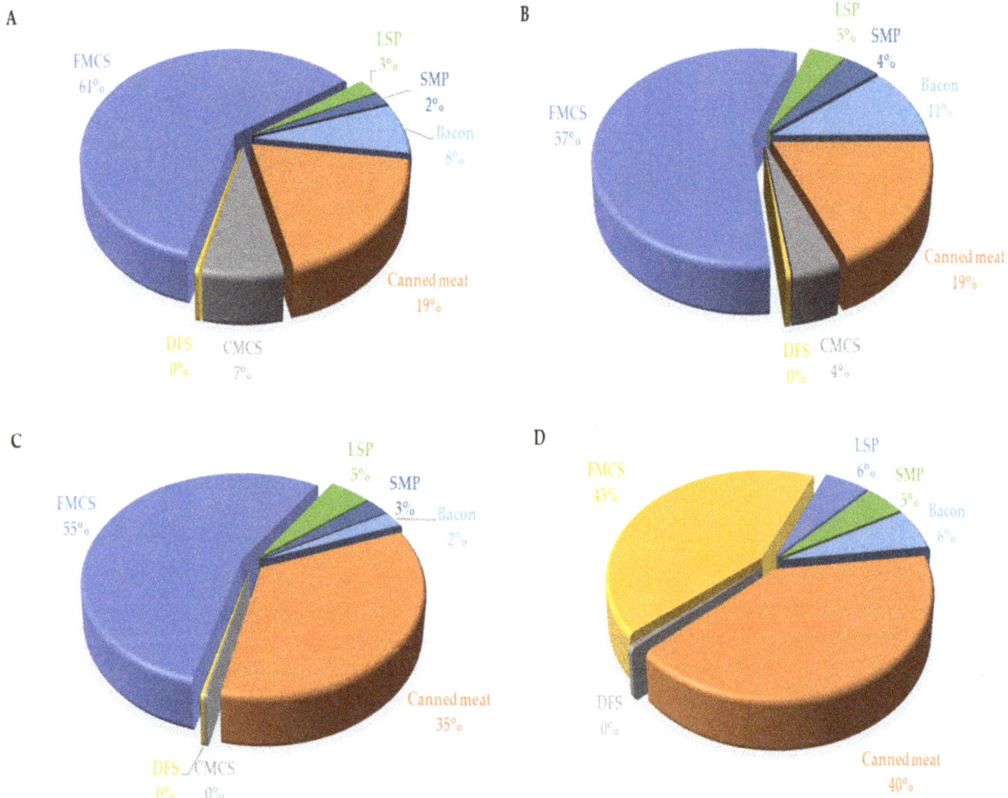

Figure 3. Contribution (%) of the most important food groups of processed meat products to dietary exposure of toddlers 1–3 years ((**A**)—male and (**B**)—female) and children 3–9 years ((**C**)—male and (**D**)—female), to nitrite ion (NO_2^-). CMCS—coarsely minced cooked sausages, DFS—dry fermented sausages, FMCS—finely minced cooked sausages, LSP—liver sausage and pate, SMP—smoked meat products.

3.4. Phosphorus Content in Processed Meat Products

Phosphorus content in different types of meat products is shown in Table 6. The results obtained are presented as mean, quartiles (Q), median values (Q2), the 95th percentile, and range. As seen in the table, the highest average and highest phosphorus content were observed in smoked meat products (6.12 ± 1.33 g/kg, 10.64 g/kg, respectively), while the lowest average values were obtained for liver sausage and pate (2.71 ± 1.05 g/kg). The average content of phosphorus (5.18 ± 1.31 g/kg) varied within a range between 0.27 and 10.64 g/kg. Analysis of the frequency distribution of phosphates contents (Figure 4) shows a normal distribution, and consequently, 95% of the data are between the average (SD ± 1.31) standard deviations. Of the 1900 retail samples we analysed, only 32 (1.7%) exceeded maximum limit for total phosphates expressed as P_2O_5 in meat products (<8 g/kg) [22]. In this study, the mean phosphate content in all groups of meat products was below the legal limit (<8 g/kg) [22]. Our results are in line with data obtained in previous studies [56,57].

Table 6. Mean levels and ranges for phosphorus (P_2O_5) in processed meat samples (g/kg).

Meat Product	*N*	Mean ± SD (g/kg)	Q1 (g/kg)	Q2 (g/kg)	Q3 (g/kg)	P95 (g/kg)	Range (g/kg)	Above MPL (%)
Bacon	230	4.41 ± 1.22	3.57	4.39	5.13	6.53	1.10–7.94	–
Canned meat	375	5.79 ± 1.01	5.14	5.76	6.43	7.40	2.37–9.83	6 (1.6)
Coarsely minced cooked sausages	353	5.21 ± 1.14	4.44	5.12	5.89	7.14	2.25–9.92	7 (2)
Finely minced cooked sausages	551	4.70 ± 0.88	4.16	4.64	5.17	6.10	1.12–9.22	4 (0.7)
Liver sausage and pate	45	2.71 ± 1.05	2.18	2.50	3.08	3.88	0.27–7.96	–
Smoked meat products	346	6.12 ± 1.33	5.35	6.02	6.97	7.98	2.11–10.64	15 (4.3)
Total	1900	5.18 ± 1.31	4.36	5.11	5.98	7.54	0.27–10.64	32 (1.7)

MPL—maximum permitted level (<8 g/kg) [22].

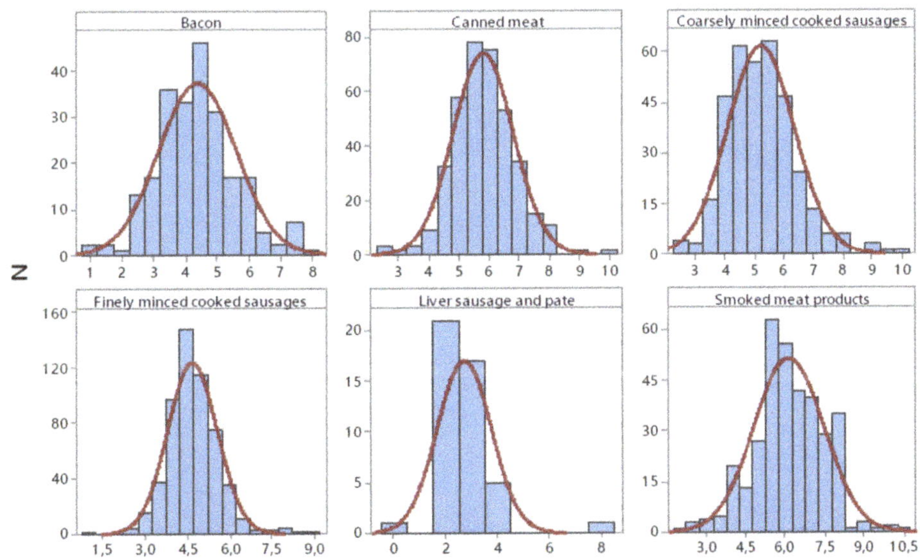

Figure 4. Distribution of phosphorus content (P_2O_5) in examined processed meat products (g kg^{-1}). N—number of samples.

Due to their a multifunctioning property, the addition of phosphates, as well as their blend, is a standard practice in the meat industry. Phosphates increase the water-holding capacity and, consequently, reduce drip loss and cooking loss [58]. The most common categories of products that use phosphates are cooked sausages, hams, and other whole-muscle products. This may be attributed to the fact that moisture retention is an important parameter of their quality. The addition of phosphates increases water holding in cooked sausages by 30–40 g per 100 g of meat [59]. On the other hand, phosphates are sometimes used in products, such as cured meat, to reduce cured colour development. In other products such as bacon, phosphates are used to improve texture during cooking by the consumer [60]. Yet, the use of phosphates at higher levels can produce a rubbery texture or impair the sensory characteristics of products such as giving the meat a soapy flavour [61]. In many countries, including Serbia, meat processing plants do not have an obligation to indicate the amount of phosphorus contained in (or added to) their products. However, comparing results observed in the present study with the information available in the literature, it could be concluded that the use of food-grade phosphates as a food additive in Serbia is generally in line with existing recommendations (up to 0.5%). Despite the fact that phosphates are widely used to improve quality characteristics of meat products, due to potential risks associated with chronic kidney disease (CKD), bone disease, and cardiovascular disease (CVD), phosphate reduction is a growing issue for meat producers.

3.5. Contribution of Processed Meat Products to the Daily Intake of Phosphorous

Daily intake of phosphorus through processed meat products in child populations is expressed as EDI: at the mean, 2.06 mg/kg bw/day, and at P95th level l–2.73 mg/kg bw/day, contributed 5.14% and 6.82% to ADI, respectively (Table 7). There was a slightly lower phosphorus intake per kilogram body weigh in the children aged 1–3 years, compared with the children aged 3–9 years, with no observable difference between males and females. Major sources of phosphorus in meat products in the observed study sample were finely minced cooked sausages, canned meat products, and bacon (Figure 5), which were, at the same time, the most consumed processed meat products by children.

Table 7. EDI of phosphorus and risk characterization of phosphorus intake (mg/kg bw/day).

Meat Product	Age Group	Gender	ADC (g/day)	Phosphorus Content		EDI (mg/kg bw/day)		Contribution to ADI (%)		ADI (mg/kg bw/day)
				Mean ± SD (g/kg)	P95th (g/kg)	Mean	P95th	Mean	P95th	
Bacon	Toddlers, 1–3 years	M	2.17	1.92 ± 0.53	2.85	0.30	0.44	0.75	1.11	40
		F	2.87			0.40	0.59	0.99	1.47	
	Children, 3–9 years	M	1.24			0.10	0.15	0.25	0.37	
		F	2.93			0.24	0.35	0.59	0.87	
Canned meat	Toddlers, 1–3 years	M	2.37	2.53 ± 0.44	3.23	0.43	0.55	1.08	1.38	
		F	2.17			0.39	0.50	0.98	1.26	
	Children, 3–9 years	M	8.01			0.84	1.08	2.11	2.70	
		F	8.32			0.88	1.12	2.20	2.81	
Coarsely minced cooked sausages	Toddlers, 1–3 years	M	0.92	2.27 ± 0.49	3.11	0.15	0.21	0.38	0.52	
		F	0.50			0.08	0.11	0.20	0.28	
	Children, 3–9 years	M	0.05			0.00	0.01	0.01	0.02	
		F	0.00			0.00	0.00	0.00	0.00	
Finely minced cooked sausages	Toddlers, 1–3 years	M	6.49	2.05 ± 0.38	2.66	0.96	1.24	2.39	3.10	
		F	5.68			0.84	1.09	2.09	2.72	
	Children, 3–9 years	M	11.11			0.95	1.23	2.37	3.08	
		F	7.66			0.66	0.85	1.64	2.13	
Liver sausage and pate	Toddlers, 1–3 years	M	0.86	1.18 ± 0.46	1.7	0.07	0.11	0.18	0.26	
		F	1.41			0.12	0.17	0.30	0.43	
	Children, 3–9 years	M	3.22			0.16	0.23	0.40	0.57	
		F	3.21			0.16	0.23	0.40	0.57	
Smoked meat products	Toddlers, 1–3 years	M	0.40	2.67 ± 0.58	3.48	0.08	0.10	0.19	0.25	
		F	0.75			0.14	0.19	0.36	0.47	
	Children, 3–9 years	M	0.95			0.11	0.14	0.26	0.34	
		F	1.54			0.17	0.22	0.43	0.56	
Average	Toddlers, 1–3 years	M	2.04	2.26 ± 0.57	3.30	0.28	0.38	0.71	0.95	
		F	2.07			0.28	0.38	0.70	0.95	
	Children, 3–9 years	M	3.76			0.31	0.40	0.77	1.01	
		F	3.76			0.30	0.40	0.75	0.99	
Total	Toddlers, 1–3 years	M	14.28	2.26 ± 0.57	3.30	1.99	2.65	4.97	6.62	
		F	14.51			1.97	2.65	4.93	6.62	
	Children, 3–9 years	M	26.33			2.16	2.83	5.40	7.07	
		F	26.30			2.10	2.78	5.26	6.95	

ADC—average daily consumption of meat products; ADI—acceptable daily intake [28].

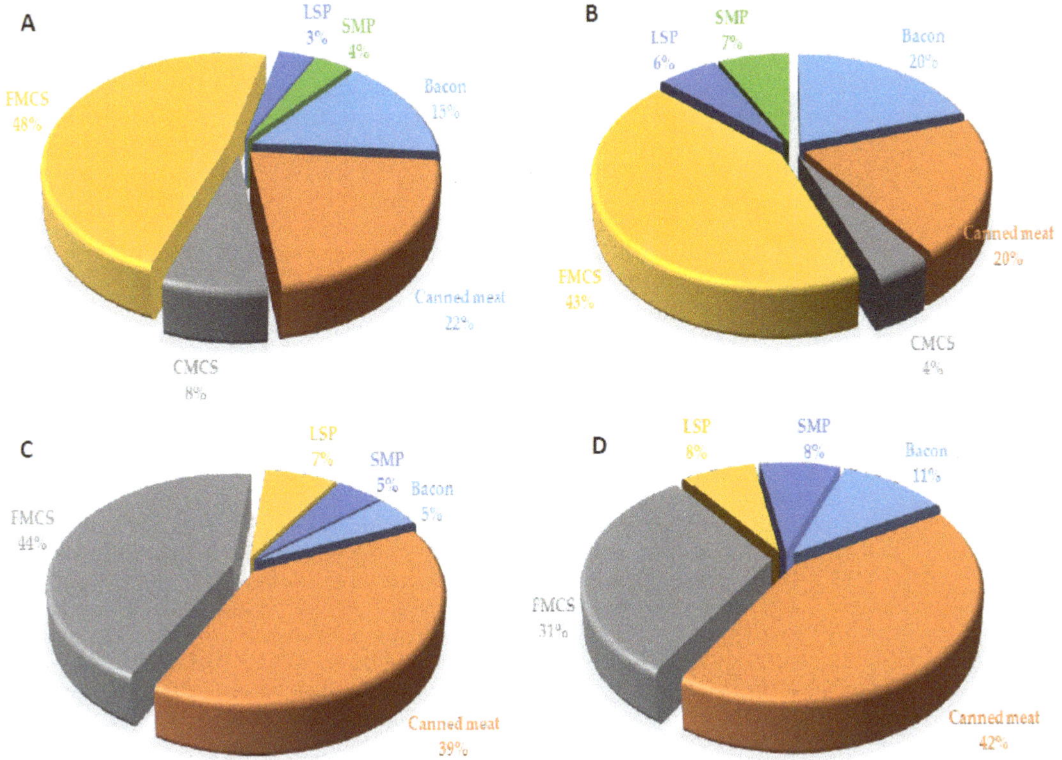

Figure 5. Contribution (%) of the most important food groups of processed meat products to dietary exposure of toddlers 1–3 years ((**A**)—male and (**B**)—female) and children 3–9 years ((**C**)—male and (**D**)—female) to phosphorus ion. CMCS—coarsely minced cooked sausages, FMCS—finely minced cooked sausages, LSP—liver sausage and pate, SMP—smoked meat products.

Even though these EDI values are far below ADI, it should be considered that other studies on the dietary assessment of phosphorus report on excessive intake of phosphorus from the whole diet [62,63], and particularly from meat and meat products [64], and that phosphorus as additive accounts only as a part of total phosphorus content in these food groups [65]. The EFSA Panel from 2019, which analysed total phosphorus exposure in children, concluded that when the whole diet is regarded, the intake of phosphorus would exceed the ADI of 40 mg/kg bw/day in children aged 1–9 years, both at the mean and high exposure levels [28]. The report represented the results from 15 EU countries (including Austria, Belgium, Bulgaria, Czech Republic, Denmark, Finland, France, Germany, Greece, Italy, Latvia, Netherlands, Spain, Sweden, and UK), involving 13637 children (the same as for the nitrite Panel). The mean phosphorus EDI from all dietary sources was 69.70 (55–74) mg/kg bw/day in the 1–3-year-old group and 49.92 (33–62) mg/kg bw/day in the 3–9 age group. The highest exposure for the 1–3-year-old group was noted in Denmark and UK and in Finland for the 3–9-year-old group.

The main contributing food categories to total phosphate EDI were bread, rolls and fine bakery wares, processed cheese, meat products, and sugars and syrups. Processed meat products contributed to the whole diet phosphorus exposure, at 6.1–14.0% in children 1–3 years old and 5.2–17.8% in children 3–9 years old [28]. According to these data, we could also predict that in Serbian children, the ADI could be exceeded if the all dietary sources of phosphates are considered. Thus, future studies should focus on determination of total phosphorus intake from the whole diet in this vulnerable population in Serbia.

The EFSA Panel from 2019 concluded that phosphates have "low acute oral toxicity and there is no concern with respect to genotoxicity and carcinogenicity", and that they "do not present any risk for reproductive or developmental toxicity" [28]. However, even though phosphorus is an essential element in the human body, excessive consumption of phosphorus can be related to several health risks, including cardiovascular complications—impaired endothelial function and hypertension, vascular calcifications, left ventricular hypertrophy, heart failure, and atrial fibrillation, as well as increased cardiovascular mortality [28]. High phosphorus intake can disrupt the hormonal regulation of phosphorus, calcium, and vitamin D and stimulate excessive secretion of parathormone and fibroblast growth factor-23 (FGF-23), leading to impaired bone turnover and bone demineralization, osteoporosis, bone fractures, and, due to higher renal tubular exposure to phosphates, tubular nephropathy (disseminated atrophy and tissue necrosis) and nefrocalcinosis, leading to renal impairment. Fibroblast growth factor-23 is directly related to cardiovascular events and has been associated with left ventricular hypertrophy, atrial fibrillation, heart failure, and cardiovascular mortality. Some studies have shown that FGF23 can directly induce left ventricular hypertrophy. However, there are insufficient data in humans to confirm all these health risks, and more research is needed [28]. Nevertheless, in cases of chronic renal disease with decreased renal function or vitamin D intoxication and low calcium dietary intake, gut absorption of phosphorus can increase. Phosphorus from food additives has the highest bioavailability because the inorganic sodium and potassium phosphate salts dissociate easily and do not require release by luminal phosphatases, making them thus more easily absorbed than phosphorus from organic phosphate in natural, unprepared animal or plant foods. Phosphorus from plants sources is less digestible and less bioavailable than phosphorus from animal sources because of complexes with phytates; bioavailability of phosphate from phytates is only 20–30%, since humans lack the enzyme phytase. Different cooking methods or industrial food processing can affect bioavailability and absorption, as well as the consumption of vitamin D and phosphate binders (calcium, magnesium, some amino acids), which can increase or decrease phosphate absorption.

Considering all these potential health risks of excessive phosphates intakes, efforts should be made to decrease exposure to phosphates through additives. The meat industry should consider other technological approaches to preservation, such as application of ultrasound (US), high-pressure processing (HPP), and pulsed electric field (PEF), that modify the protein structure and improve its functional properties, allowing for the reduction of the content of additives in meat products [66].

3.6. *Study Strengths and Limitations*

This study on nitrites and phosphorus exposure risk assessment through meat products was conducted on a representative sample of the Serbian child population using harmonized food consumption data collected according to EFSA guidance methodology (within the EU Menu project survey), which makes these data comparable with other harmonized food consumption data in the whole of Europe. These data allow longitudinal monitoring and exposure assessment.

One of the study limitations is that only 2 days 24 h food records were explored, while it would be better to have at least 3 days, ideally 7 days, to cover interday variation in the intakes [67]. However, we performed FPQ data analysis on frequency of consumption, and the results were congruent with the 24 h food records data, indicating that frequency of consumption was well presented by the twice repeated 24 h records in studied population. Additionally, the study sample was stratified to proportionally cover weekdays and weekend days, capturing variability of dietary patterns. According to EFSA EU Menu methodology, food consumption data collection studies are performed using twice repeated 24 h records [33]. Total exposure from nitrites and phosphate—total EDI—could not be extracted in this study, as only consumption of processed meat products was observed. Further research should include other food groups, i.e., perform exposure risk assessment on the total diet.

4. Conclusions

The present study has shown that the content of nitrites in meat products in Serbia is within MPL, with only one product exceeding the limit. Intake of nitrites in the child population, on average, is below ADI. However, when intake was accounted for those who consumed meat products, these values indicated that intake of nitrites only from processed meat products was alarmingly high and exceeded ADI in a substantial number of consumers, particularly among the youngest children.

The content of phosphorus in analysed meat products was within MPL, with 3.7% of all processed meat products exceeding this value. Assessed intake of phosphorus from processed meat products was far below ADI values, even when regarded for the highest exposure scenario, which does not raise a concern. However, this finding should be taken with caution, as phosphorus from processed meat products is just one part of the total phosphorus intake from various dietary sources.

Taking into account the cumulative effect of long-term consumption, these findings raise a concern of exposure risk of nitrites in the child population. The results highlighted that intake of processed meat is high and frequent among this population group, which calls for attention in the food industry to consider reducing content of these additives in meat products. As learned from other countries, such as Denmark and Korea, lowering of MPL for nitrites resulted in gradual reduction of nitrite intake in a population. Decreased usage of nitrite in processed meat products by meat industries is recommended, especially for those products which are majorly consumed in child populations—finely minced cooked sausages and canned meat.

Regulatory authorities should be informed that there is a potential risk and have their attention called to lower MPL for nitrites. The system for continual dietary exposure monitoring for these and other additives should be established in the country. On a wider perspective, these findings suggest that the food industry should move towards applying healthier, "greener" technologies for preservation, as the actual transformation of food systems towards nutrition-sensitive and sustainability require. Finally, rising awareness on harmful effect of nitrites in processed meat can bring a major shift in dietary preferences in children and their parents and caregivers.

Author Contributions: Conceptualization, J.M., D.R.M. and M.G. (Mirjana Gurinović); methodology, D.V. and M.G. (Maria Glibetić); software, J.M. and M.Z.; validation, I.Š.; formal analysis, I.Š.; investigation, D.V., V.K. and B.B.; resources, D.R.M. and M.G. (Maria Glibetić); data curation, M.Z.; writing—original draft preparation, J.M., D.R.M. and I.Š.; writing—review and editing, J.M., D.R.M., M.G. (Mirjana Gurinović) and I.Š.; visualization, I.Š.; supervision, M.G. (Maria Glibetić); project administration, M.G. (Mirjana Gurinović); funding acquisition, M.G. (Mirjana Gurinović). All authors have read and agreed to the published version of the manuscript.

Funding: This research was funded by the Ministry of Education, Science, and Technological Development of the Republic of Serbia, Agreement on realization and financing of scientific research work of NIO for 2021, no. 451-03-9/2021-14/200050 and no. 451-03-9/2021-14/200015. Research was also funded by the European Food Safety Authority OC/EFSA/DATA/2016/02-CTO1.

Institutional Review Board Statement: The study was conducted according to the guidelines of the Declaration of Helsinki and approved by the Ethical Committee of the Institute for Medical Research, University of Belgrade on 8 December 2017 (EO 123/2017).

Informed Consent Statement: Written informed consent for inclusion was obtained from all the participants.

Data Availability Statement: Not applicable.

Acknowledgments: Authors of this manuscript are grateful to Agnes Kadvan for IT support in the EU Menu National Food Consumption Survey.

Conflicts of Interest: The authors declare no conflict of interest.

References

1. GBD 2019. Risk Factors Collaborators. Global burden of 87 risk factors in 204 countries and territories, 1990–2019: A systematic analysis for the Global Burden of Disease Study 2019. *Lancet* **2020**, *396*, 1223–1249. [CrossRef]
2. World Cancer Research Fund, American Institute for Cancer Research. *Diet, Nutrition, Physical Activity and Cancer: A Global Perspective. Continuous Update Project Expert Report*; World Cancer Research Fund: London, UK, 2018.
3. Bouvard, V.; Loomis, D.; Guyton, K.Z.; Grosse, Y.; Ghissassi, F.E.; Benbrahim-Tallaa, L.; Guha, N.; Mattock, H.; Straif, K. Carcinogenicity of consumption of red and processed meat. *Lancet Oncol.* **2015**, *16*, 1599–1600. [CrossRef]
4. Honikel, K.O. The use and control of nitrate and nitrite for the processing of meat products. *Meat Sci.* **2008**, *78*, 68–76. [CrossRef] [PubMed]
5. Gøtterup, J.; Olsen, K.; Knøchel, S.; Tjener, K.; Stahnke, L.H.; Møller, J.K. Colour formation in fermented sausages by meat-associated staphylococci with different nitrite- and nitrate-reductase activities. *Meat Sci.* **2008**, *78*, 492–501. [CrossRef] [PubMed]
6. Pennington, J.A.T. Dietary exposure models for nitrates and nitrites. *Food Control* **1998**, *9*, 385–395. [CrossRef]
7. Sebranek, J.G. Basic curing Ingredients in Meat Products. In *Ingredients in Meat Products: Properties, Functionality and Applications*; Tarté, R., Ed.; Springer Science: New York, NY, USA, 2009; pp. 1–24.
8. Marco, A.; Navarro, J.L.; Flores, M. The influence of nitrite and nitrate on microbial, chemical and sensory parameters of slow dry fermented sausage. *Meat Sci.* **2006**, *73*, 660–673. [CrossRef]
9. Van den Brand, A.D.; Beukers, M.; Niekerk, M.; van Donkersgoed, G.; van der Aa, M.; van de Ven, B.; Sprong, C.R. Assessment of the combined nitrate and nitrite exposure from food and drinking water: Application of uncertainty around the nitrate to nitrite conversion factor. *Food Addit. Contam. Part A* **2020**, *37*, 568–582. [CrossRef] [PubMed]
10. Hord, N.G.; Tang, Y.; Bryan, N.S. Food sources of nitrates and nitrites: The physiologic context for potential health benefits. *Am. J. Clin. Nutr.* **2009**, *90*, 1–10. [CrossRef]
11. Chetty, A.A.; Prasad, S.; Pinho, O.C.; de Morais, C.M. Estimated dietary intake of nitrate and nitrite from meat consumed in Fiji. *Food Chem.* **2019**, *278*, 630–635. [CrossRef]
12. Tamme, T.; Reinik, M.; Rosato, M. Nitrates and Nitrites in Vegetables: Occurrence and Health Risks. In *Bioactive Foods in Promoting Health*; Wotson, R., Preedy, V., Eds.; Academic Press: London, UK; San Diego, CA, USA, 2010; p. 754.
13. Ward, M.H.; Heineman, E.F.; Markin, R.S.; Weisenburger, D.D. Adenocarcinoma of the stomach and esophagus and drinking water and dietary sources of nitrate and nitrite. *Int. J. Occup. Environ. Health* **2008**, *14*, 193–197. [CrossRef] [PubMed]
14. Tricker, A.R. N-nitroso compounds and man: Sources of exposure, endogenous formation and occurrence in body fluids. *Eur. J. Cancer Prev.* **1997**, *6*, 226–268. [CrossRef] [PubMed]
15. Cross, A.J.; Ferrucci, L.M.; Risch, A.; Graubard, B.I.; Ward, M.H.; Park, Y.; Hollenbeck, A.R.; Schatzkin, A.; Sinha, R. A large prospective study of meat consumption and colorectal cancer risk: An investigation of potential mechanisms underlying this association. *Cancer Res.* **2010**, *70*, 2406–2414. [CrossRef] [PubMed]
16. World Cancer Research Fund, American Institute for Cancer Research. *Food, Nutrition, Physical Activity, and the Prevention of Cancer: A Global Perspective*; American Institute for Cancer Research: Washington, DC, USA, 2007.
17. Ozel, M.Z.; Gogus, F.; Yagci, S.; Hamilton, J.F.; Lewis, A.C. Determination of volatile nitrosamines in various meat products using comprehensive gas chromatography-nitrogen chemiluminescence detection. *Food Chem. Toxicol.* **2010**, *48*, 3268–3273. [CrossRef]
18. Gilchrist, M.; Winyard, P.G.; Benjamin, N. Dietary nitrate—Good or bad? *Nitric Oxide* **2010**, *15*, 104–109. [CrossRef]
19. EFSA Panel on Food Additives and Nutrient Sources added to Food (EFSA ANS Panel); Mortensen, A.; Aguilar, F.; Crebelli, R.; Di Domenico, A.; Dusemund, B.; Frutos, M.J.; Galtier, P.; Gott, D.; Gundert-Remy, U.; et al. Scientific Opinion on the re-evaluation of potassium nitrite (E 249) and sodiumnitrite (E 250) as food additives. *EFSA J.* **2017**, *15*, 4786. [CrossRef]
20. Tonacchera, M.; Pinchera, A.; Dimida, A.; Ferrarini, E.; Agretti, P.; Vitti, P.; Santini, F.; Crump, K.; Gibbs, J. Relative potencies and additivity of perchlorate, thiocyanate, nitrate, and iodide on the inhibitionof radioactive iodide uptake by the humansodium iodide symporter. *Thyroid* **2004**, *14*, 1012–1019. [CrossRef]
21. De Groef, B.; Decallonne, B.R.; Van der Geyten, S.; Darras, V.M.; Bouillon, R. Perchlorate versus other environmental-sodium/iodide symporter inhibitors: Potentialthyroid-related health effects. *Eur. J. Endocrinol.* **2006**, *155*, 17–25. [CrossRef]
22. Regulation on the Quality of ground meat, meat preparations and meat products. *Official Gazette of RS No 50*, 12 July 2019.
23. Regulation on the food additives. *Official Gazette of RS No 53*, 11 July 2018.
24. Joint FAO/WHO Expert Committee on Food Additives. *World Health Organization & Food and Agriculture Organization of the United Nations. Evaluation of Certain food Additives: Fifty-Ninth Report of the Joint FAO/WHO Expert Committee on Food Additives. World Health Organization*; WHO: Geneva, Switzerland, 2002. Available online: https://apps.who.int/iris/handle/10665/42601 (accessed on 17 November 2021).
25. Habermeyer, M.; Roth, A.; Guth, S.; Diel, P.; Engel, K.H.; Epe, B.; Fürst, P.; Heinz, V.; Humpf, H.U.; Joost, H.G.; et al. Nitrate and nitrite in the diet: How to assess their benefit and risk for human health. *Mol. Nutr. Food Res.* **2015**, *59*, 106–128. [CrossRef]
26. Calvo, M.S.; Sherman, R.A.; Uribarri, J. Dietary Phosphate and the Forgotten Kidney Patient: A Critical Need for FDA Regulatory Action. *Am. J. Kidney Dis.* **2019**, *73*, 542–551. [CrossRef] [PubMed]
27. Wallace, T.C.; Bailey, R.L.; Lappe, J.; O'Brien, K.O.; Wang, D.D.; Sahni, S.; Weaver, C.M. Dairy intake and bone health across the lifespan: A systematic review and expert narrative. *Crit. Rev. Food Sci. Nutr.* **2021**, *61*, 3661–3707. [CrossRef] [PubMed]

28. EFSA Panel on Food Additives and Flavourings (EFSA FAF Panel); Younes, M.; Aquilina, G.; Castle, L.; Engel, K.-H.; Fowler, P.; Frutos Fernandez, M.J.; Fürst, P.; Gürtler, R.; Husøy, T.; et al. Scientific Opinion on the re-evaluation of phosphoric acid–phosphates–di-, tri- and polyphosphates (E 338–341, E 343, E 450–452) as food additives and the safety of proposed extension of use. *EFSA J.* **2019**, *17*, 5674. [CrossRef]
29. European Uniion. Regulation (EC) No 1333/2008 of the European Parliament and of the Council of 16 December 2008 on food additives. *Off. J. Eur. Union* **2008**, *354*, 16–31.
30. Institute of Public Health of Serbia "Dr Milan Jovanovic Batut". *Health Statistical Year Book of Republic of Serbia 2020*; Institute of Public Health of Serbia "Dr Milan Jovanovic Batut: Belgrade, Serbia, 2021. Available online: www.batut.org.rs (accessed on 17 November 2021).
31. *ISO 2918:1999*; Meat and Meat Products—Determination of Nitrite Content (Reference Method). ISO: Geneva, Switzerland, 1996.
32. *ISO 13730:1996*; Meat and Meat Products—Determination of Total Phosphorus Content—Spectrometric Method. ISO: Geneva, Switzerland, 1996.
33. Gurinović, M.; Nikolić, M.; Zeković, M.; Milešević, J.; Kadvan, A.; Ranić, M.; Glibetic, M. Implementation of Harmonized Food Consumption Data Collection in Balkan Region According to the EFSA EU Menu Methodology Standards. *Front. Nutr.* 2021. *provisionally accepted; in press*. [CrossRef]
34. Zekovic, M.; Gurinovic, M.; Milesevic, J.; Glibetic, M. National Food Consumption Survey among children from 1 to 9 years in Serbia–final deliverable. *EFSA Support. Publ.* **2021**, *18*, 27. [CrossRef]
35. Nikolic, M.; Milesevic, J.; Zekovic, M.; Gurinovic, M.; Glibetic, M. The development and validation of food atlas for portion size estimation in the Balkan region. *Front. Nutr.* **2018**, *5*, 78. [CrossRef]
36. Gurinović, M.; Milešević, J.; Kadvan, A.; Nikolić, M.; Zeković, M.; Djekić-Ivanković, M.; Glibetić, M. Development, features and application of DIET ASSESS & PLAN (DAP) software in supporting public health nutrition research in Central Eastern European Countries (CEEC). *Food Chem.* **2018**, *238*, 186–194. [CrossRef]
37. Gurinović, M.; Milešević, J.; Kadvan, A.; Đekić-Ivanković, M.; Debeljak-Martačić, J.; Takić, M.; Glibetić, M. Establishment and advances in the online Serbian food and recipe data base harmonized with EuroFIRTM standards. *Food Chem.* **2016**, *193*, 30–38. [CrossRef]
38. European Food Safety Authority (EFSA). Guidance on the EU Menu methodology. *EFSA J.* **2014**, *12*, 3944.
39. JECFA—Joint FAO/WHO Expert Committee on Food Additives. *Evaluation of Certain Contaminants in Food: Eighty-Third Report of the Joint FAO/WHO Expert Committee on Food Additives*; WHO Technical Report Series 1002; WHO: Geneva, Switzerland, 2017; pp. 1–166. Available online: https://apps.who.int/iris/handle/10665/254893 (accessed on 17 November 2021).
40. European Food Safety Authority (EFSA). Management of left-censored data in dietary exposure assessment of chemical substances. *EFSA J.* **2010**, *8*, 1557.
41. Lee, H.S. Exposure estimates of nitrite and nitrate from consumption of cured meat products by the U.S. population. *Food Addit. Contam. Part A. Chem. Anal. Control Expo. Risk Assess.* **2018**, *35*, 29–39. [CrossRef] [PubMed]
42. Temme, E.H.M.; Vandevijvere, S.; Vinkx, C.; Huybrechts, I.; Goeyens, L.; Oyen, H.V. Average daily nitrate and nitrite intake in the Belgian population older than 15 years. *Food Addit. Contam.* **2011**, *28*, 1193–1204. [CrossRef]
43. Choi, S.-H.; Suh, H.-J. Determination and estimation of daily nitrite intake from processed meats in Korea. *J. Consum. Prot. Food. Saf.* **2017**, *12*, 15–22. [CrossRef]
44. Andrée, S.; Jira, W.; Schwind, K.H.; Wagner, H.; Schwägele, F. Chemical safety of meat and meat products. *Meat Sci.* **2010**, *86*, 38–48. [CrossRef]
45. Gibson, A.; Roberts, T.A.; Robinson, A. The effect of pH, sodium chloride, sodium nitrite and storage temperature on the growth of *Clostridium perfringens* and faecal streptococci in laboratory media. *J. Food Sci. Technol.* **1984**, *3*, 195–210. [CrossRef]
46. Bajcic, A.; Petronijevic, R.; Katanic, N.; Trbovic, D.; Betic, N.; Nikolic, A.; Milojevic, L. Evaluation of the content and safety of nitrite utilisation in meat products in Serbia in the period 2016–2018. *Meat Technol.* **2018**, *59*, 102–109. [CrossRef]
47. Zivkovic, D.; Stajic, S. *Meat Preservation Procedures. Meat Technology 1*; Faculty of Agriculture, University of Belgrade: Belgrade, Serbia, 2016.
48. Elias, A.; Jalakas, S.; Roasto, M.; Reinik, M.; Nurk, E.; Kaart, T.; Tuvike, A.; Meremäe, K.; Nelis, K.; Elias, T. Nitrite and nitrate content in meat products and estimated nitrite intake by the Estonian children. *Food Addit. Contam. Part A Chem. Anal. Control Expo. Risk Assess.* **2020**, *37*, 1229–1237. [CrossRef]
49. Adam, A.H.B.; Mustafa, N.E.M.; Rietjens, I.M.C.M. Nitrite in processed meat products in Khartoum, Sudan and dietary intake. *Food Addit. Contam. Part B Surveill.* **2017**, *10*, 79–84. [CrossRef] [PubMed]
50. Lim, H.S.; Choi, E.; Lee, S.J.; Nam, H.S.; Lee, J.K. Improved spectrophotometric method for nitrite determination in processed foods and dietary exposure assessment for Korean children and adolescents. *Food Chem.* **2022**, *367*, 130628. [CrossRef]
51. Lee, D.Y.; Lee, S.Y.; Jo, C.; Yoon, Y.; Jeong, J.Y.; Hur, S.J. Effect on health from consumption of meat and meat products. *J. Anim. Sci. Technol.* **2021**, *63*, 955–976. [CrossRef]
52. IARC Working Group on the Evaluation of Carcinogenic Risks to Humans. *Ingested Nitrate and Nitrite, and Cyanobacterial Peptide Toxins*; IARC Monographs on the Evaluation of Carcinogenic Risks to Humans; International Agency for Research on Cancer: Lyon, France, 2010; Volume 94.

53. IARC Working Group on the Evaluation of Carcinogenic Risks to Humans, International Agency for Research on Cancer. *Red Meat and Processed Meat*, 6th ed.; IARC Monographs on the Evaluation of Carcinogenic Risks to Humans; International Agency for Research on Cancer: Lyon, France, 2018; Volume 114, ISBN 978-92-832-0152-6.
54. Bahadoran, Z.; Ghasemi, A.; Mirmiran, P.; Azizi, F.; Hadaegh, F. Nitrate-nitrite-nitrosamines exposure and the risk of type 1 diabetes: A review of current data. *World J. Diabetes* **2016**, *7*, 433. [CrossRef] [PubMed]
55. Bedale, W.; Sindelar, J.J.; Milkowski, A.L. Dietary nitrate and nitrite: Benefits, risks, and evolving perceptions. *Meat Sci.* **2016**, *120*, 85–92. [CrossRef]
56. Koricanac, V.; Vranic, D.; Lilic, S.; Milicevic, D.; Sobajic, S.; Zrnic, M. Total phosphorus content in various types of cooked sausages from the Serbian market. International 58th Meat Industry Conference "Meat Safety and Quality: Where it goes?" *Proc. Food Sci.* **2015**, *5*, 152–155. [CrossRef]
57. Saičić, S.; Vranić, D.; Trbović, D.; Pavlov, N. Total phosphorus content in meat products. *Tehnol. Mesa* **2008**, *49*, 147–152.
58. Polak, T.; Lušnic Polak, M.; Tomović, V.; Žlender, B.; Demšar, L. Characterization of the Kranjska klobasa, a traditional slovenian cooked, cured, and smoked sausage from coarse ground pork. *J. Food Process. Preserv.* **2017**, *41*, e13269. [CrossRef]
59. Puolanne, E.J.; Ruusunen, M.H.; Vainionpaa, J.I. Combined effects of NaCl and raw meat pH on water-holding in cooked sausage with and without added phosphate. *Meat Sci.* **2001**, *58*, 1–7. [CrossRef]
60. Polak, T.; Polak Lušnic, M.; Zahija, I.; Babič, K.; Demšar, L. Comparison of the physico-chemical parameters and sensory properties of selected pasteurized meat products on Slovenian market. *Meso* **2020**, *22*, 196–208. [CrossRef]
61. Gadekar, Y.P.; Sharma, B.D.; Shinde, A.K.; Mendiratta, S.K. Effect of Different Phosphates on Quality of Goat Meat and Restructured Goat Meat Product. *Agric. Res.* **2014**, *3*, 370–376. [CrossRef]
62. Cuadrado-Soto, E.; Lopez-Sobaler, A.M.; Jimenez-Ortega, A.I.; Aparicio, A.; Bermejo, L.M.; Hernandez-Ruiz, A.; Lara Villoslada, F.; Leis, R.; Mertinez de Victoria, E.; Moreno, J.M.; et al. Usual dietary intake, nutritional adequacy and food sources of calcium, phosphorus, magnesium and vitamin D of Spanish children aged one to <10 Years–Results from the EsNuPI Study. *Nutrients* **2020**, *12*, 1787.
63. Chouraqui, J.P.; Tavoularis, G.; Turck, D.; Ferry, C.; Feillet, F. Mineral and vitamin intake of infants and young children: The Nutri-Bébé 2013 survey. *Eur. J. Nutr.* **2020**, *59*, 2463–2480. [CrossRef]
64. Olza, J.; Aranceta-Bartrina, J.; González-Gross, M.; Ortega, R.M.; Serra-Majem, L.; Varela-Moreiras, G.; Gil, Á. Reported dietary intake, disparity between the reported consumption and the level needed for adequacy and food sources of calcium, phosphorus, magnesium and vitamin D in the Spanish population: Findings from the ANIBES study. *Nutrients* **2017**, *9*, 168. [CrossRef]
65. Uribarri, J.; Calvo, M.S. Dietary phosphorus intake and health. *Am. J. Clin. Nutr.* **2014**, *99*, 247–248. [CrossRef]
66. Pinton, M.B.; dos Santos, B.A.; Lorenzo, J.M.; Cichoski, A.J.; Boeira, C.P.; Campagnol, P.C.B. Green technologies as a strategy to reduce NaCl and phosphate in meat products: An overview. *Curr. Opin. Food Sci.* **2021**, *40*, 1–5. [CrossRef]
67. European Food Safety Authority (EFSA). Outcome of the questions for health professionals in the fields of nephrology, mineral metabolism, cardiovascular and nutrition medicine on phosphates food additives re-evaluation. *EFSA Support. Publ.* **2019**, *16*, 1624E.

nutrients

MDPI

Article

Dietary Exposure and Risk Assessment of Aflatoxin M1 for Children Aged 1 to 9 Years Old in Serbia

Dragan R. Milićević [1,*], Jelena Milešević [2], Mirjana Gurinović [2], Saša Janković [1], Jasna Đinović-Stojanović [1], Milica Zeković [2] and Maria Glibetić [2]

[1] Institute of Meat Hygiene and Technology, Kaćanskog 13, 11040 Belgrade, Serbia; sasa.jankovic@inmes.rs (S.J.); jasna.djinovic@inmes.rs (J.Đ.-S.)
[2] Centre of Research Excellence in Nutrition and Metabolism, Institute for Medical Research, National Institute of Republic of Serbia, University of Belgrade, Bulevar Oslobođenja 18, 11000 Belgrade, Serbia; jelena.milesevic@gmail.com (J.M.); mirjana.gurinovic@gmail.com (M.G.); zekovicmilica@gmail.com (M.Z.); mglibetic@gmail.com (M.G.)
* Correspondence: dragan.milicevic@inmes.rs

Abstract: The present study was conducted to estimate the exposure and characterize the risk for the child population of Serbia to Aflatoxin M1 (AFM1) from milk and milk-based food. A total of 3404 samples comprising milk and different milk-based food samples were collected from various regions of Serbia from 2017 to 2019. Evaluation of AFM1 exposure was carried out using the deterministic method, whereas risk characterization was evaluated using the margin of exposure (MOE) and the risk of hepatocellular carcinoma (HCC). Detection rates for AFM1 in milk and milk-based food samples ranged between 2% and 79%, with the highest incidence (79%) and mean level (22.34 ± 0.018 ng kg^{-1}) of AFM1 being detected in pasteurized and UHT milk. According to the three consumption estimates, the values of estimated daily intake (EDI) were higher for toddlers as compared with children aged 3–9 years. Children aged 1–3 years had the highest risk of exposure to AFM1 in milk, with an estimated daily intake of 0.164 and 0.193 ng kg^{-1} bw day^{-1} using lower bound (LB) and upper bound (UB) exposure scenarios, respectively. Such difference could result from the higher consumption to weight in younger children. Based on the estimated daily intake (EDI) found in this study, the risk of AFM1 exposure due to consumption of milk and milk-based food was low since the MOE values obtained were >10,000. In addition, the risk of HCC cases/year/10^5 individuals of different age groups showed that the value of HCC, using potency estimates of 0.0017 (mean), was maximum (0.00034) in the age group 1–3 years, which indicates no health risk for the evaluated groups. The present study revealed the importance of controlling and preventing AFM1 contamination in milk through continuous monitoring and regular inspection to reduce the risk of AFM1 exposure, especially in children.

Keywords: aflatoxin M1; milk; dairy products; risk assessment; children

Citation: Milićević, D.R.; Milešević, J.; Gurinović, M.; Janković, S.; Đinović-Stojanović, J.; Zeković, M.; Glibetić, M. Dietary Exposure and Risk Assessment of Aflatoxin M1 for Children Aged 1 to 9 Years Old in Serbia. *Nutrients* **2021**, *13*, 4450. https://doi.org/10.3390/nu13124450

Academic Editor: Rosa Casas

Received: 29 October 2021
Accepted: 12 December 2021
Published: 13 December 2021

1. Introduction

Mycotoxins are toxic compounds produced as secondary metabolites by certain groups of fungi and constitute a significant hazard to food safety and public health [1]. Under certain environmental conditions (i.e., temperature and humidity) and/or biotic stress, toxigenic fungi and their metabolites may contaminate crops and food commodities in different phases of production and processing [2]. Mycotoxins show stability against heat processes, which makes their occurrence in processed food likely expected even if toxin-producing molds are eliminated during the food preparation process [3]. Consumption of mycotoxin-contaminated food may lead to different health adverse effects, including immune suppression, target organ toxicity, genotoxicity, or carcinogenicity [4]. Moreover, when animals ingest these toxins, their metabolites or unmetabolized compounds may be transferred to products such as milk and further contaminate dairy products. Aflatoxin

M1 (AFM1) is a principal hydroxylated metabolite of Aflatoxin B1 (AFB1), which may be found in milk from lactating animals after ingesting feed contaminated with AFB1 [5]. The ubiquitous occurrence of mycotoxins in the food chain has been shown in numerous reports over the last decades [6,7]. Thus, to ensure consumer safety due to exposure through food, strict regulations and guidelines have been set by different organizations such as the World Health Organization (WHO) and the European Food Safety Authority (EFSA) to control, measure, and diminish occurrence of the major mycotoxins [8].

Long-term exposure to mycotoxins is associated with myriad health consequences that belong to non-communicable diseases (NCDs), among which are liver and renal cancers, chronic gastritis, and nervous system disorders [9]. Among mycotoxins, aflatoxins represent the major public health concern because they are hepatotoxic, teratogenic, and immunosuppressive. These secondary metabolites are produced by some *Aspergillus* species, especially *A. flavus, A. nomius,* and *A. parasiticus* [10]. The International Agency for Research on Cancer (IARC) classified aflatoxins as Class 1 carcinogenic compounds to humans [11]. According to the risks associated with mycotoxins, the European Union has established the strictest maximum levels for AFM1 (0.05 $\mu g\ kg^{-1}$) in raw milk, heat-treated milk and milk for the manufacture of milk-based products and 0.025 $\mu g\ kg^{-1}$ in infant formulae and follow-on formulae, including infant milk and follow-on milk [12]. Previous regulations have not eradicated milk AFM1 successfully, which resulted in periodic changes to the official regulations. In the meanwhile, Serbia has set standards for aflatoxins where the maximum regulatory level for AFM1 in raw milk, heat-treated milk, and milk for the manufacture of milk-based products is 0.25 $\mu g\ kg^{-1}$ [13]. Dairy products are not included in the Serbian regulation, while for infant formulae and follow-on formulae, including infant milk and follow-on milk, as well as for dietary foods for special medical purposes intended specifically for infants, the permitted level of AFM1 has been set at 0.025 $\mu g\ kg^{-1}$.

An important health effect of aflatoxins is their link with liver cancer. In 2012, about 745,000 deaths worldwide were estimated to have been caused mostly by aflatoxin-induced hepatocellular carcinoma (HCC) [14]. In the same year, a total of 782,451 new liver cancer cases and 745,533 related deaths were estimated to occur per year [15]. In addition to liver cancer and cirrhosis, aflatoxins have also been linked to growth stunting in children, malnutrition, kwashiorkor, or marasmus diseases, and the suppression of immune responses [16].

Although, the levels of mycotoxins found in the diet are often low, because of their longer life duration (from now) than adults, children are critically affected by natural contaminants such as mycotoxins and thus are prone to develop chronic syndromes in the future (e.g., mycotoxin-related cancers) [17,18]. Moreover, infants and young children are more vulnerable to the deleterious effects of mycotoxins, because of their larger intake/body weight ratio, higher metabolic rate, and lower detoxification capabilities [19,20]. Therefore, it is necessary to evaluate mycotoxin presence in foods and the level of exposure in children [21].

Food security and food safety is an important prerequisite for good health. Milk and dairy products are a source of many nutrients including proteins, fatty acids, calcium, vitamins, and minerals essential for human health, especially in infants and children [22]. However, the risk of contamination by AFM1 is an important food safety concern for milk. Despite the available data on AFM1 occurrence in milk and dairy products, information on exposure and risk assessment of infants and young children in Serbia is lacking. This is due to a combination of limited monitoring systems and a lack of food consumption data. Thus, the extent and health implications associated with mycotoxin exposure of infants and young children need to be evaluated and should be given a priority in Serbia.

Therefore, the objective of the present study was to conduct a preliminary risk assessment and evaluate the dietary exposure of the child population in Serbia to AFM1. The results of our study are helpful to risk managers in their prioritization for food monitoring programs as part of risk-based food control, as well as in the application of adequate measures to protect the health of children.

2. Materials and Methods

2.1. Sample Collection

We conducted a risk assessment for AFM1 by combining the concentration of AFM1 in food commodities from several studies published between 2017 and 2019 [5,23,24]. In brief, a total of 3404 milk and milk-based food samples was randomly collected from various regions of Serbia from 2017 to 2019. The samples consisted of different types of milk, dairy products, and infant formula. The majority of collected samples were from local dairy processing plants that manufacture fluid milk, cheese, cream, or cultured dairy products, while some of the food samples (i.e., infant formula, milk beverages) were from imported sources.

2.2. Sample Preparation and Analysis

The sample analysis was performed with an enzyme-linked immunosorbent assay (ELISA). Preparation of the samples and the ELISA test procedure for the determination of AFM1 in milk-based food samples was performed according to the manufacturer's instructions (Tecna S.r.l., Mirandola (MO), Italy). For ELISA analysis, 100 µL of diluted antibody solution was added to each well and shaken for 30 s. The plate was incubated for 45 min at room temperature (20 to 25 °C). After four washing steps, 100 µL of enzyme conjugate solution was pipetted into separate duplicate wells and the plate gently shaken to mix. After incubation of the plate for 15 min at room temperature in the dark, the liquid in the wells was discarded, and, to complete removal of the remainder of the liquid, the plate was tapped against an absorbent paper (three times). After four washing steps, 100 µL of developing solution was added to each well and the plate was shaken for 30 s. The plate was incubated for 15 min at room temperature in the dark. Then, using a multichannel pipette, 50 µL of the stop solution reagent was added to each well and mixed for several seconds.

Optical density was measured using ELISA-reader Thermo Scientific (Waltham, MA, USA, SAD) model 364, at a wavelength of 450 nm. Ascent software (v.1.0) was used for data acquisition and processing. The detection limit of the method was 0.005 µg/kg, while specificity was 100%, and 16% for AFM1 and AFM2, respectively. Relative standard deviation of reproducibility was 6%, and recovery was 110%. Quality assurance regarding the ELISA method was confirmed by participation in a proficiency testing scheme (PROGETTO TRIESTE) of lyophilized milk. The proficiency test results were satisfactory according to the calculated z-score of 0.09 and 1.27 for AFM1 and AFM2, respectively (acceptable range for z: −2 to 2).

The samples with AFM1 levels above the MRL were confirmed and also quantified by LC-MS/MS analytical techniques.

2.3. Extraction of Milk Samples for LC-MS/MS

LC-MS/MS analysis of AFM1 was carried out according to the method previously published by Milicevic et al. [5,23,24].

2.3.1. Standard Solution Preparation

AFM1 standard was purchased from Sigma Aldrich Chemical Company (St. Louis, MO, USA). Working solutions, prepared by diluting the stock solution, were used to prepare the calibration curve and to spike milk samples. The final concentrations of AFM1 used in the calibration curve were 0.2, 1.0, 2.5, 10.0 and 20.0 ng/mL. All stock and working standard solutions were stored in brown vials at −18 °C. For recovery studies, defatted milk was enriched with AFM1 working standard solution at three spiked levels: 0.025, 0.05, 0.075 µg/kg (i.e., 0.5 times MRL, MRL and 1.5 times MRL).

2.3.2. Chromatographic and MS Parameters

The instrument used for LC-MS/MS was a Waters Acquity UPLC system (Waters, Milford, MA, USA) coupled with a TQD mass spectrometer (Waters Micromass, Manchester,

UK). A Purospher Star (Merck, Darmstadt, Germany) RP-18 column (50 × 2.1 mm, 2 μm particle size) was used for the separation of AFB1. The mobile phase was 0.1% acetic acid and methanol (35:65). Isocratic flow was maintained at 0.3 mL/min. Two product ions were monitored (329 > 273 Da and 329 > 259.1 Da). Quantification ion was 273 Da. MassLynx 4.1 software was employed for data acquisition and processing. The detection limit of the method was 0.02 μg/kg, relative standard deviation of reproducibility was 5.4%, and recovery was 65–81%. Linear regression analysis was performed using JMP v.10 software.

2.4. National Food Consumption Survey on Toddlers and Children

A Serbian National Food Consumption Survey on toddlers and children was conducted between 2017 and 2021 according to the EFSA EU MENU methodology [25]. Valid data were collected from a total of 576 participants with 290 toddlers aged from one to below three years old and 286 children aged from three to nine years old. Data collection was conducted using project-specific national survey pack that included a general questionnaire, an age-appropriate food propensity questionnaire (FPQ), and a 24 h food diary. The consumed portion sizes were estimated based on natural units, household measures, packaging information and country-specific portion size measurement aid (PMSA) (i.e., previously tested Food Atlas) [26]. Following EFSA guidance on the EU Menu methodology, a previously developed and validated innovative nutritional software tool DIET ASSESS and PLAN (DAP) was used [27] for standardized and harmonized food consumption data collection and comprehensive dietary intake assessment. Basic FoodEx2 codes including implicit facets were assigned to all foods and recipes from the Serbian Food Composition Data Base (FCDB) which is integrated into the DAP platform. Weight measurements were obtained for children without shoes and jackets using a digital balance and data were recorded to the nearest 0.1 kg. For children's height measurements, portable stadiometers were applied with 0.1 cm accuracy.

2.5. Health Risk Assessment

Deterministic methods (or single point) were employed to derive a worst-case risk estimate. Assessment of cumulative risks posed to the health of children by consumption of milk and milk-based food was performed in three stages which comprised exposure assessment, risk characterization, and assessment of liver cancer risk.

2.6. Exposure Assessment

Chronic AFM1 exposure among the two age groups was estimated by the deterministic approach involving the average probable daily intake (APDI) method [28]. Exposure was calculated for all the food categories, and for both consumer groups according to their gender and age to highlight the differences in exposure. The EDI of AFM1 (expressed as ng kg^{-1} bw day^{-1}) was calculated based on the concentration of AFM1 detected and the intake rate of analyzed foods, according to Equation (1):

$$EDI = \Sigma c * C/bw \quad (1)$$

where Σc is the average concentration of AFM1 (ng kg^{-1}), C is the daily average consumption of the commodity (kg per day), and *bw* is the body weight for the male and female child populations (kg).

The mean concentrations of AFM1 in selected milk and dairy products were taken from Table 1. Within the general framework of chemical risk assessment, a difficult step in dietary exposure evaluation is handling concentration data reported to be below the limit of detection (LOD). These data are known as non-detects and the resulting occurrence distribution is left-censored. The left-censored data (data below LOD and LOQ) were processed by applying EFSA's substitution method [29]. According to this guidance, for dietary exposure assessments, three exposure scenarios were considered. Middle bound (MB), assuming that the not detected results correspond to half of the LOD

(ND = 2.5 ng kg^{-1}) was used for all AFM1 when a finding with a value <LOD was in $\leq$60% of samples. In contrast, when a large percentage of the results were below the LOD (>60 but $\leq$80% non-quantified and with at least 25 results quantified), two estimates used a lower bound (LB) scenario, in which zero was assigned to samples showing AFM1 concentration below LOD/LOQ, and the upper bound (UB) was obtained assuming the value for the LOD of AFM1 (5.0 ng kg^{-1}) for the results of AFM1 reported as not detected (ND = LOD). Furthermore, following EFSA recommendations, exposure calculations at the 95th percentile (P95) of AFM1 concentration (P95) were performed to evaluate the worst-case scenarios [30]. The daily average consumption of these products and mean body weights were obtained from the data provided in the food frequency questionnaire by age (Tables 2 and 3). The different food commodities were grouped within each food category to better explain their contribution to the total dietary exposure to AFM1.

Table 1. Aflatoxin M1 incidence and concentration in milk and dairy product samples included in the study.

Type of Sample	*n*/*N* (%)	Mean (ng kg^{-1} ± SD) of All Samples				Mean Positives (ng kg^{-1} ± SD)	Median Positives (ng kg^{-1})	Q1 Positives (ng kg^{-1})	Q3 Positives (ng kg^{-1})	Range (ng kg^{-1})
		LB	MB	UB	P95					
Infant formula	14/92 (15.2)	1.6 ± 0.004	3.76 ± 0.003	5.88 ± 0.002	12.5	10.0 ± 0.002	11.0	8.00	13.00	8.0–14.0
Fermented milk products	158/775 (20.3)	9.58 ± 0.02	11.5 ± 0.02	13.56 ± 0.02	57.0	47.0 ± 0.022	38.0	34.0	56.25	25.0–174.0
Clotted cream	0/48	-	-	-		-	-	-	-	<5.0
Butter	14/143 (10)	5.20 ± 0.01	7.44 ± 0.016	9.70 ± 0.01	47.0	53.0 ± 0.016	47.0	41.75	58.0	39.0–92.0
Milk beverages	145/714 (20)	4.22 ± 0.01	6.21 ± 0.011	8.20 ± 0.01	23.0	20.77 ± 0.02	14.0	10.50	22.50	5.0–117.0
Sour cream	19/132 (14)	6.90 ± 0.02	9.04 ± 0.018	11.18 ± 0.02	48.0	47.95 ± 0.002	39.0	31.0	60.0	25.0–103.0
Cheese	7/404 (2)	1.36 ± 0.01	3.82 ± 0.014	6.28 ± 0.01	5.0	7.89 ± 0.08	49.0	40.0	58.0	39.0–276.0
Pasteurized and UHT milk	574/725 (79)	22.34 ± 0.02	22.87 ± 0.018	23.40 ± 0.01	53.0	28.22 ± 0.016	25.00	19.00	35.0	5.0–132.0
Milk powder	67/201 (33)	9.12 ± 0.02	10.79 ± 0.020	12.46 ± 0.02	47.0	27.37 ± 0.03	16.00	9.00	36.00	5.0–155.0
Whey liquid	13/90 (14)	14.82 ± 0.05	16.96 ± 0.05	19.10 ± 0.05	70.0	102.6 ± 0.10	70.0	20.50	211.0	5.0–278.0
Total	**1012/3404 (29.7)**	**9.47 ± 0.02**	**11.23 ± 0.02**	**12.99 ± 0.02**	**44.0**	**31.86 ± 0.026**	**27.00**	**16.00**	**38.00**	**5.0–278.0**

N = number of analyzed samples. *n* = number of positive samples (AFM1 > LOD). % = percentage of positive samples. The limit of detection (LOD) for AFM1 is 5.0 ng kg^{-1}. Lower bound (LB) = assuming that the not detected results are equal to 0 (ND = 0). Middle bound (MB) = assuming that the not detected results correspond to half of the LOD (ND = 2.5 ng kg^{-1}). Upper bound (UB) = assuming that the not detected results correspond to the LOD (ND = 5.0 ng kg^{-1}). P95 = 95th percentile. First quartile (Q1) 25% of the data are less than or equal to this value. Second quartile (Q2) = the median. A total of 50% of the data are less than or equal to this value. Third quartile (Q3) = 75% of the data are less than or equal to this value.

Table 2. Characteristics of the study sample [25].

Age Group	Body Weight (kg)		*N*	
	Male	Female	Male	Female
Toddlers, 1–3 years	14	13	98	91
Children, 3–9 years	24	24	159	150

N = number of participants.

Table 3. The average intake of food groups by child population (g/day) [25].

Age Group		Infant Formula	Fermented Milk Products	Clotted Cream	Butter	Milk Beverages	Sour Cream	Cheese	Pasteurized and UHT Milk	Whey Liquid
Toddlers. 1–3 years	M	31.51	133.40	9.94	4.88	230.00	15.00	22.50	102.87	-
	F	29.93	115.76	8.72	5.30	213.30	17.47	24.40	112.05	-
Children. 3–9 years	M	-	153.24	13.31	7.95	220.83	16.58	27.19	99.55	250.0
	F	-	150.50	13.60	8.02	199.75	15.05	26.94	93.58	-
Average		**30.72**	**138.23**	**11.41**	**6.54**	**215.97**	**16.02**	**25.25**	**102.01**	**250.0**

M = male. F = female. Food groups were categorized according to a national survey.

2.7. Risk Characterization

Since AFM1 is considered carcinogenic, there is no TDI based on a dose of no observable effect (NOEL). Therefore, risk characterization originating from the oral exposure to aflatoxins was calculated using two approaches; the qualitative margin of exposure (MOE) approach established by EFSA [30] for substances that are both genotoxic and carcinogenic and the quantitative approach to liver cancer risk estimation proposed by the FAO/WHO [31].

The MOE value was calculated using Equation (2):

$$MOE = BMDL_{10}/EDI \tag{2}$$

where BMDL10 is the benchmark dose lower confidence limit (BMDL10) for 10% increased cancer risk. Based on animal data, EFSA concluded that AFM1 induces liver cancer with a potency one-tenth that of AFB1 (for AFB1 0.4 μg kg^{-1} bw per day^{-1}), so hence, a potency factor of 0.1 for the AFM1 risk assessment was used in this study. EDI is the average daily intake used to estimate chronic dietary exposure to AFB1, as calculated in Equation (1). A calculated MOE value lower than 10,000 implies that exposure to a carcinogenic and genotoxic substance contributes to the risk of HCC and is of concern to public health [30].

2.8. Assessment of Liver Cancer Risk—The Carcinogenic Potency

Most health concerns for aflatoxins are related to primary liver cancer burden, as the ingestion of these toxins has been directly linked to HCC development, particularly in individuals infected with hepatitis virus. To estimate the risk of cancer posed by dietary exposure to AFM1, we used the following equation:

$$\text{Population risk} = \text{EDI} \times \text{Average potency} \tag{3}$$

Regarding the differences in carcinogenic potency, for AFM1, according to JECFA [28], AFM1 induces liver cancer with one-tenth of the potency of AFB1. Therefore, the carcinogenic potency (CP) of AFM1 was calculated to be 0.0562 additional cancer cases per 100,000/year per 1 ng kg^{-1} bw day^{-1} for hepatitis B virus (HBV) surface antigen positive ($HBsAg^{+}$) populations and 0.0049 additional cancer cases per 100,000/year per 1 ng kg^{-1} bw day^{-1} for $HBsAg^{-}$ populations. The prevalence used of HBV-infected individuals in Serbia was 1%, based on earlier studies [5]. Thus, the CP of 1 ng AFM1 kg^{-1} bw day^{-1} in a population with a 1% prevalence of HBV infection would be 0.005413 cases per year per 100,000 people according to Equation (4):

$$\text{Average cancer potency} = (0.0049 \times 0.99 + 0.0562 \times 0.01) \tag{4}$$

2.9. Statistical Analysis

Data were analyzed using Minitab statistical software version 17 (Minitab Ink., Coventry, UK). AFM1 concentrations for the studied samples were expressed in the form of descriptive statistics and presented in Tables 1 and 4–7. A Shapiro–Wilk test of normality was run to check the normality of data and after recording the data as normal, a further test was used for statistical evaluation of the data.

Table 4. Estimated daily intake (ng kg^{-1} bw day^{-1}) of AFM1 in selected food products for two age categories.

Food Group	Exposure (ng kg^{-1} bw day^{-1})											
	Toddlers, 1–3 Years						Children, 3–9 Years					
	Male			Female			Male			Female		
	LB	UB	P95	LB	UB	P95	LB	UB	P95	LB	UB	P95
Infant formula	0.004	0.014	0.029	0.004	0.014	0.029						
Fermented milk products	0.091	0.129	0.543	0.085	0.121	0.508	0.061	0.087	0.364	0.060	0.085	0.358
Butter	0.002	0.003	0.016	0.002	0.004	0.019	0.002	0.003	0.016	0.002	0.003	0.016
Milk beverages	0.069	0.135	0.378	0.069	0.134	0.377	0.039	0.075	0.212	0.035	0.068	0.191
Sour cream	0.007	0.012	0.051	0.009	0.015	0.065	0.005	0.008	0.033	0.004	0.007	0.030
Cheese	0.002	0.010	0.008	0.003	0.012	0.009	0.002	0.007	0.006	0.001	0.007	0.005
Pasteurized and UHT milk	0.164	0.172	0.389	0.193	0.202	0.457	0.093	0.097	0.220	0.087	0.091	0.207
Whey liquid							0.154	0.177	0.199			
Total	**0.340**	**0.475**	**1.415**	**0.365**	**0.501**	**1.463**	**0.201**	**0.277**	**0.850**	**0.190**	**0.262**	**0.807**

Lower bound (LB) = assuming that the not detected results are equal to 0 (ND = 0). Middle bound (MB) = assuming that the not detected results correspond to half the LOD (ND = 2.5 ng kg^{-1}). Upper bound (UB) = assuming that the not detected results correspond to the LOD (ND = 5.0 ng kg^{-1}). P95 = 95th percentile. M = male. F = female.

Table 5. The margin of exposure (MOE) values based on dietary exposure to AFM1 for two age categories.

Food Group	MOE											
	Toddlers, 1–3 Years						Children, 3–9 Years					
	Male			Female			Male			Female		
	LB	UB	P95	LB	UB	P95	LB	UB	P95	LB	UB	P95
Infant formula	1,076,492	292,923	137,791	1,085,686	295,425	138,968						
Fermented milk products	43,819	30,958	7365	46,890	33,127	7881	65,393	46,200	10,991	66,566	47,028	11,188
Butter	2,206,810	1,183,032	244,158	1,886,792	1,011,476	208,752	2,322,206	1,244,894	256,925	2,301,937	1,234,028	254,682
Milk beverages	57,696	29,692	10,586	57,851	29,772	10,614	103,015	53,015	18,901	113,886	58,610	20,896
Sour cream	541,063	333,930	77,778	431,381	266,237	62,011	839,146	517,899	120,627	924,455	570,549	132,890
Cheese	1,830,065	396,320	497,778	1,567,020	339,355	426,230	2,596,110	562,215	706,142	2,679,887	580,358	728,929
Pasteurized and UHT milk	24,368	23,264	10,271	20,773	19,832	8756	43,166	41,211	18,195	45,920	43,840	19,356
Whey liquid							25,911	22,642	20,105			
Average	**825,759**	**327,160**	**140,818**	**728,056**	**285,032**	**123,316**	**856,421**	**355,439**	**164,555**	**1,022,109**	**422,402**	**194,657**

MOE calculations were based on benchmark dose ($BMDL_{10}$) for AFB1 of 0.4 μg kg^{-1} bw day^{-1} and potency factor for AFM1 of 0.1 [30]. Lower bound (LB) = assuming that the not detected results are equal to 0 (ND = 0). Middle bound (MB) = assuming that the not detected results correspond to half the LOD (ND = 2.5 ng kg^{-1}). Upper bound (UB) = assuming that the not detected results correspond to the LOD (ND = 5.0 ng kg^{-1}). P95 = 95th percentile.

Table 6. Cancer risk estimates calculated from chronic dietary exposure to AFM1. Scenario 1 (mean).

Food Group	Liver Cancer Risk (Case/100,000 Persons)											
	Toddlers, 1–3 Years						Children, 3–9 Years					
	Male			Female			Male			Female		
	LB	UB	P95	LB	UB	P95	LB	UB	P95	LB	UB	P95
Infant formula	0.00001	0.00003	0.00006	0.00001	0.00003	0.00006						
Fermented milk products	0.00018	0.00025	0.00106	0.00017	0.00024	0.00099	0.00012	0.00017	0.00071	0.00012	0.00017	0.00070
Butter	0.00000	0.00001	0.00003	0.00000	0.00001	0.00004	0.00000	0.00001	0.00003	0.00000	0.00001	0.00003
Milk beverages	0.00014	0.00026	0.00074	0.00013	0.00026	0.00074	0.00008	0.00015	0.00041	0.00007	0.00013	0.00037
Sour cream	0.00001	0.00002	0.00010	0.00002	0.00003	0.00013	0.00001	0.00002	0.00006	0.00001	0.00001	0.00006
Cheese	0.00000	0.00002	0.00002	0.00000	0.00002	0.00002	0.00000	0.00001	0.00001	0.00000	0.00001	0.00001
Pasteurized and UHT milk	0.00032	0.00034	0.00076	0.00038	0.00039	0.00089	0.00018	0.00019	0.00043	0.00017	0.00018	0.00040
Whey liquid							0.00030	0.00034	0.00039			
Total	**0.00066**	**0.00093**	**0.00276**	**0.00071**	**0.00098**	**0.00286**	**0.00069**	**0.00089**	**0.00166**	**0.00037**	**0.00037**	**0.00158**

Potency estimates of 0.0017 (mean) per 100,000 person-years per ng kg^{-1} bw day^{-1} were calculated for HBsAg-negative individuals. For HBsAg-positive individuals, potency estimates of 0.0269 (mean) per 100,000 person-years per ng kg^{-1} bw day^{-1} were calculated [31]. The risk of liver cancer was estimated as new cancer cases $year^{-1}$ per 100,000 population by multiplying the AFM1 EDI by the average HCC potency 0.001952 (mean) based on 1% prevalence of HBV infection in Serbia.

Table 7. Cancer risk estimates calculated from the chronic dietary exposure to AFM1. Scenario 2 (95% upper bound (UB)).

Food Group	Liver Cancer Risk (Case/100,000 Persons)											
	Toddlers, 1–3 Years						Children, 3–9 Years					
	Male			Female			Male			Female		
	LB	UB	P95	LB	UB	P95	LB	UB	P95	LB	UB	P95
Infant formula	0.00002	0.00007	0.00016	0.00002	0.00007	0.00016						
Fermented milk products	0.00049	0.00070	0.00294	0.00046	0.00065	0.00275	0.00033	0.00047	0.00197	0.00033	0.00046	0.00194
Butter	0.00001	0.00002	0.00009	0.00001	0.00002	0.00010	0.00001	0.00002	0.00008	0.00001	0.00002	0.00009
Milk beverages	0.00038	0.00073	0.00205	0.00037	0.00073	0.00204	0.00021	0.00041	0.00115	0.00019	0.00037	0.00104
Sour cream	0.00004	0.00006	0.00028	0.00005	0.00008	0.00035	0.00003	0.00004	0.00018	0.00002	0.00004	0.00016
Cheese	0.00001	0.00005	0.00004	0.00001	0.00006	0.00005	0.00001	0.00004	0.00003	0.00001	0.00004	0.00003
Pasteurized and UHT milk	0.00089	0.00093	0.00211	0.00104	0.00109	0.00247	0.00050	0.00053	0.00119	0.00047	0.00049	0.00112
Whey liquid							0.00084	0.00096	0.00108			
Total	**0.00184**	**0.00257**	**0.00766**	**0.00197**	**0.00271**	**0.00792**	**0.00109**	**0.00150**	**0.00460**	**0.00103**	**0.00142**	**0.00437**

Potency estimates of 0.0049 (95% upper bound (UB)) per 100,000 person-years per ng kg^{-1} bw day^{-1} were calculated for HBsAg-negative individuals. For HBsAg-positive individuals, potency estimates of 0.0562 (95% UB) per 100,000 person-years per ng kg^{-1} bw day^{-1} were calculated [31]. The risk of liver cancer was estimated as new cancer cases $year^{-1}$ per 100,000 population by multiplying the AFM1 EDI by the average HCC potency 0.005413 (UB) based on 1% prevalence of HBV infection in Serbia.

3. Results

3.1. Prevalence of AFM1 in Milk and Milk-Based Food

The prevalence of AFM1 in milk and milk product samples collected from various regions of Serbia from 2017 to 2019 is presented in Table 1. Amongst the collected samples, 574/725 pasteurized and UHT milk, 67/201 milk powder, 158/775 fermented milk products, 145/714 milk beverages, 14/92 infant formula, 19/132 sour cream, 13/90 whey, 14/143 butter, and 7/404 cheese were contaminated with AFM1. Overall, the mean levels (ng kg^{-1}) of AFM1 based on the LB mean ranked as follows: pasteurized and UHT milk > whey > fermented milk products > milk powder > sour cream > butter > milk beverages > infant formula > cheese. As expected, the highest incidence of contamination (79%) and the greatest mean concentration of AFM1 were observed in pasteurized and UHT milk (22.34 $\pm$ 0.02 ng kg^{-1}), while cheese with 1.36 $\pm$ 0.01 ng kg^{-1} showed the lowest mean concentration. Among the different milk products, the maximum AFM1 level found in this study was registered in a whey sample, reaching a contamination level of 278 ng kg^{-1}, followed by cheese (276 ng kg^{-1}) and a fermented milk product (174 ng kg^{-1}).

The mean concentration of AFM1 in the present study is slightly lower compared to the previous studies from Serbia [32–34]. In addition, the results of this study are in agreement with the reported AFM1 concentrations in milk and dairy products from global studies, where the prevalence of AFM1 in milk worldwide was 79.1% [30]. This could be explained by the fact that preventive and control activities during harvest, processing, and storage of dairy feeds, combined with the improvement of risk management actions in dairy processing industries have been improved in recent years to a considerable extent. The variation in the mean AFM1 contamination in milk previously reported may be attributed to differentiation in carry-over rates of AFB1 in milk, which depends on the animal species, but these rates can also vary greatly depending upon nutritional, environmental, and physiological factors such as stage of lactation, systemic diseases, local (mammary) infections, level of AFB1 in feed, rate of feed ingestion, and geographical and seasonal conditions [35]. It is also important to highlight that many of the data have been obtained using different methodologies, with a consequence of different sensitivity and precision.

Aflatoxin contamination of foods of animal origin (milk, dairy products, eggs, and edible animal products) is a global public health and economic concern. The presence of AFM1 in milk and milk products is most probably the consequence of feeding dairy cows a diet contaminated with AFB1. Since their presence has been responsible for significant adverse health and economic issues affecting consumers and farmers worldwide, the formulation of regulations to control their presence in animal feed has been triggered [8]. Various investigations conducted in Serbia in the last decade have revealed a significant presence of aflatoxins in maize [3,36]. In general, the reported concentration of AFB1 in maize and consequently the presence of AFM1 in milk showed year-to-year variations in

AFM1 prevalence [24]. Therefore, to estimate the risk of illness in the Serbian population exposed to aflatoxins, a series of survey studies have been conducted to monitor the incidence of AFM1 contamination, particularly in raw milk in Serbia.

Albeit mean concentrations of AFM1 in our study were lower in whey and cheese than in the milk samples, the highest concentrations of AFM1, which we measured in whey and cheese, was also observed in these two products in previous studies, which concluded that during cheese production, 60% of the initial content of AFM1 accumulates in the whey, while 40% of the AFM1 remains in the curd or fresh cheese [23,37,38]. This might be due to the water-soluble nature of AFM1 and its affinity to form a hydrophobic bond with the hydrophobic part of casein that is subsequently concentrated in cheese [39]. Furthermore, the AFM1 concentration in soft cheeses was generally 2.5–3.3 times higher, and in hard cheeses, 3.9–5.8 times higher, than in the milk from which the cheeses were made [34,37,40]. Most studies have reported that AFM1 concentrations in milk products are strongly dependent on the AFM1 concentrations in milk, and hence milk concentrations could be a good predictor of the AFM1 concentration in cheese and whey. Notably, in the present study, AFM1 was found in 20% (158/775) of fermented milk products (ranging from 25 to 174 ng kg^{-1}, mean 9.58 $\pm$ 0.02 ng kg^{-1}). The presence of AFM1 in fermented milk products may be due to manufacturers usually using imported dry milk for producing dairy products that were contaminated with AFM1. However, this low level of AFM1 in fermented milk products could also be attributed to the function of lactic acid bacteria, during fermentation [41]. Results indicated that the incidence and mean AFM1 values obtained in the present study are low to moderate. Hence the risk of AFM1 exposure could not be a public health concern for the general population. However, as children use milk and dairy products in their diets frequently and are more sensitive to the adverse effects of aflatoxins compared to adults, ingestion of low doses of AFM1 in milk over long periods must be considered a risk, and should not be underestimated or neglected.

3.2. Dietary Exposure Assessment

Risk assessment through dietary exposure is the process of estimating the magnitude and the probability of a harmful effect on individuals or populations from specified agents or activities. Per definition, exposure assessment, as one component of risk assessment methodology, combines mycotoxin levels in food with consumption patterns, and therefore, provides valuable information for risk management if mycotoxins compromise food safety and health hazards, at either an individual or a population level [9]. Following the recommendations of EFSA [29], the current study utilized the most comprehensive (chronic) exposure scenario to assess the EDI of AFM1 by Serbian children, taking into account a range of LB and UB values.

Based on the data described before (Section 2.4), the EDI of AFM1 (ng kg^{-1} bw day^{-1}) through milk and dairy product consumption in different age categories was calculated and is presented in Table 4. It is widely considered that the LB scenario generally underestimates contamination and exposure levels and that the UB scenario overestimates them [29]. As can be seen from Table 4, the exposure of AFM1 differs from product to product, and a significant difference ($p < 0.05$) between the exposure values assessed, considering the UB, and LB scenarios, was found within products. In consequence, for these purposes, the middle-bound approach should be applied. In this study, the highest EDIs of AFM1, i.e., 0.164–0.172 ng kg^{-1} bw day^{-1} (LB-UB) and 0.193–0.202 ng kg^{-1} bw day^{-1}, were found for the consumption of pasteurized and UHT milk by male and female toddlers, respectively.

The AFM1 exposures were ranked for all the food types: pasteurized and UHT milk > whey > fermented milk products > milk beverages > sour cream > infant formula > butter > cheese. The food categories pasteurized and UHT milk (46 to 48%) and fermented milk products (27 to 31%) were the main contributors to the overall AFM1 mean exposure throughout both age groups (Figure 1). Due to the limited number of consumption and concentration data for milk powder and clotted cream, these food categories were not taken into account for risk assessment.

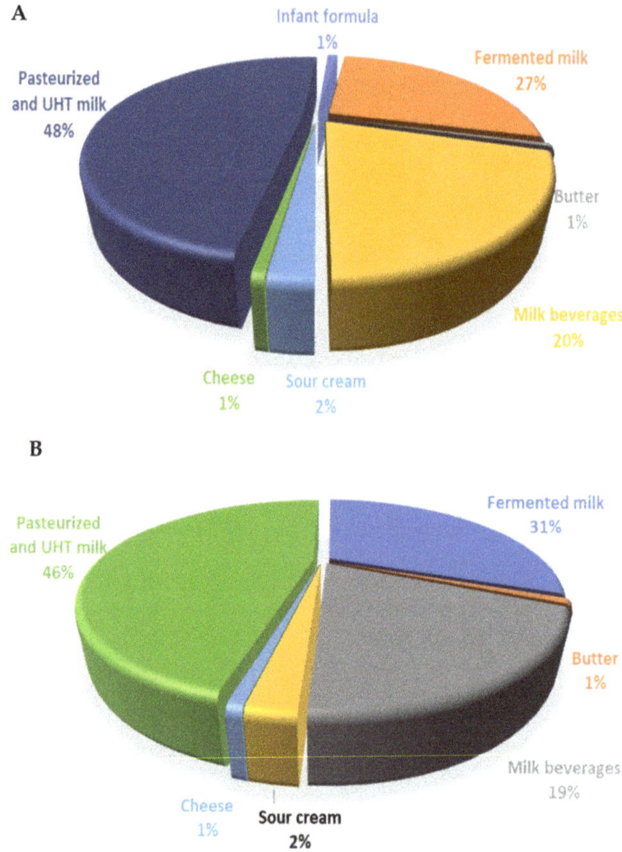

Figure 1. Contribution (%) of the most important food groups to the long-term dietary exposure of children aged 2 to 6 years (**A**) and toddlers, 1–3 years (**B**), to AFM1.

The findings obtained in this study showed a remarkably lower exposure of the Serbian child population in comparison with the estimates of AFM1 intake reported by Kos et al. [32] and Milićević et al. [5]. These contradictory results regarding the EDI could be attributed to some uncertainties, including occurrence data (sampling strategy, low number of samples, seasonal effects, lack of sensitivity of some analytical methods) and exposure modeling. In addition, data from a health survey of the population of Serbia indicate a negative trend in the consumption of milk and dairy products. At least 41.8% of the population consumed milk and dairy products on a daily basis in 2021, which is significantly less than in 2013, when 51.7% of the population reported daily consumption.

Although aflatoxin contamination in food occurs in many countries around the world, the nations that have been identified to be substantially exposed to AFM1 (sometimes dramatically) are primarily in sub- Saharan Africa and South Asia (Iran and Pakistan), and of particular concern are populations of children [42]. Generally, the mean dietary AFM1 exposure from milk and dairy product consumption in European populations is comparatively low, which may be the result of strict regulations on mycotoxins in feed and milk products and from the adoption of an integrated food safety management system. In comparison to international studies, our results were lower than the results of several studies. In addition, the current EDI values did not exceed the previously established international TDI limit (0.2 ng kg^{-1} bw day^{-1}) [43]. Although the estimated AFM1 exposure levels for milk and dairy product consumers in the present study are relatively low, owing to the genotoxic and carcinogen nature of aflatoxins, the approach of "as low as reasonably achievable" (ALARA) could be adopted in forthcoming regulations to protect Serbian consumers against the health effects caused by AFM1.

3.3. Risk Characterization/Cancer Risk Attributable to AFM1

The risk of exposure to AFM1 through milk and dairy product consumption was characterized using MoE (Table 5), and the liver cancer risk approach (Tables 6 and 7). According to the EFSA scientific committee guidance [30], when the MoE value is $\geq$ 10,000, it is considered that there is a low risk of a negative impact on public health. Our results showed that MoE values for LB and UB exposure scenarios to AFM1 were far higher than 10,000 in toddlers and other children, which indicates no health concern due to exposure to AFM1 through consumption of milk and dairy products. However, as children consume more milk relative to their body weight, children's exposure risk to AFM1 in milk and dairy products should be a continuous focus of attention.

The results of the characterization of HCC risk (cases per 100,000 individuals per year) for different age groups due to AFM1 exposure based on the calculation of the risk by P-cancer and EDI, are presented in Tables 6 and 7. The additional cancer risk due to mean exposure to AFM1 associated with milk and dairy product consumption in toddlers using potency estimates of 0.0017 (mean) for the LB scenario ranged from 0.00032 to 0.00001 and from 0.00038 to 0.00001 cases per 100,000 individuals per year for males and females, respectively. For other children, the mean estimated number of liver cancer cases for the LB scenario ranged from 0.00030 to 0.00012 cases per 100,000 individuals per year for males and from 0.00017 to 0.00001 cases per 100,000 individuals per year for females. The main contribution of HCC risk due to AFM1 exposure was caused by the consumption of pasteurized and UHT milk, estimated at 0.00038 and 0.00039 cases per 100,000 individuals per year for the LB and UB scenarios, respectively. Our results are considerably lower than those reported in an assessment by EFSA [30] where the estimated cancer risk (mean and UB) ranged between 0.002–0.035, 0.008–0.032, 0.003–0.018, 0.001–0.006, 0.001–0.004, and 0.001–0.003 aflatoxin-induced cancers per 100,000 person-years for infants, toddlers, other children, adolescents, adults, and the elderly, respectively. Globally the standardized annual incidence rate for liver cancer is 15.3 per 100,000 among men and 5.4 per 100,000 among women [44]. Several studies conducted in African and South Asian countries have investigated the health impacts of early dietary exposure to aflatoxins. Prolonged exposure to aflatoxin might be the underlying cause of congenital disabilities and child growth impairment. Most of the studies have reported that exposure to aflatoxins might be the underlying cause of child growth impairment [45,46]. Nonetheless, the possible association between chronic exposure to aflatoxins in early life and the early onset of hepatic cancer has been explored by several studies. AFB1 is the most potent human hepatocarcinogen, accounting for around 4.6–28.2% of the total HCC cases worldwide [47]. Further, there is a strong synergistic association between AFB1 and HBV infection in the etiology of HCC. Recent results from a national study in Serbia [48] revealed that the rate of acute cases of HBV infection continued to decline (incidence of 1.25/100,000 inhabitants) over the last few years (2010–2019), which is following global trends and most likely reflects the impact of national vaccination programs. On the other hand, there is an increasing trend in the numbers of registered cases of chronic HBV and hepatitis C infections. Improving the health and well-being of children are priority health policies of many countries. It is necessary to provide children with stability and an environment for growth and development that includes good health and proper nutrition. Numerous epidemiological studies link childhood health with health outcomes in adults, and investing in children's health is one of the most important measures that society can take to improve the health of the entire population. In summary, future work in this area would focus on the survey of occurrence and exposure to AFM1 to identify geographic regions where AFB1 levels in staple food are high enough to cause concern for human populations.

4. Strengths and Limitations

This is the first ever conducted study on AFM1 exposure risk assessment of children population through milk and milk products in Serbia using harmonized food consumption data collected within the EU Menu project survey according to EFSA guidance methodology, which makes this data comparable with other harmonized food consumption data in whole Europe. A limitation of this study is that relatively small number of infant formula, clotted cream, butter and sour cream have been considered, leading to underestimated health risk associated with exposure to AFM1 from milk products among the children. Furthermore, limitation of the study lies in the fact that total exposure to AFM1 was not assessed for the whole diet, i.e., from other food groups that certainly contain AFM1, and will be the area of research in further studies.

5. Conclusions

Considering the present evidence on the negative health effects of AFM1, this study through the MOE approach and the population risk assessment method suggests that milk and dairy products had negligible health risk to the child population due to AFM1 exposure. Despite current AFM1 concentrations being not high enough to elicit toxic effects, risk data should be interpreted carefully due to the present study investigating only the consumption of milk and dairy products. Thus, the focus of future studies should be on exposure from complete diets commonly consumed by Serbian children to estimate cumulative exposure from all sources of aflatoxins. In addition, further research is advisable, in particular related to the association of liver cancer with AF intake and HBV infection. Since the contamination of feedstuffs with AFB1 plays a major role in the contamination of milk, the government and all stakeholders involved in the milk supply chain should pay more attention to implementing an integrated food safety management system to prevent the production of mycotoxins in dairy cattle feed and to reduce AFM1 residues in milk.

Author Contributions: Conceptualization, D.R.M., J.M. and M.G. (Mirjana Gurinović); methodology, D.R.M., J.M., M.G. (Mirjana Gurinović) and S.J.; validation, S.J.; formal analysis, J.Đ.-S. and J.M.; investigation, M.Z.; writing—original draft preparation, D.R.M. and J.M.; writing—review and editing, D.R.M., J.M. and M.G. (Mirjana Gurinović); supervision, M.G. (Maria Glibetić). All authors have read and agreed to the published version of the manuscript.

Funding: This research was funded by the Ministry of Education, Science and Technological Development of the Republic of Serbia, Agreement on Realization and Financing of Scientific Research work of NIO for 2021, No. 451-03-9/2021-14/200050 and No. 451-03-9/2021-14/200015. In addition, research was funded by the European Food Safety Authority OC/EFSA/DATA/2016/02-CTO1.

Institutional Review Board Statement: The study was conducted in accordance with the guidelines laid down in the Declaration of Helsinki and all procedures involving human subjects were approved by the Institute for Medical Research Ethics Committee in Serbia on 8 December 2017 (EO 123/2017).

Informed Consent Statement: Written informed consent for inclusion was obtained from all the participants.

Data Availability Statement: Results attained in this study are included in the manuscript. Individual data are not publicly available due to ethical restrictions.

Acknowledgments: The authors of this manuscript are grateful to Agnes Kadvan for IT support in the EU Menu National Food Consumption Survey.

Conflicts of Interest: The authors declare no conflict of interest.

References

1. Medina, A.; González, J.M.; Sainz, M.J. Impact of global warming on mycotoxins. *Curr. Opin. Food Sci.* **2017**, *18*, 76–81. [CrossRef]
2. Raffa, N.; Keller, N.P. A call to arms: Mustering secondary metabolites for success and survival of an opportunistic pathogen. *PLoS Pathog.* **2019**, *15*, e1007606. [CrossRef] [PubMed]
3. Udovicki, B.; Audenaert, K.; De Saeger, S.; Rajkovic, A. Overview on the mycotoxins incidence in Serbia in the period 2004–2016. *Toxins* **2018**, *10*, 279. [CrossRef] [PubMed]
4. Milićević, D.; Škrinjar, M.; Baltić, T. Real and perceived risks for mycotoxin contamination in foods and feeds: Challenges for food safety control. *Toxins* **2010**, *2*, 572–592. [CrossRef] [PubMed]
5. Milićević, D.; Spirić, D.; Radičević, T.; Velebit, B.; Stefanović, S.; Milojević, L.; Janković, S. A review of the current situation of aflatoxin M 1 in cow's milk in Serbia: Risk assessment and regulatory aspects. *Food Addit. Contam. A* **2017**, *34*, 1617–1631. [CrossRef]
6. Milićević, D.; Nastasijevic, I.; Petrovic, Z. Mycotoxin in the food supply chain—Implications for public health program. *J. Environ. Sci. Health C* **2016**, *34*, 293–319. [CrossRef]
7. Pleadin, J.; Frece, J.; Markov, K. Mycotoxins in food and feed. *Adv. Food Nutr. Res.* **2019**, *89*, 297–345.
8. Milicević, D.; Udovički, B.; Petrović, Z.; Janković, S.; Radulović, S.; Gurinović, M.; Rajković, A. Current status of mycotoxin contamination of food and feeds and associated public health risk in Serbia. *Meat Technol.* **2020**, *61*, 1–36. [CrossRef]
9. Wild, C.P.; Miller, J.D.; Groopman, J.D. (Eds.) *Mycotoxin Control in Low and Middle-Income Countries*; (IARC Working Group Reports, No. 9.); International Agency for Research on Cancer: Lyon, France, 2015. Available online: https://www.ncbi.nlm.nih.gov/books/NBK350558/2015 (accessed on 11 October 2021).
10. De Ruyck, K.; De Boevre, M.; Huybrechts, I.; De Saeger, S. Dietary mycotoxins, co-exposure, and carcinogenesis in humans: Short review. *Mutat. Res. Rev. Mutat. Res.* **2015**, *766*, 32–41. [CrossRef]
11. IARC (International Agency for Research on Cancer). *Chemical Agents and Related Occupations*; WHO: Geneva, Switzerland, 2012; pp. 225–248.
12. European Commission. Regulation (EC) No 165/2010 of the European Parliament and of the Council of 26 February 2010 amending Regulation (EC) No 1881/2006 setting maximum levels for certain contaminants in foodstuffs as regards aflatoxins. *Off. J. Eur. Union* **2010**, *L50/8*, 12. Available online: https://eur-lex.europa.eu (accessed on 11 October 2021).
13. Serbian Regulation. Maximum allowed contents of contaminants in food and feed. *Off. Bull. Repub. Serbia* **2020**, *132/2020*. Available online: https://www.paragraf.rs (accessed on 11 October 2021).
14. Ferlay, J.; Soerjomataram, I.; Dikshit, R.; Eser, S.; Mathers, C.; Rebelo, M.; Parkin, D.M.; Forman, D.; Bray, F. Cancer incidence and mortality worldwide: Sources, methods and major patterns in GLOBOCAN 2012. *Int. J. Cancer* **2015**, *136*, 359–386. [CrossRef]
15. Wong, M.C.S.; Jiang, J.Y.; Goggins, W.B.; Liang, M.; Fang, Y.; Fung, F.D.H.; Leung, C.; Wang, H.H.X.; Wong, G.L.H.; Wong, V.W.S.; et al. International incidence and mortality trends of liver cancer: A global profile. *Sci. Rep.* **2017**, *7*, 45846. [CrossRef]
16. Tesfamariam, K.; De Boevre, M.; Kolsteren, P.; Belachew, T.; Mesfin, A.; De Saeger, S.; Lachat, C. Dietary mycotoxins exposure and child growth. immune system. morbidity. and mortality: A systematic literature review. *Crit. Rev. Food Sci. Nutr.* **2020**, *60*, 3321–3341. [CrossRef]
17. Raiola, A.; Tenore, G.C.; Manyes, L.; Meca, G.; Ritieni, A. Risk analysis of main mycotoxins occurring in food for children: An overview. *Food Chem. Toxicol.* **2015**, *84*, 169–180. [CrossRef]
18. Peraica, M.; Richter, D.; Rašić, D. Mycotoxicoses in children. *Arch. Ind. Hyg. Toxicol.* **2014**, *65*, 347–363. [CrossRef]
19. Assuncao, R.; Vasco, E.; Nunes, B.; Loureiro, S.; Martins, C.; Alvito, P. Single-compound and cumulative risk assessment of mycotoxins present in breakfast cereals consumed by children from Lisbon region. *Food Chem. Toxicol.* **2015**, *86*, 274–281. [CrossRef]
20. Hulin, M.; Bemrah, N.; Nougadere, A.; Volatier, J.L.; Sirot, V.; Leblanc, J.C. Assessment of infant exposure to food chemicals: The French total diet study design. *Food Addit. Contam. A* **2014**, *31*, 1226–1239. [CrossRef]
21. Juan, C.; Raiola, A.; Mañes, J.; Ritieni, A. Presence of mycotoxin in commercial infant formulas and baby foods from Italian market. *Food Control* **2014**, *39*, 227–236. [CrossRef]
22. Pereira, P.C. Milk nutritional composition and its role in human health. *Nutrition* **2014**, *30*, 619–627. [CrossRef]
23. Milićević, D.; Spirić, D.; Janković, S.; Velebit, B.; Radičević, T.; Petrović, Z.; Stefanović, S. Aflatoxin M1 in processed milk: Occurrence and seasonal variation with an emphasis on risk assessment of human exposure in Serbia. In Proceedings of the 59th International Meat Industry Conference MEATCON2017, Zlatibor, Serbia, 1–4 October 2017. [CrossRef]
24. Milićević, D.; Petronijević, R.; Petrović, Z.; Đjinović-Stojanović, J.; Jovanović, J.; Baltić, T.; Janković, S. Impact of climate change on aflatoxin M1 contamination of raw milk with special focus on climate conditions in Serbia. *J. Sci. Food Agric.* **2019**, *99*, 5202–5210. [CrossRef]
25. Zekovic, M.; Gurinovic, M.; Milesevic, J.; Glibetic, M. National Food Consumption Survey among children from 1–9 years in Serbia–final deliverable. *EFSA Supporting Publ.* **2021**, *18*, 6994E. [CrossRef]
26. Nikolic, M.; Milesevic, J.; Zekovic, M.; Gurinovic, M.; Glibetic, M. The development and validation of food atlas for portion size estimation in the Balkan region. *Front. Nutr.* **2018**, *5*, 78. [CrossRef]
27. Gurinović, M.; Milešević, J.; Kadvan, A.; Nikolić, M.; Zeković, M.; Djekić-Ivanković, M.; Glibetić, M. Development, features and application of DIET ASSESS & PLAN (DAP) software in supporting public health nutrition research in Central Eastern European Countries (CEEC). *Food Chem.* **2018**, *238*, 186–194. [CrossRef]

28. JECFA—Joint FAO/WHO Expert Committee on Food Additives. Evaluation of certain contaminants in food: Eighty-third report of the joint FAO/WHO expert committee on food additives. *WHO Tech. Rep. Ser.* **2017**, *1002*, 1–166. Available online: https://apps.who.int/iris/handle/10665/254893 (accessed on 12 October 2021).
29. EFSA. Management of left-censored data in dietary exposure assessment of chemical substances. *EFSA J.* **2010**, *8*, 1557.
30. Schrenk, D.; Bignami, M.; Bodin, L.; Chipman, J.K.; del Mazo, J.; Grasl-Kraupp, B.; Hogstrand, C.; Hoogenboom, L.R.; Leblanc, J.C.; Nebbia, C.S.; et al. (EFSA Panel on Contaminants in the Food Chain) Scientific opinion—Risk assessment of aflatoxins in food. *EFSA J.* **2020**, *18*, 112.
31. FAO/WHO/UN. *Safety Evaluation of Certain Contaminants in Food: Prepared by the Eighty-Third Meeting of the Joint FAO/WHO Expert Committee on Food Additives (JECFA)*; WHO Food Additives Series, No. 74; FAO JECFA Monographs 19 bis; World Health Organization; Food and Agriculture Organization of the United Nations: Geneva, Switzerland, 2018.
32. Kos, J.; Lević, J.; Đuragić, O.; Kokić, B.; Miladinović, I. Occurrence and estimation of aflatoxin M1 exposure in milk in Serbia. *Food Control* **2014**, *38*, 41–46. [CrossRef]
33. Škrbić, B.; Živančev, J.; Antić, I.; Godula, M. Levels of aflatoxin M1 in different types of milk collected in Serbia: Assessment of human and animal exposure. *Food Control* **2014**, *40*, 113–119. [CrossRef]
34. Tomašević, I.; Petrović, J.; Jovetić, M.; Raičcević, S.; Milojević, M.; Miočinović, J. Two year survey on the occurrence and seasonal variation of aflatoxin M1 in milk and milk products in Serbia. *Food Control* **2015**, *56*, 64–70. [CrossRef]
35. Walte, H.G.; Schwake-Anduschus, C.; Geisen, R.; Fritsche, J. Aflatoxin: Food chain transfer from feed to milk. *J. Consum. Protect. Food Saf.* **2016**, *11*, 295–297. [CrossRef]
36. Kos, J.; Janić Hajnal, E.; Šarić, B.; Jovanov, P.; Mandić, A.; Đuragić, O.; Kokić, B. Aflatoxins in maize harvested in the Republic of Serbia over the period 2012–2016. *Food Addit. Contam. B* **2018**, *11*, 246–255. [CrossRef]
37. Škrbić, B.; Antić, I.; Živančev, J. Presence of aflatoxin M1 in white and hard cheese samples from Serbia. *Food Control* **2015**, *50*, 111–117. [CrossRef]
38. Milicevic, D.; Jankovic, S.; Muric, K.; Petrovic, Z.; Petronijevic, R.; Raseta, M.; Djinovic-Stojanovic, J. Safety of milk and whey from Zlatibor region in relation to aflatoxin M1 contamination: A seasonal study. In Proceedings of the 60th International Meat Industry Conference MEATCON2019, Kopaonik, Serbia, 22–25 September 2019. [CrossRef]
39. Costamagna, D.; Gaggiotti, M.; Chiericatti, C.A.; Costabel, L.; Audero, G.M.L.; Taverna, M.; Signorini, M.L. Quantification of aflatoxin M1 carry-over rate from feed to soft cheese. *Toxicol. Rep.* **2019**, *6*, 782–787. [CrossRef] [PubMed]
40. Torović, L.; Popović, N.; Živkov-Baloš, M.; Jakšić, S. Risk estimates of hepatocellular carcinoma in Vojvodina (Serbia) related to aflatoxin M1 contaminated cheese. *J. Food Compos. Anal.* **2021**, *103*, 104122. [CrossRef]
41. Iqbal, S.Z.; Asi, M.R. Assessment of aflatoxin M1 in milk and milk products from Punjab, Pakistan. *Food Control* **2013**, *30*, 235–239. [CrossRef]
42. Turnaa, N.S.; Wu, F. Aflatoxin M1 in milk: A global occurrence, intake & exposure assessment. *Trends Food Sci. Tech.* **2021**, *110*, 183–192.
43. Kuiper-Goodman, T. Uncertainties in the risk assessment of three mycotoxins: Aflatoxin, ochratoxin, and zearalenone. *Canad. J. Physiol. Pharmacol.* **1990**, *68*, 1017–1024. [CrossRef]
44. Pitt, J.I.I.; Wild, C.P.; Baan, R.A.; Gelderblom, W.C.A.; Miller, J.D.; Riley, R.T.; Wu, F. *Improving Public Health through Mycotoxin Control*; IARC WHO: Lyon, France, 2012; pp. 225–248, ISBN 978-92-832-2158-6.
45. Makori, N.; Matemu, A.; Kimanya, M.; Kassim, N. Inadequate management of complementary foods contributes to the risk of aflatoxin exposure and low nutrition status among children. *World Mycotoxin J.* **2019**, *12*, 67–76. [CrossRef]
46. Ismail, A.; Naeem, I.; Gong, Y.Y.; Routledge, M.N.; Akhtar, S.; Riaz, M.; Ramalho, L.N.Z.; de Oliveira, C.A.F.; Ismail, Z. Early life exposure to dietary aflatoxins, health impact and control perspectives: A review. *Trends Food Sci. Tech.* **2021**, *112*, 212–224. [CrossRef]
47. Liu, Y.; Wu, F. Global burden of aflatoxin-induced hepatocellular carcinoma: A risk assessment. *Environ. Health Perspect.* **2010**, *118*, 818–824. [CrossRef]
48. Institute of Public Health of Serbia "Dr Milan Jovanovic Batut" The Report on Infectious Diseases in the Republic of Serbia for 2019. 2020. Available online: https://www.batut.org.rs/download/izvestaji/Godisnji%20izvestaj%20o%20zaraznim%20bolestima%202019.pdf (accessed on 26 October 2021).

nutrients

MDPI

Article

Dietary Intake of Adult Residents in Luxembourg Taking Part in Two Cross-Sectional Studies—ORISCAV-LUX (2007–2008) and ORISCAV-LUX 2 (2016–2017)

Farhad Vahid [1], Alex Brito [2,3], Gwenaëlle Le Coroller [4], Michel Vaillant [4], Hanen Samouda [1,†], Torsten Bohn [1,*,†] and on behalf of ORISCAV Working Group [‡]

1 Nutrition and Health Research Group, Population Health Department, Luxembourg Institute of Health, 1445 Strassen, Luxembourg; Farhad.vahid@lih.lu (F.V.); hanene.samouda@lih.lu (H.S.)
2 Laboratory of Pharmacokinetics and Metabolomics Analysis, Institute of Translational Medicine and Biotechnology, I.M. Sechenov First Moscow State Medical University, 119991 Moscow, Russia; abrito@labworks.ru
3 World-Class Research Center "Digital Biodesign and Personalized Healthcare", I.M. Sechenov First Moscow State Medical University, 119991 Moscow, Russia
4 Competence Center in Methodology and Statistics, Luxembourg Institute of Health, 1445 Strassen, Luxembourg; Gwenaelle.LeCoroller@lih.lu (G.L.C.); michel.vaillant@lih.lu (M.V.)
* Correspondence: Torsten.bohn@lih.lu; Tel.: +352-621-216-637
† These authors contributed equally to the study.
‡ Membership of ORISCAV Working Group is provided in the Acknowledgments.

Abstract: Background: A balanced diet is an important lifestyle component and has been associated with a reduced risk of chronic diseases. Objectives: To assess dietary intake of adult residents in Luxembourg taking part in two population-based cross-sectional studies (ORISCAV-LUX, 2007–2008 and ORISCAV-LUX 2, 2016–2017). Methods: Dietary intake of the study participants (1242 in 2007/08 and 1326 in 2016/17), 25–69 years old, were evaluated using food-frequency questionnaires (134 items in 2007/2008 and 174 items in 2016/2017) according to the French ANSES-CIQUAL food composition database. Both food-group- and nutrient-based analyses were conducted. Results: Dietary patterns in ORISCAV-LUX 2, 2016–2017, were characterized by an increase in the estimated marginal means (EMM) of the intake of energy, total fat, saturated fatty acids, alcohol, and decreased EMM of total carbohydrates, magnesium, and calcium compared to 2007/08. We also observed an increased EMM of the intake of protein-rich food items and ready-to-eat foods/fast foods, together with a decreased intake of grains, dairy products, and vegetables (all *p*-values <0.05, linear mixed models). The intake of most micronutrients was stable or slightly increased in ORISCAV-LUX 2 vs. ORISCAV-LUX, except for the drop in magnesium and calcium, and generally met recommendations, in particular, EFSA population reference intakes (PRI), except for vitamin D. Conclusions: Though most micronutrient recommendations were met, nutrient consumption in terms of high energy, total fat, and sodium, as well as low carbohydrates, were not aligned with recommendations for balanced eating.

Keywords: dietary habits; food groups; calorie intake; vitamins; minerals; beta-carotene; sugar-sweetened beverages; exploratory factor analysis

Citation: Vahid, F.; Brito, A.; Le Coroller, G.; Vaillant, M.; Samouda, H.; Bohn, T.; on behalf of ORISCAV Working Group. Dietary Intake of Adult Residents in Luxembourg Taking Part in Two Cross-Sectional Studies—ORISCAV-LUX (2007–2008) and ORISCAV-LUX 2 (2016–2017). *Nutrients* **2021**, *13*, 4382. https://doi.org/10.3390/nu13124382

Academic Editor: Ana Baylin

Received: 16 November 2021
Accepted: 1 December 2021
Published: 7 December 2021

1. Introduction

Dietary patterns are an important nutritional and lifestyle component [1]. Poor dietary habits, such as over- or under-consumption of calories and macronutrients and a low intake of certain micronutrients or nonessential constituents, such as dietary fiber and secondary plant compounds, including carotenoids and polyphenols, have been related to several chronic noncommunicable diseases, such as cardiometabolic, neurodegenerative, and autoimmune diseases, as well as mental illnesses [1] and cancer [2]. Dietary patterns and food choices are influenced by various factors, such as age, gender, socioeconomic factors, and neurophysiological variations [3]. However, far from being fixed, these patterns might change over time at the population level due to changes in the population composition, sociodemographic factors, altered nutritional knowledge, and possibly food unavailability or inaccessibility, among others [4]. As improvements in dietary habits have been shown to be an important factor influencing overall health and all-cause mortality [5–7], monitoring dietary patterns is an important aspect for studying changes related to population health.

Luxembourg is characterized by a large percentage (about 50% of a total of ca. 650,000) of residents with foreign origin, possibly contributing to the diversity of eating habits [8]. Approximately 14.3% [8] of the population is 65 years old or older, a slightly higher fraction than in other European countries (12%) [9]. The country is also characterized, like many other Westernized countries, by a high prevalence of noncommunicable chronic diseases and related health conditions, potentially, although not exclusively, associated with a poor nutritional status, including prediabetes (25.6%) and diabetes (6.5%), overweight (37.3%) and obesity (20.6%), depression (21.6%), hypertension (31%), dyslipidemia (30.4%), cancer (3.6%), dementia (3.8%), cognitive impairment (26.1%), and Parkinson's disease (0.2%) [10–13].

In Luxembourg, the dietary patterns, similar as to other Western-type diets [14], have been relatively high in saturated fat (18.4%), ready-made meals (7%), and alcohol consumption (13 L annual per capita) [15], all of which have been associated with cardiometabolic disturbances, such as a high incidence of metabolic syndrome, although this has been investigated in a cross-sectional study not adjusted for additional exposure [16]. In contrast, the Mediterranean dietary pattern, rich in unsaturated fats, whole grains, green leafy vegetables, fruits, and legumes, has been less frequently associated with various types of cancer and cardiometabolic diseases [17–19]. In general, dietary guidelines recommend choosing various vegetables, fruits, pulses, whole grains, and consuming a minimum amount of free sugars, processed/smoked meat, salt, and trans- and saturated fats, especially prevalent in ready-to-eat meals [20,21]. However, in Luxembourg, according to an earlier study in 2006/07, about 65% of the individuals did not reach the recommendations for dietary fiber intake [22], and about 50% did not consume five portions of 80–100 g of fruits and vegetables per day [22]. In addition, Luxembourg residents were ranked the highest meat consumers worldwide (136 kg per capita) in 2007 [23], with a large number of individuals regularly consuming ready-to-eat meals [16]. General dietary trends in the past years in most Westernized countries included tendencies for lower carbohydrate consumption [24], higher meat and processed food intake [25], but also leaning toward more organic/bio-foods [26].

Food choices, as a critical component in the overall dietary patterns, are complex and influenced by various factors [27]. One of the crucial goals of population-based longitudinal or repetitive cross-sectional studies is to monitor dietary changes to investigate food intake trends [28]. Such investigations are an important base for developing improved public health policy approaches [29]. This study was designed to investigate the changes in dietary patterns and habits concerning the intake of food groups, macro- and micronutrients, as well as non-nutrient compounds, in adult residents in Luxembourg taking part in two population-based, cross-sectional studies over the past decade (ORISCAV-LUX, 2007–2008 [8] and ORISCAV-LUX 2, 2016–2017 [30]).

2. Participants and Methods

2.1. Study Population and Design

The complete description of the study population and methods has previously been published in 2010 and 2019 [30]. Briefly, the Observation of Cardiovascular Risk Factors in Luxembourg (ORISCAV-LUX) surveys included two cross-sectional studies in adults residing in Luxembourg. In the original ORISCAV-LUX survey (2007–2008) [8], N = 1432 participants were included by a systematic random sampling procedure. In the original ORISCAV-LUX 2 survey (2016–2017) [30], N = 1558 participants were included by an initial baseline sampling plus complimentary sampling. A total of 660 individuals participated in both studies. The age ranges of the study participants in the original surveys were 18–69 years for ORISCAV-LUX and 25–79 years for ORISCAV-LUX 2.

In the present analysis, the same age ranges, 25–69 years, were retained to enable an accurate comparison between the two surveys (ORISCAV-LUX: N = 1242; ORISCAV-LUX 2: N = 1326). The participants were randomly selected based on sociodemographic attributes, including the district of residence, age, and gender. After a telephone appointment, the study participants were invited to take part in the surveys in the nearest study center from their domicile. During the study appointment, the study investigator gave the participants all the information related to the study, including the aim of the research project and the study protocol. The study participants received comprehensive guidance on the survey, including the general information questionnaire and the food frequency questionnaires (FFQ). In addition to the FFQ, selected parameters on anthropometric, demographic, and socioeconomic factors were collected. All the participants were duly informed and consented to take part in the study. The study design and information collected were approved by the National Research Ethics Committee (CNER) and the National Commission for Private Data Protection (CNPD).

2.2. Assessment of Dietary Intake

The dietary intake data were extracted from a validated quantitative FFQ [31]. In the ORISCAV-LUX study, a 134-item FFQ was used [31]. The FFQ was divided into nine food groups: 14 carbohydrate-related questions, 13 related to fruits, 13 to vegetables, 18 to meat–poultry–fish items, 11 to ready-made meals (prepared dishes), 22 to dairy products, 16 to fats (for spreading, cooking, and seasoning), 14 to drinks and beverages, and 13 to miscellaneous items. Miscellaneous items included jam, chocolate, peanut butter, dry biscuits, ice cream products, jellified desserts, sugar, and cocoa. The study participants indicated the portion size and frequency of all consumed beverages and food items on a scale ranging from "never or rarely", "two or more times/day", "once a day", "3 to 5 times/week", "1 to 2 times/week", and "1 to 3 times/month". The macro- and micronutrient intake was calculated by multiplying each food item's consumption frequency by the specific nutrient content of each portion. Portion size images were used to accurately identify the portion sizes of all the consumed food and beverage items.

Similarly, in ORISCAV-LUX 2, a validated quantitative 174-item FFQ was used [32]. In fact, more questions about certain food items were asked in order to increase the accuracy in the second wave. For example, in the first wave, the question was about the total amount of "butter" consumed, while, in the second wave, the question was divided into two parts: "unsalted butter" and "lightly salted or salted butter". The FFQ in the second wave comprised nine food groups, including 16 carbohydrate-related items, 12 fruit items, 13 vegetable items, 26 meat–poultry–fish items, 17 ready-made meal items, 22 dairy product items, 28 fat items, 21 drink and beverage items, and 18 miscellaneous items. The methods for completing the FFQ and extracting the data and the food database used [33] were similar for the two waves of the survey [33].

The amount of macro- and micronutrient intake was converted into a daily consumption and reported as median and interquartile range. For this purpose, the macro- and micronutrient intake amounts were obtained by linking the consumed food/beverage items with the ANSES-CIQUAL French Food Composition Table database [33]. The total energy was obtained as the sum of 37 kJ/g (9 kcal/g) for fat, 29 kJ/g (7 kcal/g) for alcohol, 17 kJ/g (4 kcal/g) for protein, 17 kJ/g (4 kcal/g) for carbohydrates (except for polyols), 13 kJ/g (3 kcal/g) for organic acids, 10 kJ/g (2.4 kcal/g) for polyols, and 8 kJ/g (2 kcal/g) for dietary fiber.

2.3. Anthropometric Measures

A trained research nurse performed the anthropometric measures of weight and height. Body mass index (BMI) was calculated. The height (cm) and body weight (kg) were measured in a slight dress without shoes. The participants' BMI was estimated as weight in kg divided by the square of height in meters (kg/m^2).

2.4. Demographic and Socioeconomic Factors

Age, gender, marital status, education, job, income, and the number of persons living in the same household were obtained from the General Information Questionnaire.

2.5. Data Management

From all the enrolled participants (ORISCAV-LUX = 1432, ORISCAV-LUX 2 = 1558), only the data of the participants who had completed the FFQ were considered in the present analysis. In this regard, 80 participants from ORISCAV-LUX and 127 participants from ORISCAV-LUX 2 were excluded due to the lack of FFQ data. From the 1352 participants who completed the FFQ in ORISCAV-LUX and the 1431 participants who completed the FFQ in ORISCAV-LUX 2, 110 participants under 25 years (ORISCAV-LUX) and 105 participants over 69 years (ORISCAV-LUX 2) were excluded to obtain the same age-range groups in the present paper.

Finally, the data of 1242 participants from ORISCAV-LUX and 1326 participants from ORISCAV-LUX 2 who completed the FFQ and were in the same age ranges were included in our analyses (see the flowchart of participants, Figure 1).

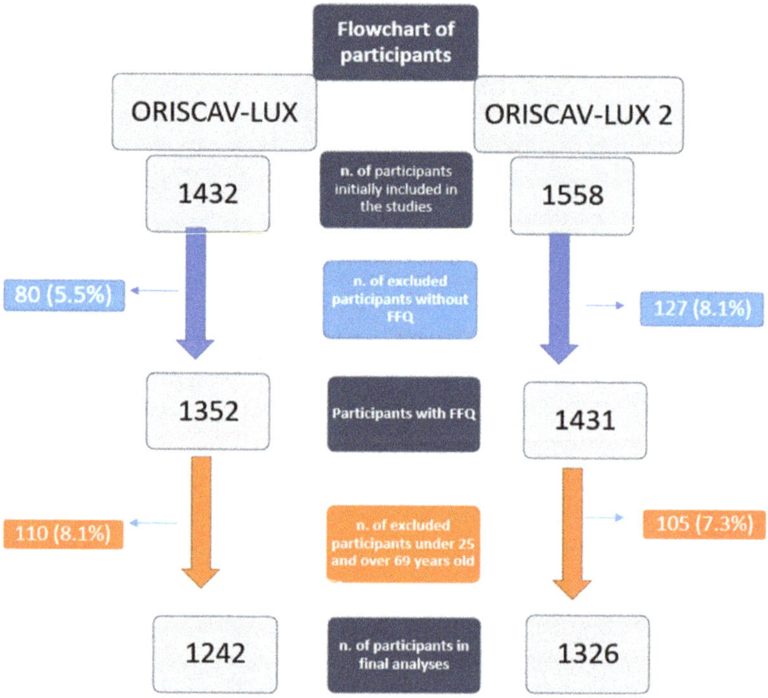

Figure 1. Flowchart of participants in ORISCAV-LUX and ORISCAV-LUX 2.

Missing data: For the variable "income", we grouped the missing data into "did not answer". Therefore, no other participants were excluded from the analysis due to missing data in sociodemographic variables (Table 1). The replacement of missing data by the means or their imputation was not considered as a suitable strategy as data were not to be missing at random.

Table 1. Distribution of demographic, anthropometric, and socioeconomic characteristics of participants.

Variables	Mean ± SD (Minimum–Maximum) or Number (%)	
	ORISCAV-LUX (N = 1242)	**ORISCAV-LUX 2 (N = 1326)**
Age (year)	46.3 ± 11.6 (25.2–69.9)	49.5 ± 10.0 (25.2–69.9)
BMI (kg/m^2)	26.9 ± 5.0 (16.1–51.2)	26.1 ± 4.7 (12.9–50.4)
Gender		
- Women	634 (51.1%)	709 (53.4%)
- Men	608 (48.9%)	617 (46.6%)
Marital status		
- Single [a]	155 (12.5%)	150 (11.3%)
- Married	944 (76.0%)	984 (74.2%)
- Widow(er)	32 (2.6%)	162 (12.2%)
- Divorced or separated [b]	111 (8.9%)	150 (11.3%)
Education		
- * No diploma	292 (23.5%)	169 (12.7%)
- Secondary education **	478 (38.5%)	463 (34.9%)
- Post-secondary education ***	364 (29.3%)	587 (44.3%)
- Did not answer	108 (8.7%)	107 (8.1%)
Occupation (Job)		
- Employed	835 (67.2%)	917 (69.1%)
- Unemployed [c]	216 (17.4%)	141 (10.6%)
- Leave [d]	170 (13.8%)	251 (19.0%)
- Did not answer	20 (1.6%)	17 (1.3%)
Income (EUR)		
- Less than 750	13 (1.0%)	4 (0.3%)
- 750 to 1499	49 (3.9%)	22 (1.7%)
- 1500 to 2249	143 (11.5%)	45 (3.4%)
- 2250 to 2999	195 (15.7%)	74 (5.6%)
- 3000 to 4999	381 (30.7%)	306 (23.1%)
- 5000 to 10,000	277 (22.3%)	466 (35.1%)
- More than 10,000	57 (4.6%)	111 (8.4%)
- Did not answer	127 (10.2%)	298 (22.5%)
Country of birth		
- Luxembourg	738 (59.4%)	778 (58.7%)
- Portugal	149 (12.0%)	110 (8.3%)
- Other European countries	246 (19.8%)	315 (23.3%)
- Non-European countries	109 (8.8%)	123 (9.3%)

Standard deviation = SD, body mass index = BMI. * Pre-primary and primary education. ** CATP—Certificate of Technical and Professional Aptitude, CITP—Certificate of Technical and Professional Initiation, CCM—Certificate of Manual Capability, Diploma for Completion of Secondary Technical Studies, Diploma for Completion of Secondary General Studies. *** Technician diploma, Bac +2 (BTS), Bac +3 (Bachelors/Degree), Bac +4 (Masters), Bac +5 and more (3rd Cycle, DEA, DESS, MBA, Masters, Ph.D., etc.), Diploma from a Grande Ecole, an Engineering School. [a] Single, never married, and never in a registered partnership. [b] Divorced, separated, separated but still legally married. [c] In school, university or in training, at home, unemployed or in search of employment. [d] Retired or in early retirement, on long-term leave.

2.6. Statistical Analyses

The normality of the data distribution and equality of variance were measured by Q–Q normality plots and the Kolmogorov–Smirnov test and box plots, respectively. A log transformation was performed for the non-normally distributed data.

Since about 45% of the participants in the second survey also participated in the first survey, linear mixed model (LMM) analyses were performed on log-transformed values to compare the estimated marginal means (EMMs) of the energy and macro- and micronutrient intake between the 2 surveys. LMMs included random intercepts for subjects and fixed effects for ORISCAV-LUX vs. ORISCAV-LUX 2, and adjustment for age at baseline, gender, marital status, education, job, income, and number of persons living in the same household. The LMMs, using an unstructured variance–covariance matrix, enabled the post hoc comparison of the estimated marginal means (EMMs) of the energy and macro- and micronutrient intake between the two surveys, and also according to the gender. A post hoc test (Tukey's) was used. In order to decrease the false discovery rate due to the LMMS performed for each dietary parameter, we applied the Benjamini–Hochberg adjustment for multiple comparisons. The EMMs adjusted for age, gender, education, occupation (job), marital status, number of persons living in the household, and income were reported. In addition to the EMM and 95% confidence interval (95% CI), the raw data were reported as median and interquartile ranges. A *p*-value of 0.05 was considered as significant. The EMMs were also reported, adjusted only for age and gender, as a supplementary analysis. We also used the exploratory factor analysis (EFA) method to identify dietary patterns (2 major components), using the data from the FFQ, organized into 12 major food groups (Figure 2). Absolute values > 0.30 were considered to have a significant role in the components. Small coefficients below this value were suppressed.

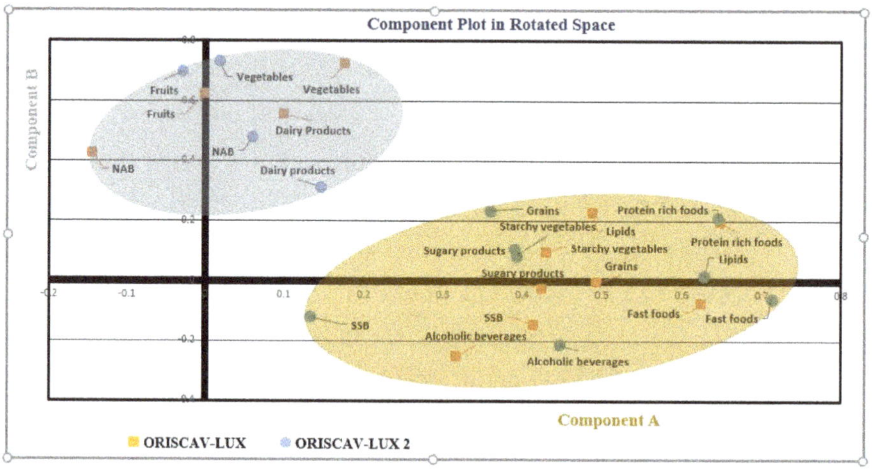

Figure 2. The exploratory factor analysis (EFA) was used to identify two major components, using the data from the FFQ organized into 12 major food groups. A component plot in rotated space for both ORISCAV-LUX and ORISCAV-LUX 2 is represented. Component A included fast foods (ready-to-eat meals), lipids, grains, starchy vegetables, alcoholic beverages, SSB (sugar-sweetened beverages), sugary products, and protein-rich foods. Component B included fruits, vegetables, NAB (nonalcoholic beverages), and dairy products.

A comparison of the average intake of macro- and micronutrients of the study participants in the two surveys with the recommended values published by the World Health Organization (WHO), European Food Safety Authority (EFSA, PRI), United States Department of Agriculture (USDA, RDA) dietary guidance, British Nutrition Foundation (BNF), and German-(D), Austrian-(A), and Swiss (CH) (DACH) reference values was also carried out. The SPSS statistical software (IBM SPSS statistics 25.0, IBM Corp., Armonk, NY, USA) was used for the statistical analyses.

3. Results

Overall, 51% of the participants in the ORISCAV-LUX and 53.4% of the participants in the ORISCAV-LUX 2 were women. The mean age was 46.3 ± 11.6 years in the ORISCAV-LUX and 49.5 ± 10.0 years in the ORISCAV-LUX 2. The general characteristics of the study participants are presented in Table 1.

The EMM obtained from the linear mixed models, as well as the median and interquartile ranges of total energy, water, alcohol, and macronutrient intake of participants, are presented in Table 2. There was a significant increase in the EMMs of total energy intake, total water, total protein, animal protein, total fat, cholesterol, polyunsaturated fatty acids (PUFA), saturated fatty acids (SFA), monounsaturated fatty acids (MUFA), linoleic acid, alpha-linolenic acid, and alcohol, and a significant decrease in the EMMs of total carbohydrate, simple sugar, and added sugar intake in ORISCAV-LUX 2 compared to ORISCAV-LUX.

Table 2. Median (interquartile range) and estimated marginal means of participants' total energy, alcohol, and macronutrient intake (*p*-values are based on linear mixed models, as further detailed in footnotes).

Parameter	Raw Values Median (IQR) ORISCAV-LUX	Raw Values Median (IQR) ORISCAV-LUX 2	Modeled Values Estimated Marginal Means (95% CI) * ORISCAV-LUX	Modeled Values Estimated Marginal Means (95% CI) * ORISCAV-LUX 2	*p*-Value **
Total energy intake (kcal/day)	2213 (1142)	2374 (1136)	3.349 (3.340, 3.358)	3.379 (3.371, 3.387)	<0.001
Total water (g/day)	2920 (1353)	3083 (1250)	3.469 (3.461, 3.478)	3.482 (3.475, 3.490)	0.018
Total protein (g/day)	88.3 (46.4)	89.6 (45.3)	1.939 (1.929, 1.948)	1.954 (1.946, 1.963)	0.017
VSP (g/day)	26.6 (14.8)	26.7 (14.2)	1.427 (1.417, 1.437)	1.432 (1.422, 1.441)	0.474
Animal source protein (g/day)	56.8 (36.2)	60.6 (37.4)	1.746 (1.734, 1.757)	1.772 (1.760, 1.783)	<0.001
Total fat (g/day)	93.4 (57.1)	116.7 (64.7)	1.975 (1.964, 1.986)	2.066 (2.056, 2.075)	<0.001
SFA (g/day)	32.2 (21.4)	39.8 (23.1)	1.511 (1.499, 1.522)	1.594 (1.583, 1.604)	<0.001
MUFA (g/day)	39.2 (24.5)	46.8 (26.2)	1.598 (1.587, 1.610)	1.674 (1.664, 1.684)	<0.001
PUFA (g/day)	14.8 (10.0)	20.7 (13.4)	1.180 (1.168, 1.193)	1.323 (1.311, 1.335)	<0.001
Linoleic acid (g/day)	12.2 (8.5)	17.3 (11.7)	1.092 (1.079, 1.105)	1.240 (1.228, 1.252)	<0.001
Alpha-linoleic acid (g/day)	1.04 (0.74)	1.79 (1.40)	0.045 (0.031, 0.058)	0.265 (0.251, 0.279)	<0.001
Arachidonic acid (g/day)	0.15 (0.11)	0.19 (0.13)	−0.821 (−0.834, −0.808)	−0.720 (−0.733, −0.707)	<0.001
EPA (g/day)	0.12 (0.13)	0.20 (0.23)	−0.934 (−0.955, −0.912)	−0.769 (−0.794, −0.743)	<0.001
DPA (g/day)	0.06 (0.05)	0.08 (0.07)	−1.230 (−1.245, −1.214)	−1.133 (−1.151, −1.115)	<0.001
DHA (g/day)	0.18 (0.19)	0.28 (0.31)	−0.746 (−0.767, −0.726)	−0.591 (−0.614, −0.569)	<0.001
Cholesterol (mg/day)	310.5 (196.1)	356.5 (204.5)	2.485 (2.474, 2.497)	2.554 (2.543, 2.565)	<0.001
Total carbohydrates (g/day)	229.4 (125.8)	217.6 (110.1)	2.361 (2.351, 2.371)	2.341 (2.332, 2.350)	0.002
Simple sugars (g/day)	108.3 (71.5)	100.1 (59.6)	2.038 (2.025, 2.051)	1.995 (1.983, 2.006)	<0.001
Added sugars (g/day)	31.7 (32.9)	28.1 (27.6)	1.504 (1.485, 1.523)	1.433 (1.414, 1.451)	<0.001
Starch (g/day)	108.5 (68.0)	102.9 (61.7)	2.029 (2.018, 2.040)	2.019 (2.008, 2.030)	0.220
Total fiber (g/day)	22.9 (12.8)	23.1 (12.0)	1.367 (1.356, 1.377)	1.357 (1.347, 1.367)	0.178
Soluble fiber (g/day)	4.6 (2.6)	4.7 (2.4)	0.667 (0.666, 0.687)	0.662 (0.652, 0.673)	0.049
Alcohol (g/day)	4.1 (11.5)	5.6 (11.2)	0.641 (0.601, 0.682)	0.762 (0.729, 0.794)	<0.001

* Linear mixed model (based on log-transformed data) adjusted for age, gender, marital status, education, job, income, number of persons living in the same household. ** Benjamini–Hochberg correction was applied to all *p*-values: all *p*-values are displayed after this correction. Vegetable source protein = VSP, interquartile range = IQR, saturated fatty acids = SFA, monounsaturated fatty acids = MUFA, polyunsaturated fatty acids = PUFA, eicosapentaenoic acid = EPA, docosapentaenoic acid = DPA, docosahexaenoic acid = DHA.

Regarding micronutrient intake (Table 3), a significant reduction in the EMMs of magnesium and calcium intake was observed in ORISCAV-LUX 2 compared to ORISCAV-LUX.

The distribution of the participants' food group intake is shown in Table 4. A significant decrease in the EMM of grains, vegetables, starchy vegetables, dairy products, and sugary products intake was found when comparing ORISCAV-LUX to ORISCAV-LUX 2, along with a significant increase in the EMM of protein-rich foods, ready-to-eat and fast food, lipids, noncaloric beverages, and alcoholic beverages. The same models for macro-, micro-nutrients and food groups adjusted only for age at baseline and gender are presented in Supplementary Tables S1–S3.

The within- and between-group comparisons of macronutrient (Table 5) and micronutrient (Table 6) intakes based on gender groups showed that men consumed significantly more energy, fat, proteins, total carbohydrates, cholesterol, total fiber, and alcohol compared to women in both ORISCAV-LUX and ORISCAV-LUX 2. The intake of most micronutrients (except folate, vitamin E, vitamin C, and calcium) were also lower in women compared to men in both waves of ORISCAV. In addition, there was a significant increase in the intake of total energy, total fat, and total alcohol in men participating in ORISCAV-LUX2 compared to men in ORISCAV-LUX. Similar significant increases were seen in women in ORISCAV-LUX 2 compared to ORISCAV-LUX.

On the other hand, there was a significant decrease in consumed total carbohydrates in both men and women in ORISCAV-LUX 2, compared to ORISCAV-LUX (Table 5). In parallel with the total increased fat intake, there was a significantly higher intake of fat-soluble vitamins (A, D, and E) in both men and women participating in ORISCAV-LUX 2 compared to

ORISCAV-LUX. In accordance with the reduction in grains and dairy product consumption in ORISCAV-LUX 2 compared to the ORISCAV-LUX, the intake of calcium and magnesium showed a significant decrease in both genders in ORISCAV-LUX 2 (Table 6).

Table 3. Median (interquartile range) and estimated marginal means of micronutrient intake of participants (*p*-values are based on linear mixed models, as explained in footnotes).

Parameter	Raw Values		Modeled Values		*p*-Value **
	Median (IQR)		Estimated Marginal Means (95% CI) *		
	ORISCAV-LUX	ORISCAV-LUX 2	ORISCAV-LUX	ORISCAV-LUX 2	
Vitamin A (μg/day)	364.8 (399.7)	475.5 (340.3)	2.595 (2.578, 2.612)	2.670 (2.657, 2.684)	<0.001
Beta-carotene (μg/day)	4121 (3690)	4973 (4094)	3.637 (3.620, 3.654)	3.689 (3.673, 3.705)	<0.001
Vitamin D (μg/day)	2.6 (3.2)	5.1 (4.6)	0.410 (0.389, 0.430)	0.693 (0.677, 0.710)	<0.001
Vitamin E (mg/day)	13.8 (8.8)	18.4 (11.6)	1.151 (1.139, 1.163)	1.269 (1.258, 1.280)	<0.001
Vitamin C (mg/day)	135.1 (109.1)	145.1 (101.7)	2.129 (2.114, 2.144)	2.150 (2.136, 2.163)	0.049
Thiamine (mg/day)	1.55 (0.85)	1.53 (0.81)	0.184 (0.174, 0.194)	0.192 (0.183, 0.201)	0.221
Riboflavin (mg/day)	1.86 (1.04)	1.83 (0.96)	0.271 (0.261, 0.281)	0.269 (0.260, 0.278)	0.794
Niacin (mg/day)	21.2 (11.4)	23.0 (12.0)	1.325 (1.315, 1.334)	1.361 (1.352, 1.370)	<0.001
Pantothenic acid (mg/day)	5.29 (2.71)	5.85 (2.91)	0.720 (0.711, 0.730)	0.771 (0.762, 0.780)	<0.001
Pyridoxine (mg/day)	2.18 (1.18)	2.39 (1.22)	0.340 (0.330, 0.349)	0.381 (0.372, 0.390)	<0.001
Folate (μg/day)	351.0 (196.8)	349.6 (172.3)	2.546 (2.536, 2.557)	2.538 (2.529, 2.547)	0.220
Vitamin B12 (μg/day)	5.34 (4.35)	6.30 (4.54)	0.723 (0.709, 0.737)	0.798 (0.785, 0.812)	<0.001
Calcium (mg/day)	1047 (518.3)	933.6 (454.3)	3.022 (3.013, 3.032)	2.969 (2.961, 2.978)	<0.001
Iron (mg/day)	13.9 (7.1)	14.3 (6.7)	1.143 (1.133, 1.152)	1.154 (1.145, 1.162)	0.082
Iodide (μg/day)	143.8 (82.6)	154.9 (78.8)	2.157 (2.147, 2.167)	2.193 (2.183, 2.202)	<0.001
Magnesium (mg/day)	411.0 (177.2)	373.5 (161.0)	2.620 (2.612, 2.628)	2.574 (2.566, 2.581)	<0.001
Potassium (mg/day)	3575 (1638)	3526 (1547)	3.550 (3.541, 3.559)	3.543 (3.535, 3.551)	0.220
Phosphorus (mg/day)	1354 (686.6)	1330 (612.4)	3.134 (3.124, 3.143)	3.126 (3.118, 3.134)	0.220
Sodium (mg/day)	2332 (1878)	3333 (1957)	3.497 (3.487, 3.508)	3.531 (3.521, 3.541)	<0.001

* Linear mixed model (based on log-transformed data) adjusted for age, gender, marital status, education, job, income, number of persons living in the same household. ** Benjamini–Hochberg correction was applied to all *p*-values: all *p*-values are displayed after this correction.

Table 4. Median (interquartile range) and estimated marginal means of food group intake of participants (*p*-values are based on linear mixed models, as explained in footnotes).

Parameter	Raw Values		Modeled Values		*p*-Value **
	Median (IQR)		Estimated Marginal Means (95% CI) *		
	ORISCAV-LUX	ORISCAV-LUX 2	ORISCAV-LUX	ORISCAV-LUX 2	
Grains (g/day)	196.7 (140.0)	119.1 (101.1)	2.275 (2.260, 2.289)	2.075 (2.085, 2.092)	<0.001
Fruits (g/day)	289.8 (315.5)	286.6 (268.2)	2.419 (2.395, 2.442)	2.414 (2.395, 2.434)	0.779
Vegetables (g/day)	261.6 (232.6)	216.4 (171.8)	2.427 (2.410, 2.444)	2.302 (2.286, 2.318)	<0.001
Starchy vegetables (g/day)	57.1 (82.8)	56.7 (60.7)	1.765 (1.743, 1.786)	1.725 (1.704, 1.745)	0.008
Protein-rich foods (g/day)	161.0 (118.1)	213.7 (147.9)	2.181 (2.166, 2.196)	2.322 (2.309, 2.335)	<0.001
Ready-to-eat and fast foods (g/day)	83.3 (87.9)	95.7 (103.8)	1.879 (1.858, 1.899)	1.948 (1.927, 1.969)	<0.001
Dairy products (g/day)	233.8 (254.0)	178.4 (199.7)	2.322 (2.299, 2.346)	2.163 (2.139, 2.188)	<0.001
Lipids (fats and oils) (g/day)	40.8 (37.4)	61.4 (51.5)	1.606 (1.588, 1.623)	1.768 (1.752, 1.784)	<0.001
Sugary products (g/day)	38.0 (46.7)	33.6 (41.4)	1.545 (1.519, 1.572)	1.495 (1.472, 1.518)	0.006
- NCB (g/day)	1515 (989.3)	1698 (1011)	3.131 (3.114, 3.148)	3.198 (3.186, 3.210)	<0.001
- SSB (g/day)	53.5 (237.2)	70.7 (233.3)	2.045 (2.005, 2.086)	2.074 (2.039, 2.109)	0.333
- Alcoholic beverages (g/day)	58.6 (172.3)	76.2 (157.4)	1.908 (1.874, 1.941)	1.955 (1.927, 1.983)	0.019

* Linear mixed model (based on log-transformed data) adjusted for age, gender, marital status, education, job, income, number of persons living in the same household. ** Benjamini–Hochberg correction was applied to all *p*-values: all *p*-values are displayed after this correction. Interquartile range = IQR, noncaloric beverages = NCB, sugar-sweetened beverages = SSB.

Finally, Table 7 displays the comparison of the average intake of macro- and micronutrients of the study participants in the two surveys with the recommended values published by the World Health Organization (WHO), European Food Safety Authority (EFSA), United States Department of Agriculture (USDA) dietary guidance, British Nutrition Foundation (BNF), and German-(D), Austrian-(A), and Swiss (CH) (DACH) reference values.

Table 5. Within- and between-group comparisons * of macronutrients based on gender groups (*p*-values are based on linear mixed model) [e].

	ORISCAV-LUX					ORISCAV-LUX 2					Men W 1 vs. W 2	Women W 1 vs. W 2
	Men (n = 608)		Women (n = 634)		*p*-Value [a]	Men (n = 617)		Women (n = 709)		*p*-Value [a]		
	Median (IQR)	EMM (95% CI)	Median (IQR)	EMM (95% CI)		Median (IQR)	EMM (95% CI)	Median (IQR)	EMM (95% CI)		*p*-Value [c]	*p*-Value [d]
Total energy intake (kcal/day)	2435 (1242)	3.395 (3.375, 3.415)	2015 (1005)	3.313 (3.293, 3.333)	<0.001	2684 (1187)	3.432 (3.413, 3.452)	2133 (922)	3.336 (3.317, 3.354)	<0.001	<0.001	<0.001
Total water (g/day)	2944 (1370)	3.468 (3.451, 3.485)	2901 (1325)	3.451 (3.434, 3.469)	0.421	3162 (1389)	3.483 (3.465, 3.501)	3006 (1118)	3.466 (3.449, 3.483)	0.126	0.035	0.040
Total protein intake (g/day)	96.5 (50.3)	1.994 (1.974, 2.014)	80.1 (40.7)	1.894 (1.874, 1.914)	<0.001	102 (48.6)	2.006 (1.986, 2.026)	80.1 (39.1)	1.900 (1.880, 1.919)	<0.001	0.126	0.521
Vegetable protein (g/day)	29.1 (15.8)	1.489 (1.467, 1.510)	24.4 (13.3)	1.406 (1.384, 1.427)	<0.001	29.2 (15.5)	1.490 (1.468, 1.511)	24.7 (12.3)	1.410 (1.389, 1.431)	<0.001	0.906	0.656
Animal source protein (g/day)	64.1 (37.8)	1.787 (1.762, 1.812)	51.6 (32.0)	1.684 (1.659, 1.709)	<0.001	70.1 (40.4)	1.822 (1.795, 1.848)	53.4 (32.1)	1.694 (1.668, 1.720)	<0.001	0.001	0.890
Total fat (g/day)	99.8 (60.5)	2.000 (1.976, 2.024)	88.7 (52.5)	1.950 (1.926, 1.974)	<0.001	128 (66.7)	2.101 (2.078, 2.124)	108.9 (56.8)	2.028 (2.005, 2.050)	<0.001	<0.001	<0.001
SFA (g/day)	35.2 (24.3)	1.536 (1.511, 1.562)	30.0 (19.2)	1.476(1.451, 1.502)	<0.001	44.5 (25.1)	1.627 (1.601, 1.652)	36.2 (20.0)	1.545 (1.521, 1.569)	<0.001	<0.001	<0.001
MUFA (d/day)	41.7 (24.8)	1.619 (1.595, 1.644)	37.3 (24.0)	1.575 (1.550, 1.600)	<0.001	52.3 (27.7)	1.707 (1.683, 1.732)	43.2 (22.4)	1.633 (1.610, 1.657)	<0.001	<0.001	<0.001
PUFA (g/day)	15.4 (10.0)	1.196 (1.169, 1.223)	14.4 (9.6)	1.156 (1.128, 1.183)	<0.001	22.6 (13.9)	1.354 (1.326, 1.382)	19.6 (12.5)	1.287 (1.260, 1.314)	<0.001	<0.001	<0.001
- Linoleic acid (g/day)	12.5 (8.6)	1.105 (1.077, 1.133)	12.0 (8.3)	1.065 (1.036, 1.094)	<0.001	18.8 (12.0)	1.271 (1.242, 1.300)	16.2 (11.1)	1.200 (1.171, 1.228)	<0.001	<0.001	<0.001
- Alpha-linoleic acid (g/day)	1.08 (0.73)	0.065 (0.036, 0.095)	1.02 (0.77)	0.039 (0.009, 0.070)	0.295	1.89 (1.29)	0.303 (0.272, 0.334)	1.68 (1.45)	0.016 (0.228, 0.290)	0.003	<0.001	<0.001
- Arachidonic acid (g/day)	0.17 (0.13)	−0.746 (−0.775, −0.718)	0.13 (0.09)	−0.894 (−0.923, −0.864)	<0.001	0.22 (0.15)	−0.657 (−0.687, −0.628)	0.17 (0.11)	−0.792 (−0.821, −0.762)	<0.001	<0.001	<0.001
- EPA (g/day)	0.13 (0.13)	−0.868 (−0.917, −0.819	0.10 (0.13)	−0.988 (−1.038, −0.937)	<0.001	0.21 (0.24)	−0.705 (−0.758, −0.653)	0.18 (0.23)	−0.808 (−0.863, −0.753)	<0.001	<0.001	<0.001
- DPA (g/day)	0.07 (0.06)	−1.150 (−1.185, −1.115)	0.05 (0.04)	−1.298 (−1.334, −1.265)	<0.001	0.08 (0.08)	−1.064 (−1.101, −1.027)	0.07 (0.07)	−1.196 (−1.235, −1.157)	<0.001	<0.001	<0.001
- DHA (g/day)	0.20 (0.20)	−0.680 (−0.727, −0.634)	0.16 (0.18)	−0.792 (−0.840, −0.745)	<0.001	0.29 (0.33)	−0.537 (−0.585, −0.488)	0.26 (0.31)	−0.619 (−0.668, −0.570)	<0.001	<0.001	<0.001
Cholesterol intake (mg/day)	352 (220)	2.541 (2.516, 2.566)	279 (173)	2.433 (2.408, 2.459)	<0.001	395 (225)	2.600 (2.574, 2.625)	323 (169)	2.502 (2.477, 2.527)	<0.001	<0.001	<0.001
Total carbohydrates (g/day)	250 (134)	2.421 (2.399, 2.442)	210 (105)	2.337 (2.315, 2.359)	<0.001	240 (121)	2.406 (2.384, 2.428)	197 (92.2)	2.310 (2.289, 2.331)	<0.001	0.084	<0.001
Total fiber (g/day)	23.4 (13.1)	1.390 (1.367, 1.412)	22.6 (12.6)	1.364 (1.341, 1.387)	0.030	23.6 (12.4)	1.384 (1.361, 1.408)	22.8 (11.7)	1.360 (1.337, 1.383)	0.015	0.858	0.565
- Soluble fiber (g/day)	4.5 (2.5)	0.686 (0.662, 0.709)	4.7 (2.6)	0.679 (0.655, 0.703)	0.830	4.6 (2.4)	0.676 (0.652, 0.701)	4.7 (2.4)	0.673 (0.649, 0.969)	0.664	0.662	0.516
Alcohol (g/day)	8.2 (18.1)	0.728 (0.640, 0.816)	2.0 (5.7)	0.303 (0.209, 0.396)	<0.001	9.3 (15.6)	0.837 (0.750, 0.925)	3.3 (7.5)	0.466 (0.379, 0.552)	<0.001	<0.001	<0.001

[a] Within-group comparison. [c] Between-group comparison, men. [d] Between-group comparison, women. [e] Linear mixed model adjusted for age, marital status, education, job, income, number of persons living in the same household. * Benjamini–Hochberg correction was applied to all *p*-values: all *p*-values are displayed after this correction. Saturated fatty acids = SFA, monounsaturated fatty acids = MUFA, polyunsaturated fatty acids = PUFA, eicosapentaenoic acid = EPA, docosapentaenoic acid = DPA, docosahexaenoic acid = DHA, interquartile range = IQR, marginal means = EMM based on log-transformed data, wave = W.

Table 6. Within- and between-group comparisons* of micronutrients based on gender groups (p-values are based on linear mixed model) [e].

	ORISCAV-LUX					ORISCAV-LUX 2					Men W 1 vs. W 2	Women W 1 vs. W 2
	Men (n = 608)		Women (n = 634)			Men (n= 617)		Women (n = 709)				
	Median (IQR)	EMM (95% CI)	Median (IQR)	EMM (95% CI)	p-Value [a]	Median (IQR)	EMM (95% CI)	Median (IQR)	EMM (95% CI)	p-Value [a]	p-Value [c]	p-Value [d]
Vitamin A (µg/day)	412 (491)	2.662 (2.625, 2.698)	337 (306)	2.562 (2.526, 2.597)	<0.001	541 (375)	2.737 (2.703, 2.771)	425 (276)	2.640 (2.607, 2.674)	<0.001	<0.001	<0.001
Beta-carotene (µg/day)	3789 (3241)	3.609 (3.572, 3.645)	4427 (4195)	3.669 (3.632, 3.707)	<0.001	4693 (3733)	3.669 (3.630, 3.707)	5108 (4515)	3.727 (3.690, 3.763)	<0.001	<0.001	<0.001
Vitamin D (µg/day)	2.9 (3.4)	0.468 (0.425, 0.511)	2.3 (2.7)	0.362 (0.319, 0.406)	<0.001	5.6 (4.7)	0.733 (0.692, 0.774)	4.8 (4.4)	0.664 (0.623, 0.704)	<0.001	<0.001	<0.001
Vitamin E (mg/day)	13.7 (8.3)	1.137 (1.111, 1.163)	14.1 (9.0)	1.143 (1.117, 1.170)	0.208	20.3 (13.5)	1.295 (1.269, 1.321)	16.7 (9.5)	1.225 (1.200, 1.250)	<0.001	<0.001	<0.001
Vitamin C (mg/day)	128 (102)	2.130 (2.098, 2.162)	141 (120)	2.152 (2.119, 2.186)	0.038	141 (99.1)	2.161 (2.129, 2.194)	150 (103)	2.183 (2.152, 2.215)	0.093	0.009	0.086
Thiamine (mg/day)	1.7 (0.9)	0.234 (0.212, 0.255)	1.4 (0.7)	0.151 (0.129, 0.173)	<0.001	1.7 (0.9)	0.245 (0.223, 0.267)	1.4 (0.6)	0.154 (0.133, 0.175)	<0.001	0.110	0.469
Riboflavin (mg/day)	1.9 (1.1)	0.309 (0.287, 0.330)	1.7 (0.9)	0.250 (0.228, 0.272)	<0.001	2.0 (1.0)	0.315 (0.293, 0.337)	1.6 (0.8)	0.238 (0.217, 0.259)	<0.001	0.279	0.038
Niacin (mg/day)	23.8 (12.3)	1.373 (1.352, 1.393)	19.4 (9.6)	1.273 (1.253, 1.294)	<0.001	26.3 (13.2)	1.414 (1.393, 1.435)	20.2 (9.8)	1.302 (1.282, 1.322)	<0.001	<0.001	<0.001
Pantothenic acid (mg/day)	5.6 (2.8)	0.765 (0.744, 0.785)	5.0 (2.4)	0.699 (0.678, 0.720)	<0.001	6.4 (3.3)	0.819 (0.798, 0.840)	5.4 (2.4)	0.747 (0.727, 0.767)	<0.001	<0.001	<0.001
Pyridoxine (mg/day)	2.3 (1.2)	0.386 (0.365, 0.407)	2.0 (1.0)	0.305 (0.284, 0.327)	<0.001	2.6 (1.3)	0.434 (0.412, 0.455)	2.2 (1.0)	0.342 (0.321, 0.363)	<0.001	<0.001	<0.001
Folate (µg/day)	354 (203)	2.563 (2.541, 2.585)	347 (194)	2.547 (2.525, 2.570)	0.418	358 (179)	2.567 (2.545, 2.589)	340 (162)	2.541 (2.520, 2.563)	0.004	0.311	0.284
Vitamin B12 (µg/day)	6.0 (4.8)	0.795 (0.764, 0.826)	4.8 (3.8)	0.678 (0.647, 0.709)	<0.001	7.4 (4.9)	0.875 (0.844, 0.906)	5.5 (4.0)	0.746 (0.715, 0.777)	<0.001	<0.001	<0.001
Calcium (mg/day)	1043 (520)	3.029 (3.008, 3.050)	1048 (507)	3.024 (3.004, 3.045)	0.605	990 (473)	2.983 (2.963, 3.004)	904 (433)	2.960 (2.940, 2.981)	<0.001	<0.001	<0.001
Iron (mg/day)	15.4 (7.7)	1.192 (1.172, 1.212)	12.8 (6.5)	1.108 (1.088, 1.129)	<0.001	15.7 (7.5)	1.203 (1.182, 1.223)	13.3 (5.9)	1.120 (1.101, 1.140)	<0.001	0.104	0.174
Iodide (µg/day)	151 (80.9)	2.194 (2.171, 2.216)	134 (83.5)	2.143 (2.121, 2.166)	<0.001	166 (84.3)	2.235 (2.213, 2.257)	145 (71.7)	2.174 (2.152, 2.195)	<0.001	<0.001	<0.001
Magnesium (mg/day)	430 (192)	2.649 (2.631, 2.666)	391 (164)	2.604 (2.586, 2.621)	<0.001	404 (171)	2.611 (2.593, 2.629)	353 (146)	2.554 (2.536, 2.571)	<0.001	<0.001	<0.001
Potassium (mg/day)	3683 (1701)	3.583 (3.564, 3.602)	3378 (1531)	3.542 (3.522, 3.562)	<0.001	3721 (1635)	3.582 (3.562, 3.601)	3370 (1447)	3.540 (3.521, 3.559)	<0.001	0.928	0.354
Phosphorus (mg/day)	1483 (763)	3.178 (3.158, 3.198)	1280 (637)	3.104 (3.083, 3.124)	<0.001	1504 (694)	3.174 (3.154, 3.194)	1215 (544)	3.090 (3.071, 3.110)	<0.001	0.699	0.016
Sodium (mg/day)	3703 (2143)	3.549 (3.526, 3.572)	2895 (1524)	3.444 (3.421, 3.467)	<0.001	3894 (2143)	3.586 (3.562, 3.609)	2952 (1529)	3.462 (3.439, 3.484)	<0.001	<0.001	0.310

[a] Within-group comparison. [c] Between-group comparison, men. [d] Between-group comparison, women. [e] Linear mixed model adjusted for age, marital status, education, job, income, number of persons living in the same household. * Benjamini–Hochberg correction was applied to all p-values: all p-values are displayed after this correction. Interquartile range = IQR, marginal means = EMM based on log-transformed data, wave = W.

Table 7. Mean intake of macro- and micronutrients of participants in ORISCAV-LUX and ORISCAV-LUX 2 compared to recommended values.

	Wave 1		Wave 2		WHO [a]		USDA [ʃ]		BNF [a]		DACH		EFSA [Δ]	
	M	W	M	W	M	W	M	W	M	W	M	W	M	W
Total energy intake (kcal/day) [d]	2660	2191	2785	2264	2500	2000	ND	ND	2500 [e]	2000 [e]	2600	2000 [e]	2000 [e]	1800 [e]
Total water (g/day)	3057	2995	3242	3103	3700 [c]	2700 [c]	3700	2700	ND	ND	2700	2700	2500 [c]	2000 [c]
Total protein (g/day) (%) [t]	104 (15%)	84.4 (15%)	106 (23%)	84.1 (14%)	0.66 [b]	0.66 [b]	56	46	0.75 [b]	0.75 [b]	56	50	0.83 [b]	0.83 [b]
Total fat (g/day) (%) [t]	109 (37%)	97.0 (40%)	135 (43%)	115 (46%)	20–35%	20–35%	ND	ND	35%	35%	30%	30%	30%	30
SFA (g/day) (%) [t]	39.0 (13%)	33.4 (14%)	46.8 (15%)	38.8 (15%)	10%	10%	ND	ND	11%	11%	ND	ND	ND	ND
MUFA (g/day) (%) [t]	45.8 (15%)	41.0 (17%)	55.2 (18%)	47.0 (18%)	15–20%	15–20%	ND	ND	ND	ND	ND	ND	ND	ND
PUFA (g/day) (%)t	17.4 (11%)	16.2 (7%)	25.0 (8%)	22.4 (9%)	6–11%	6–11%	ND	ND	ND	ND	ND	ND	ND	ND
Carbohydrate (g/day) (%) [t]	282 (42%)	227 (41%)	257 (37%)	210 (37%)	55–75%	55–75%	13D	130	50%	50%	50%	50%	50%	50%
Total fiber (g/day)	25.3	24.2	25.4	24.3	38	25	38	25	30	30	30	30	25 [c]	25 [c]
Alcohol (g/day)	12.5	4.8	14.8	6.3	ND	ND	ND	ND	ND	ND	20	10	ND	ND
Vitamin A (μg/day)	549	437	603	490	600	500	900	700	700	600	1.0 [g]	0.8 [g]	750	650
Vitamin D (μg/day)	3.6	3.0	6.6	5.7	5 *	5 *	5 *	5 *	10	10	20 [s]	20 [s]	15 [c]	15 [c]
Vitamin E (mg/day)	15.4	15.6	21.7	18.8	10	7.5	15	15	ND	ND	14	12	13 [c]	11 [c]
Vitamin C (mg/day)	151	162	161	170	45	45	90	75	40	40	110	95	110	95
Thiamine (mg/day)	1.8	1.5	1.8	1.5	1.2	1.1	1.2	1.1	1.0	0.8	1.2	1.0	1.0	1.0
Riboflavin (mg/day)	2.1	1.9	2.1	1.8	1.3	1.1	1.3	1.1	1.3	1.1	1.3	1.1	1.6	1.6
Niacin (mg/day)	24.8	20.1	27.3	21.4	16	14	16	14	16	14	15	11	16	16
Pantothenic acid (mg/day)	6.0	5.2	6.7	5.7	5	5	5	5	ND	ND	6	6	5	5
Pyridoxine (mg/day)	2.5	2.1	2.8	2.3	1.7	1.5	1.3 *	1.3 *	1.4	1.2	1.6	1.4	1.7	1.6
Folate (μg/day)	374	372	383	363	400	400	400	400	200	200	300	300	330	330
Vitamin B12 (μg/day)	6.8	5.3	8.1	6.3	2.4	2.4	2.4	2.4	1.5	1.5	4.0	4.0	4.0 [c]	4.0 [c]
Calcium (mg/day)	1136	1111	1023	969	1000 *	1000 *	1000 *	1000 *	700	700	1000	1000	1000 [c]	1000 [c]
Iron (mg/day)	16.1	13.4	16.4	13.8	8	18 *	8	18 *	8.7	14.8 *	10	15/10 [p]	11	16/11 [p]
Iodide (μg/day)	162	147	177	155	200	150	150	150	140	140	190	150	150 [c]	150 [c]
Magnesium (mg/day)	453	410	419	371	260	220	420 [l]	320 [l]	300	270	350	300	350	300
Potassium (mg/day)	3886	3584	3860	3543	3400 [c]	2600 [c]	4700	4700	3500	3500	4000	4000	3500	3500
Phosphorus (mg/day)	1571	1350	1542	1284	700	700	700	700	550	550	700	700	550	550
Sodium (mg/day)	3937	3092	4152	3154	2000	2000	2300	2300	1600	1600	1500	1500	2000	2000

ORISCAV = Observation of Cardiovascular Risk Factors; WHO = World Health Organization; USDA = United States Department of Agriculture; BNF = British Nutrition Foundation; DACH = Germany (D), Austria (A), and Switzerland (CH) Reference Values; EFSA = European Food Safety Authority; [ʃ] Dietary Reference Intakes (DRIs): recommended dietary allowances (RDAs), if unavailable, acceptable intake (AI) was used; [a] recommended nutrient intake (RNI) is the daily intake that meets the nutrient requirements of almost all (97.5%) individuals of the general population (in the respective age- and gender-specific group); [b] g/kg body weight per day; [c] Adequate Intakes (AI); [d] kilojoule/day according to moderate physical activity; * Upper 65 year: calcium = 1200–1300 mg/day, vitamin D = 10 μg/day, iron = 8 mg/day, pyridoxine = 1.7 (female), 1.5 (male) mg/day recommended; [e] energy requirements are based on the average energy required for people of a healthy weight who are moderately active; [g] mg equivalent/day of retinol; [s] vitamin D in the absence of endogenous synthesis μg/day; [p] premenopausal women/postmenopausal women; [t] percentage of total energy intake; [l] above 30 years of age; [Δ] = population reference intake (PRI), if unavailable, acceptable intake (AI) was used; ND = not determined, M = men, W = women.

Adults residing in Luxembourg and participating in ORISCAV-LUX and ORISCAV-LUX 2 showed an intake of vitamin D, fiber, and folate below several recommendations, and a higher than recommended sodium and total fat intake.

4. Discussion

The present work used data from two large cross-sectional studies, ORISCAV-LUX (2007/2008) and ORISCAV-LUX 2 (2016/2017). We described the dietary patterns and their changes during the last decade (2007–2017) in adult residents in Luxembourg having taken part in the surveys. Our results highlight dietary changes over approximately 10 years amongst the study participants, with significant differences in the amount of some consumed macronutrients and micronutrients and underlying food groups. Most notably, total fat intake, MUFA, SFA, PUFA (including eicosapentaenoic acid, docosahexaenoic acid, and docosapentaenoic acid), cholesterol, alcohol, and total energy intake in men and women did increase significantly over the past decade (Table 2). In contrast, total carbohydrate, magnesium, and calcium intakes were significantly reduced (Table 3).

Regarding major food groups, there was a decrease over time in the intake of grains, vegetables, and dairy products. In contrast, the intake of protein-rich foods, ready-to-eat meals, fats, noncaloric beverages, and alcoholic beverages increased during the studied period (Table 4).

The strongest intake increases during this approximately 10-year period were seen for alcohol, ready-made meals, total fat, and SFA (Tables 2 and 3). Compared to dietary recommendations (Table 7), the intake of total fat, energy, SFA, and sodium appeared relatively high. Especially in conjunction with the increased intake of ready-to-eat foods/fast foods that also included processed foods, such as meat products, such trends have been associated in the literature with a high incidence of type 2 diabetes and other cardiometabolic diseases [34–37]. However, these associations are typically based on cross-sectional studies, and other lifestyle factors could confound such relations [38]. Despite the fact that Luxembourg, similar to other countries, is engaged in health promotion programs to stimulate healthy eating [39], it appears that health-promotion-oriented measures were insufficient to turn the tide of poor dietary patterns. In line with these findings, when using the exploratory factor analysis to determine the main dietary components, two dietary patterns were obtained, which were either characterized by a rather Mediterranean pattern, rich in whole grains, fruits, vegetables, and dairy products, or a rather westernized pattern rich in starchy vegetables, animal-based proteins, fast foods, and fats (Figure 2). The first pattern would be in line with diets that have been associated with generally favorable health outcomes [40].

The estimated intake of most micronutrients appeared comparable over the years, or even increased (Table 3), with the exception of calcium and magnesium intake undergoing a significant decline during the past decade. Magnesium is an essential macro-mineral and a reduction in dietary intake of this mineral over the past decades has been reported for other countries, such as the US [41]. The intake of this micronutrient has been related in meta-analyses to the decreased incidence of type 2 diabetes, cardiovascular diseases, and all-cause mortality [42]. As magnesium is consumed partly within the grain/carbohydrate group and within vegetables, it is possible that its decline was related to the reduced intake of these food groups observed in the present study. Moreover, due to the lower consumption of food items from the dairy group, the decrease in calcium intake is conceivable and predictable. In addition to its importance in bone mass density [43], numerous studies have examined the association between low calcium intake and an increased risk of CVD. A population-based study for instance concluded that dietary calcium intake is associated with a decreased CVD risk [44].

Despite a significant reduction in the intake of grains and vegetables, dietary fiber intake did not significantly change. In a Europe-wide cohort study [45], including over half a million participants, researchers reported that fiber intake was associated with various types of cancers, with reduced fiber intake from fruit and vegetable sources as a major possible cause [45]. In addition, fiber and associated phytochemicals originating from a diverse intake of plant-based food items might positively affect gut microbiota diversity [46], which has also been inversely associated with several chronic diseases, including diabetes [47]. Despite dietary fiber intake being marginal compared to some intake recommendations, it was very close to reaching the 25 g/day as stipulated by EFSA (Table 7).

Despite these findings, the intake of a number of micronutrients has been increasing over the past decade. In parallel with a generally higher fat intake, the intake of multi-unsaturated fatty acids, including omega-3 fatty acids (EPA, DHA, and linolenic acid), also increased. These fatty acids have generally been related to anti-inflammatory processes [48] and have been correlated, e.g., to a lower incidence of coronary heart diseases [48]. Another positive aspect associated with the intake of higher amounts of dietary lipids is the increased intake of fat-soluble vitamins A, E, and D (Table 3). This increase has resulted in almost reaching the respective intake recommendations set by several health and nutrition-related organizations (Table 7). Moreover, despite the decreased vegetable intake, beta-carotene consumption slightly increased from 2007 to 2017. This contradictory result could be due to the intake of other beta-carotene sources, such as carrots, cabbages, and avocado, which, in our study, showed an increase in their intake (results not shown). Furthermore, possibly due to increased total energy and protein intake from animal sources, some water-soluble vitamins, such as niacin and pyridoxine, also increased significantly (Table 3), and their intake was generally in line with dietary recommendations (Table 7).

However, our study results showed that the dietary patterns in ORISCAV-LUX 2 are moving to a more Westernized-type diet, characterized by a higher intake of fat and alcohol and a lower total carbohydrate intake. In line with our findings, Marques-Vidal et al. reported similar results in the French-speaking part of Switzerland [49]. These results indicate that minor changes in dietary intakes and choices over time can significantly affect overall dietary patterns.

Differences between genders were also observed. Our study results showed that men consumed significantly more total energy, protein, total fat, cholesterol, and alcohol than women. To some extent, the higher intake might be attributed to the higher energy needs of men compared to women [50]). It was also found that men in the ORISCAV-LUX 2 consumed significantly higher amounts of total energy, animal-based protein, total fat, fat-soluble vitamins, SFA, MUFA, PUFA, cholesterol, sodium, and alcohol than men in the ORISCAV-LUX survey, but lower amounts of calcium and magnesium. Similar results were observed when comparing women in the two waves, except for sodium, where no significant difference was observed. According to other studies, in line with our findings, dietary patterns in men and women have changed in the last decade or so in other westernized countries [51]. Bédard et al. showed that men especially consumed more high-fat, high-protein, and ready-to-eat foods than women [52]. Somewhat contrarily, Macdiarmid et al., in a study in the UK, reported that associations between sugar and fat intake and BMI were different between men and women. They concluded that the consumption of products rich in fats and sugars might partly explain the higher BMI in women than men [53].

Moreover, in line with our results and with worldwide trends, Sánchez-Villegas et al., in a cohort study, showed that Spanish adult residents tended to consume more ready-to-eat foods and meals and increased their intake of processed foods, and thus consumed more saturated fat and more sodium but fewer vegetables, low-fat dairy products, and fruits over the past years [54]. As observed for ORISCAV-LUX 2, Bamia et al. reported that western dietary patterns, mainly including fat, animal-based protein, and fast foods, are also rising in the elderly Europeans [55]. They emphasized that these dietary patterns could be an essential contributing factor for various diet-related diseases, such as diabetes [55] and other inflammatory diseases. For instance, Harding et al., in the EPIC Cancer-Norfolk study, found a significant association between dietary fat and cholesterol intake and diabetes [56], though, in a pooled meta-analysis of cohort studies, only saturated fat intake was related to some types of cancer, not total fat and cholesterol intake [57]. Our results also indicate that grain intake, a rich source of soluble and insoluble fiber, has rather decreased over the past decade. According to the "Dietary Patterns Amongst Older Europeans" survey, fiber, one of the important characteristics of the Mediterranean dietary pattern, has been shown to be associated with a reduction in metabolic diseases, such as diabetes and hypertension [55], possibly due to their positive influence on the gut flora and increased formation of anti-inflammatory short-chain fatty acids [58]. A review article by Matthias et al. following a Mediterranean dietary pattern with recommended amounts of whole grains, fruits, and vegetables highlighted that these patterns were associated with a significant reduction in a number of cardiometabolic diseases, such as diabetes and hypertension [59]. Therefore, a higher intake of dietary fiber, as opposed to the 25 g/day in the present study, is desired.

Furthermore, Michelle et al. reported that Mediterranean dietary patterns were associated with a lower likelihood of developing obesity in people that are overweight, suggesting that improving the nutritional status might be part of the solution in tackling obesity or overweightness [60]. Both metabolically unhealthy obesity and metabolically unhealthy normal weight have been on the rise in Luxembourg [61] and other European countries in the past decades and were associated with inflammatory and oxidative stress processes. Similarly, Buckland et al. of the EPIC cohort survey reported that the Mediterranean dietary pattern was associated with reduced breast cancer and coronary heart disease [62].

Several factors have been highlighted as contributing to the changes in dietary patterns over time. The most important ones in literature were economic/social status, education, age, and gender [63–65]. However, what remains to be more fully explored is why dietary patterns have shifted toward rather less healthy attributes in westernized countries, if not globally. In general, it is believed that factors such as the globalization of the economy and food production, widespread advertising of fast-food companies, and a lack of physical activity play important roles [66–69]. In addition, with globalization, staple foods have been shifting from local to industrial products, which entail, to a large extent, low-cost and highly processed foods and, consequently, result in a deterioration in healthy dietary patterns [70]. On the other hand, increasing working hours (together with less time eating at home), and easy access to cheap and ready-to-eat meals have been highlighted as individual factors in the westernization of dietary patterns over time [66–69].

One of our study's strengths is that it is the first survey to examine an example of a European country with a diverse demographic composition [71]. Another advantage of our study was using a validated FFQ, which allowed us to have a comprehensive interpretation of the study participants' dietary intake, together with a geographically appropriate food database. Our study has a relatively high sample size given the total population of Luxembourg, which allowed us to perform analyses for different age and gender groups. However, ORISCAV-LUX 2 is not fully representative of the general adult population residing in Luxembourg. For example, the number of Portuguese participants in the second survey was lower than the number of Portuguese participants in the EHES (14.5%) study, which is considered representative of the general population and was conducted almost at the same time as ORISCAV LUX 2 [72]. Contrarily, ORISCAV-LUX (as ORISCAV-LUX 2), when comparing respondents vs. nonrespondents at baseline, can be considered representative regarding the place of residence [30], though not for other variables, such as age or education level.

As for other observational studies, our study had some limitations. One of the shortcomings was related to the use of two FFQ, with the second one being slightly more detailed. Recall bias is considered inevitable, as the FFQ inquired about food intake in the past 3 months. However, it seems that employing trained personnel might significantly reduce this bias [73]. As for all population-based longitudinal and cohort studies, another concern in our surveys is the quality of collected data in the two study waves, such as sample measurement by an accredited laboratory. Due to the generally small number of missing data (except for income) and percentage of completed questions, this limitation does not seem detrimental to the results of this study, despite nutrition playing a crucial role in these groups. Different dietary assessment tools (e.g., multiple 24-h recall methods or food records) may be more recommendable for those groups than FFQ [73]. Therefore, it is proposed that future studies focus on these groups to obtain a more comprehensive overview and formulate more targeted nutritional interventions, starting early in life.

5. Conclusions

As for other Westernized countries, adults in Luxembourg taking part in ORISCAV-LUX 2 have been consuming relatively high amounts of processed foods, animal-based products, and thus proteins, and also fat and sodium. Concomitantly, a trend appears to consume slightly fewer vegetables, below the recommended intake. It is acknowledged that, in addition to physiological needs, an array of other factors, such as access to food, taste, the influence of peers, neurophysiological pathways to food intake, and socioeconomic factors, along with health promotion and public health actions are essential for improving dietary habits and patterns and deserve more investigation. Meanwhile, the State of Luxembourg has taken further steps to improve population health by fostering a healthier diet, including efforts such as introducing the Nutri-Score labeling [74], and has also provided community-based training based on age and gender. These measures remain to be awaiting their efficiency; further large-scale efforts and interventions to produce more substantial and lasting effects are desired. Additional monitoring of dietary patterns, including the very young, is paramount to monitor population-based efforts to steer lifestyle patterns toward healthy directions and reduce possible associated diseases.

Supplementary Materials: The following are available online at https://www.mdpi.com/article/10.3390/nu13124382/s1, Table S1: Estimated marginal means of participants' total energy, alcohol, and macronutrient intake, Table S2: Estimated marginal means of micronutrient intake of participants, Table S3: Estimated marginal means of food groups' intake of participants.

Author Contributions: F.V. and T.B. performed the statistical analyses, interpreted the data, and drafted the manuscript. H.S. provided substantial contributions to the idea, conception, and design of the work, interpretation of the data, and guidance on the intellectual content and critical revision of the manuscript. M.V. and G.L.C. guided the statistical analyses, interpretation of the data, and critical revision of the manuscript. A.B. contributed to the revision of the manuscript. T.B. provided expertise and oversight on the intellectual content. All authors have read and agreed to the published version of the manuscript.

Funding: The ORISCAV-LUX and ORISCAV-LUX 2 data collection was funded by the LIH (MESR funding). A.B. was supported by the Ministry of Science and Higher Education of the Russian Federation within the framework of state support for creating and developing World-Class Research Centers "Digital biodesign and personalized healthcare No. 075-15-2020-926".

Institutional Review Board Statement: The study was conducted according to the guidelines of the Declaration of Helsinki, and approved by the National Research Ethics Committee of Luxembourg (CNER, N° 201,505/12 and N° 200609/03).

Informed Consent Statement: Written informed consent was obtained by all participants.

Data Availability Statement: Not applicable.

Acknowledgments: We thank all the study participants of the ORISCAV-LUX and ORISCAV-LUX 2 surveys. We thank Ala'a Alkerwi, who was involved in the general conception and design of the ORISCAV-LUX survey. We also extend our gratitude to the ORISCAV-LUX researcher's working group that has been fully involved in the conception and design of the ORISCAV-LUX 2 survey. ORISCAV Working Group: Gloria Aguayo, Ala'a Alkerwi, Brice Appenzeller, Jean Beissel, Dritan Bejko, Anna Chioti, Sophie Couffignal, Charles Delagardelle, Yvan Devaux, Isabelle Ernens, Guy Fagherazzi, Manon Gantenbein, Laetitia Huiart, Marylène D'Incau, Marie-Lise Lair, Stephanie Noppe, Laurent Malisoux, Jessica Pastore, Magali Perquin, Maria Ruiz-Castell, Jean-Claude Schmit and Saverio Stranges. We thank the research nurses involved in the ORISCAV-LUX and ORISCAV-LUX 2 surveys. Moreover, we are grateful to the Department of Public Health, University of Liège (Belgium), for allowing us to use their Food Frequency Questionnaire (https://www.dssp-uliege.be/FFQ (accessed on 22 May 2021)).

Conflicts of Interest: The authors declare no conflict of interest.

References

1. Engeset, D.; Alsaker, E.; Ciampi, A.; Lund, E. Dietary patterns and lifestyle factors in the Norwegian EPIC cohort: The Norwegian Women and Cancer (NOWAC) study. *Eur. J. Clin. Nutr.* **2005**, *59*, 675–684. [CrossRef]
2. Grosso, G.; Bella, F.; Godos, J.; Sciacca, S.; Del Rio, D.; Ray, S.; Galvano, F.; Giovannucci, E.L. Possible role of diet in cancer: Systematic review and multiple meta-analyses of dietary patterns, lifestyle factors, and cancer risk. *Nutr. Rev.* **2017**, *75*, 405–419. [CrossRef]
3. Sadeghi, M.; Vahid, F.; Rahmani, D.; Akbari, M.E.; Davoodi, S.H. The Association between Dietary Patterns and Breast Cancer Pathobiological Factors Progesterone Receptor (PR) and Estrogen Receptors (ER): New Findings from Iranian Case-Control Study. *Nutr. Cancer* **2019**, *71*, 1290–1298. [CrossRef]
4. Johansson, L.; Thelle, D.S.; Solvoll, K.; Bjørneboe, G.-E.A.; Drevon, C.A. Healthy dietary habits in relation to social determinants and lifestyle factors. *Br. J. Nutr.* **1999**, *81*, 211–220. [CrossRef] [PubMed]
5. Pänkäläinen, M.; Fogelholm, M.; Valve, R.; Kampman, O.; Kauppi, M.; Lappalainen, E.; Hintikka, J. Pessimism, diet, and the ability to improve dietary habits: A three-year follow-up study among middle-aged and older Finnish men and women. *Nutr. J.* **2018**, *17*, 92. [CrossRef] [PubMed]
6. Tapsell, L.C. Dietary behaviour changes to improve nutritional quality and health outcomes. *Chronic. Dis. Transl. Med.* **2017**, *3*, 154–158. [CrossRef] [PubMed]
7. Johansson, S.E.; Sundquist, J. Change in lifestyle factors and their influence on health status and all-cause mortality. *Int. J. Epidemiol.* **1999**, *28*, 1073–1080. [CrossRef]
8. Alkerwi, A.a.; Sauvageot, N.; Donneau, A.-F.; Lair, M.-L.; Couffignal, S.; Beissel, J.; Delagardelle, C.; Wagener, Y.; Albert, A.; Guillaume, M. First nationwide survey on cardiovascular risk factors in Grand-Duchy of Luxembourg (ORISCAV-LUX). *BMC Public Health.* **2010**, *10*, 468. [CrossRef]
9. Suetens, C. Healthcare-associated infections in European long-term care facilities: How big is the challenge? *Eurosurveillance* **2012**, *17*, 20259. [CrossRef] [PubMed]
10. Bocquet, V.; Ruiz-Castell, M.; de Beaufort, C.; Barre, J.; de Rekeneire, N.; Michel, G.; Donahue, R.P.; Kuemmerle, A.; Stranges, S. Public health burden of pre-diabetes and diabetes in Luxembourg: Finding from the 2013-2015 European Health Examination Survey. *BMJ Open* **2019**, *9*, e022206. [CrossRef]
11. Perquin, M.; Diederich, N.; Pastore, J.; Lair, M.L.; Stranges, S.; Vaillant, M.; MemoVie, G. Prevalence of Dementia and Cognitive Complaints in the Context of High Cognitive Reserve: A Population-Based Study. *PLoS ONE* **2015**, *10*, e0138818.
12. Hipp, G.; Vaillant, M.; Diederich, N.J.; Roomp, K.; Satagopam, V.P.; Banda, P.; Sandt, E.; Mommaerts, K.; Schmitz, S.K.; Longhino, L.; et al. The Luxembourg Parkinson's Study: A Comprehensive Approach for Stratification and Early Diagnosis. *Front. Aging Neurosci.* **2018**, *10*, 326. [CrossRef] [PubMed]
13. Alkerwi, A.; Bahi, I.E.; Stranges, S.; Beissel, J.; Delagardelle, C.; Noppe, S.; Kandala, N.-B. Geographic variations in cardiometabolic risk factors in Luxembourg. *Int. J. Environ. Res.* **2017**, *14*, 648. [CrossRef]
14. Naska, A.; Fouskakis, D.; Oikonomou, E.; Almeida, M.D.V.; Berg, M.A.; Gedrich, K.; Moreiras, O.; Nelson, M.; Trygg, K.; Turrini, A.; et al. Dietary patterns and their socio-demographic determinants in 10 European countries: Data from the DAFNE databank. *Eur. J. Clin. Nutr.* **2006**, *60*, 181–190. [CrossRef] [PubMed]
15. World Health Organization. *Global Status Report on Alcohol and Health 2018*; World Health Organization: Geneva, Switzerland, 2019.
16. Alkerwi, A.; Crichton, G.E.; Hébert, J.R. Consumption of ready-made meals and increased risk of obesity: Findings from the Observation of Cardiovascular Risk Factors in Luxembourg (ORISCAV-LUX) study. *Br. J. Nutr.* **2015**, *113*, 270–277. [CrossRef]
17. Esposito, K.; Maiorino, M.I.; Bellastella, G.; Panagiotakos, D.B.; Giugliano, D. Mediterranean diet for type 2 diabetes: Cardiometabolic benefits. *Endocrine* **2017**, *56*, 27–32. [CrossRef] [PubMed]
18. Shen, J.; Wilmot, K.A.; Ghasemzadeh, N.; Molloy, D.L.; Burkman, G.; Mekonnen, G.; Gongora, M.C.; Quyyumi, A.A.; Sperling, L.S. Mediterranean Dietary Patterns and Cardiovascular Health. *Annu. Rev. Nutr.* **2015**, *35*, 425–449. [CrossRef] [PubMed]
19. Schwingshackl, L.; Hoffmann, G. Adherence to Mediterranean diet and risk of cancer: An updated systematic review and meta-analysis of observational studies. *Cancer Med.* **2015**, *4*, 1933–1947. [CrossRef]
20. Herforth, A.; Arimond, M.; Álvarez-Sánchez, C.; Coates, J.; Christianson, K.; Muehlhoff, E. A Global Review of Food-Based Dietary Guidelines. *Adv. Nutr.* **2019**, *10*, 590–605. [CrossRef]
21. Elmadfa, I. *European Nutrition and Health Report 2009*; Karger Medical and Scientific Publishers: Basel, Switzerland, 2009; Volume 62.
22. Alkerwi, A.a.; Sauvageot, N.; Nau, A.; Lair, M.-L.; Donneau, A.-F.; Albert, A.; Guillaume, M. Population compliance with national dietary recommendations and its determinants: Findings from the ORISCAV-LUX study. *Br. J. Nutr.* **2012**, *108*, 2083–2092. [CrossRef] [PubMed]
23. FAOSTAT 2007. Livestock and Fish Primary Equivalent, Food and Agriculture Organization of the United Nations, Viewed 31 March 2007. Available online: http://faostat.fao.org/site/610/DesktopDefault.aspx?PageID=610#ancor (accessed on 21 January 2021).

24. McEvoy, C.T.; Temple, N.; Woodside, J.V. Vegetarian diets, low-meat diets and health: A review. *Public Health Nutr.* **2012**, *15*, 2287–2294. [CrossRef]
25. McNeill, S.H. Inclusion of red meat in healthful dietary patterns. *Meat. Sci.* **2014**, *98*, 452–460. [CrossRef] [PubMed]
26. Kearney, J. Food consumption trends and drivers. *Philos. Trans. R. Soc. B Biol. Sci.* **2010**, *365*, 2793–2807. [CrossRef] [PubMed]
27. Vilaro, M.J.; Colby, S.E.; Riggsbee, K.; Zhou, W.; Byrd-Bredbenner, C.; Olfert, M.D.; Barnett, T.E.; Horacek, T.; Sowers, M.; Mathews, A.E. Food Choice Priorities Change Over Time and Predict Dietary Intake at the End of the First Year of College Among Students in the U.S. *Nutrients* **2018**, *10*, 1296. [CrossRef] [PubMed]
28. Caruana, E.J.; Roman, M.; Hernández-Sánchez, J.; Solli, P. Longitudinal studies. *J. Thorac. Dis.* **2015**, *7*, E537–E540. [PubMed]
29. Karageorgou, D.; Magriplis, E.; Mitsopoulou, A.V.; Dimakopoulos, I.; Bakogianni, I.; Micha, R.; Michas, G.; Chourdakis, M.; Ntouroupi, T.; Tsaniklidou, S.M.; et al. Dietary patterns and lifestyle characteristics in adults: Results from the Hellenic National Nutrition and Health Survey (HNNHS). *Public Health.* **2019**, *171*, 76–88. [CrossRef]
30. Alkerwi, A.a.; Pastore, J.; Sauvageot, N.; Coroller, G.L.; Bocquet, V.; d'Incau, M.; Aguayo, G.; Appenzeller, B.; Bejko, D.; Bohn, T.; et al. Challenges and benefits of integrating diverse sampling strategies in the observation of cardiovascular risk factors (ORISCAV-LUX 2) study. *BMC Med. Res. Methodol.* **2019**, *19*, 27. [CrossRef] [PubMed]
31. Sauvageot, N.; Guillaume, M.; Albert, A. Validation of the food frequency questionnaire used to assess the association between dietary habits and cardiovascular risk factors in the NESCAV study. *J. Food Sci.* **2013**, *3*, 3. [CrossRef]
32. Sauvageot, N.; Alkerwi, A.; Albert, A.; Guillaume, M. Use of food frequency questionnaire to assess relationships between dietary habits and cardiovascular risk factors in NESCAV study: Validation with biomarkers. *Nutr. J.* **2013**, *12*, 143. [CrossRef] [PubMed]
33. French Agency for Food, Environmental and Occupational Health & Safety. ANSES-CIQUAL French Food Composition Table for Nutritional Intakes Calculation CALNUT. Available online: https://ciqual.anses.fr/ (accessed on 21 January 2021).
34. Barnard, N.; Levin, S.; Trapp, C. Meat consumption as a risk factor for type 2 diabetes. *Nutrients* **2014**, *6*, 897–910. [CrossRef]
35. Micha, R.; Wallace, S.K.; Mozaffarian, D. Red and processed meat consumption and risk of incident coronary heart disease, stroke, and diabetes mellitus: A systematic review and meta-analysis. *Circulation* **2010**, *121*, 2271–2283. [CrossRef]
36. Ericson, U.; Sonestedt, E.; Gullberg, B.; Hellstrand, S.; Hindy, G.; Wirfält, E.; Orho-Melander, M. High intakes of protein and processed meat associate with increased incidence of type 2 diabetes. *Br. J. Nutr.* **2013**, *109*, 1143–1153. [CrossRef]
37. Ibsen, D.B.; Steur, M.; Imamura, F.; Overvad, K.; Schulze, M.B.; Bendinelli, B.; Guevara, M.; Agudo, A.; Amiano, P.; Aune, D. Replacement of red and processed meat with other food sources of protein and the risk of type 2 diabetes in European populations: The EPIC-InterAct Study. *Diab. Care* **2020**, *43*, 2660–2667. [CrossRef]
38. Arsenault, B.J.; Lamarche, B.; Després, J.-P. Targeting Overconsumption of Sugar-Sweetened Beverages vs. Overall Poor Diet Quality for Cardiometabolic Diseases Risk Prevention: Place Your Bets! *Nutrients* **2017**, *9*, 600. [CrossRef] [PubMed]
39. Brambila-Macias, J.; Shankar, B.; Capacci, S.; Mazzocchi, M.; Perez-Cueto, F.J.; Verbeke, W.; Traill, W.B. Policy interventions to promote healthy eating: A review of what works, what does not, and what is promising. *Food Nutr. Bull.* **2011**, *32*, 365–375. [CrossRef] [PubMed]
40. Trichopoulou, A.; Martínez-González, M.A.; Tong, T.Y.; Forouhi, N.G.; Khandelwal, S.; Prabhakaran, D.; Mozaffarian, D.; de Lorgeril, M. Definitions and potential health benefits of the Mediterranean diet: Views from experts around the world. *BMC Med.* **2014**, *12*, 112. [CrossRef]
41. Ford, E.S.; Mokdad, A.H. Dietary Magnesium Intake in a National Sample of U.S. Adults. *J. Nutr.* **2003**, *133*, 2879–2882. [CrossRef] [PubMed]
42. Fang, X.; Wang, K.; Han, D.; He, X.; Wei, J.; Zhao, L.; Imam, M.U.; Ping, Z.; Li, Y.; Xu, Y.; et al. Dietary magnesium intake and the risk of cardiovascular disease, type 2 diabetes, and all-cause mortality: A dose-response meta-analysis of prospective cohort studies. *BMC Med.* **2016**, *14*, 210. [CrossRef] [PubMed]
43. Tai, V.; Leung, W.; Grey, A.; Reid, I.R.; Bolland, M.J. Calcium intake and bone mineral density: Systematic review and meta-analysis. *BMJ* **2015**, *351*, h4183. [CrossRef] [PubMed]
44. Kong, S.H.; Kim, J.H.; Hong, A.R.; Cho, N.H.; Shin, C.S. Dietary calcium intake and risk of cardiovascular disease, stroke, and fracture in a population with low calcium intake. *Am. J. Clin. Nutr.* **2017**, *106*, 27–34. [CrossRef] [PubMed]
45. Bradbury, K.E.; Appleby, P.N.; Key, T.J. Fruit, vegetable, and fiber intake in relation to cancer risk: Findings from the European Prospective Investigation into Cancer and Nutrition (EPIC). *Am. J. Clin. Nutr.* **2014**, *100* (Suppl. 1), 394s–398s. [CrossRef]
46. Dingeo, G.; Brito, A.; Samouda, H.; Iddir, M.; La Frano, M.R.; Bohn, T. Phytochemicals as modifiers of gut microbial communities. *Food Funct.* **2020**, *11*, 8444–8471. [CrossRef] [PubMed]
47. Post, R.E.; Mainous, A.G., 3rd; King, D.E.; Simpson, K.N. Dietary fiber for the treatment of type 2 diabetes mellitus: A meta-analysis. *J. Am. Board. Fam. Med.* **2012**, *25*, 16–23. [CrossRef] [PubMed]
48. Vahid, F.; Goodarzi, R.; Shivappa, N.; Hébert, J.R.; Fazeli Moghadam, E. Dietary Inflammatory Index (DII®): A significant association between coronary heart disease and DII® in Armenian adults. *Eur. J. Prev. Cardiol.* **2020**, *27*, 2235–2237. [CrossRef]
49. Marques-Vidal, P.; Gaspoz, J.-M.; Theler, J.-M.; Guessous, I. Twenty-year trends in dietary patterns in French-speaking Switzerland: Toward healthier eating. *Am. J. Clin. Nutr.* **2017**, *106*, 217–224. [CrossRef]
50. CALERIE StudyGroup. Energy requirements in nonobese men and women: Results from CALERIE. *Am. J. Clin. Nutr.* **2013**, *99*, 71–78.
51. Harrington, J.M.; Dahly, D.L.; Fitzgerald, A.P.; Gilthorpe, M.S.; Perry, I.J. Capturing changes in dietary patterns among older adults: A latent class analysis of an ageing Irish cohort. *Public Health Nutr.* **2014**, *17*, 2674–2686. [CrossRef] [PubMed]

52. Bédard, A.; Riverin, M.; Dodin, S.; Corneau, L.; Lemieux, S. Sex differences in the impact of the Mediterranean diet on cardiovascular risk profile. *Br. J. Nutr.* **2012**, *108*, 1428–1434. [CrossRef] [PubMed]
53. Macdiarmid, J.I.; Vail, A.; Cade, J.E.; Blundell, J.E. The sugar–fat relationship revisited: Differences in consumption between men and women of varying BMI. *Int. J. Obes.* **1998**, *22*, 1053–1061. [CrossRef]
54. Sánchez-Villegas, A.; Delgado-Rodríguez, M.; Martínez-González, M.Á.; de Irala-Estévez, J.; for the Seguimiento Universidad de Navarra group. Gender, age, socio-demographic and lifestyle factors associated with major dietary patterns in the Spanish Project SUN (Seguimiento Universidad de Navarra). *Eur. J. Clin. Nutr.* **2003**, *57*, 285–292. [CrossRef]
55. Bamia, C.; Orfanos, P.; Ferrari, P.; Overvad, K.; Hundborg, H.H.; Tjønneland, A.; Olsen, A.; Kesse, E.; Boutron-Ruault, M.-C.; Clavel-Chapelon, F.; et al. Dietary patterns among older Europeans: The EPIC-Elderly study. *Br. J. Nutr.* **2005**, *94*, 100–113. [CrossRef]
56. Harding, A.-H.; Day, N.E.; Khaw, K.-T.; Bingham, S.; Luben, R.; Welsh, A.; Wareham, N.J. Dietary Fat and the Risk of Clinical Type 2 Diabetes: The European Prospective Investigation of Cancer-Norfolk Study. *Am. J. Epidemiol.* **2004**, *159*, 73–82. [CrossRef]
57. Genkinger, J.M.; Hunter, D.J.; Spiegelman, D.; Anderson, K.E.; Beeson, W.L.; Buring, J.E.; Colditz, G.A.; Fraser, G.E.; Freudenheim, J.L.; Goldbohm, R.A.; et al. A pooled analysis of 12 cohort studies of dietary fat, cholesterol and egg intake and ovarian cancer. *Cancer. Causes. Control.* **2006**, *17*, 273–285. [CrossRef] [PubMed]
58. Morrison, D.J.; Preston, T. Formation of short chain fatty acids by the gut microbiota and their impact on human metabolism. *Gut Microbes* **2016**, *7*, 189–200. [CrossRef]
59. Schulze, M.B.; Hu, F.B. Dietary patterns and risk of hypertension, type 2 diabetes mellitus, and coronary heart disease. *Curr. Atheroscler. Rep.* **2002**, *4*, 462–467. [CrossRef]
60. Mendez, M.A.; Popkin, B.M.; Jakszyn, P.; Berenguer, A.; Tormo, M.J.; Sanchéz, M.J.; Quirós, J.R.; Pera, G.; Navarro, C.; Martinez, C.; et al. Adherence to a Mediterranean Diet Is Associated with Reduced 3-Year Incidence of Obesity. *J. Nutr.* **2006**, *136*, 2934–2938. [CrossRef] [PubMed]
61. Europepan Commission. Report from the Commission to the European Parliament and the Council Regarding the Use of Additional Forms of Expression and Presentation of the Nutrition Declaration. 2020. Available online: https://eur-lex.europa.eu/legal-content/GA/TXT/?uri=CELEX:52020DC0207 (accessed on 25 February 2021).
62. Buckland, G.; Travier, N.; Cottet, V.; González, C.A.; Luján-Barroso, L.; Agudo, A.; Trichopoulou, A.; Lagiou, P.; Trichopoulos, D.; Peeters, P.H.; et al. Adherence to the mediterranean diet and risk of breast cancer in the European prospective investigation into cancer and nutrition cohort study. *Int. J. Cancer Res.* **2013**, *132*, 2918–2927. [CrossRef]
63. Mishra, G.; McNaughton, S.; Bramwell, G.a.; Wadsworth, M. Longitudinal changes in dietary patterns during adult life. *Br. J. Nutr.* **2006**, *96*, 735–744.
64. Waid, J.L.; Sinharoy, S.S.; Ali, M.; Stormer, A.E.; Thilsted, S.H.; Gabrysch, S. Dietary patterns and determinants of changing diets in Bangladesh from 1985 to 2010. *Curr. Dev. Nutr.* **2019**, *3*, nzy091. [CrossRef]
65. Mazzocchi, M.; Brasili, C.; Sandri, E. Trends in dietary patterns and compliance with World Health Organization recommendations: A cross-country analysis. *Public Health Nutr.* **2008**, *11*, 535–540. [CrossRef]
66. Uusitalo, U.; Sobal, J.; Moothoosamy, L.; Chitson, P.; Shaw, J.; Zimmet, P.; Tuomilehto, J. Dietary Westernisation: Conceptualisation and measurement in Mauritius. *Public Health Nutr.* **2005**, *8*, 608–619. [CrossRef] [PubMed]
67. Wakimoto, P.; Block, G. Dietary intake, dietary patterns, and changes with age: An epidemiological perspective. *J. Gerontol. Ser. A Biol. Sci. Med. Sci.* **2001**, *56*, 65–80. [CrossRef] [PubMed]
68. Lake, A.A.; Adamson, A.J.; Hyland, R.M.; Mathers, J.C. Dietary change and perceptions of change over time. *Nutr. Bull.* **2004**, *29*, 199–203. [CrossRef]
69. Jiang, L.; Audouze, K.; Romero Herrera, J.A.; Ängquist, L.H.; Kjærulff, S.K.; Izarzugaza, J.M.G.; Tjønneland, A.; Halkjær, J.; Overvad, K.; Sørensen, T.I.A.; et al. Conflicting associations between dietary patterns and changes of anthropometric traits across subgroups of middle-aged women and men. *Clin. Nutr.* **2020**, *39*, 265–275. [CrossRef] [PubMed]
70. Vega Mejía, N.; Ponce Reyes, R.; Martinez, Y.; Carrasco, O.; Cerritos, R. Implications of the Western Diet for Agricultural Production, Health and Climate Change. *Front. Sustain. Food Syst.* **2018**, *2*, 88. [CrossRef]
71. European Commission, Eurydice Network. Population: Demographic Situation, Languages and Religions of Luxembourg. 2020. Available online: https://eacea.ec.europa.eu/national-policies/eurydice/content/population-demographic-situation-languages-and-religions-46_en (accessed on 2 March 2021).
72. Bocquet, V.; Barré, J.; Couffignal, S.; d'Incau, M.; Delagardelle, C.; Michel, G.; Schlesser, M.; Stranges, S.; Kuemmerle, A.; Ruiz-Castell, M. Study design and characteristics of the Luxembourg European Health Examination Survey (EHES-LUX). *BMC Public Health.* **2018**, *18*, 1169. [CrossRef] [PubMed]
73. Shim, J.-S.; Oh, K.; Kim, H.C. Dietary assessment methods in epidemiologic studies. *Epidemiol. Health.* **2014**, *36*, e2014009. [CrossRef] [PubMed]
74. Hercberg, S.; Touvier, M.; Salas-Salvado, J.; on behalf of the Group of European scientists supporting the implementation of Nutri-Score in Europe. The Nutri-Score nutrition label. *Int. J. Vitam. Nutr. Res.* **2021**. [CrossRef] [PubMed]

nutrients

MDPI

Article

Dietary Intake of Folate and Assessment of the Folate Deficiency Prevalence in Slovenia Using Serum Biomarkers

Igor Pravst [1,2,3,*], Živa Lavriša [1], Maša Hribar [1,2], Hristo Hristov [1], Naska Kvarantan [1,4], Barbara Koroušić Seljak [5], Matej Gregorič [6], Urška Blaznik [6], Nadan Gregorič [7], Katja Zaletel [7,8], Adrijana Oblak [7], Joško Osredkar [7,9], Katja Žmitek [1,3] and Anita Kušar [1]

1 Nutrition Institute, Tržaška cesta 40, SI-1000 Ljubljana, Slovenia; ziva.lavrisa@nutris.org (Ž.L.); masa.hribar@nutris.org (M.H.); hristo.hristov@nutris.org (H.H.); naska.kvarantan@gmail.com (N.K.); katja.zmitek@vist.si (K.Ž.); anita.kusar@nutris.org (A.K.)
2 Biotechnical Faculty, University of Ljubljana, Jamnikarjeva 101, SI-1000 Ljubljana, Slovenia
3 VIST–Higher School of Applied Sciences, Gerbičeva cesta 51A, SI-1000 Ljubljana, Slovenia
4 Division of Human Nutrition and Health, Wageningen University & Research, NL-6708 Wageningen, The Netherlands
5 Computer Systems Department, Jožef Stefan Institute, SI-1000 Ljubljana, Slovenia; barbara.korousic@ijs.si
6 National Institute of Public Health, Trubarjeva 2, SI-1000 Ljubljana, Slovenia; matej.gregoric@nijz.si (M.G.); urska.blaznik@nijz.si (U.B.)
7 University Medical Centre Ljubljana, Zaloška cesta 7, SI-1000 Ljubljana, Slovenia; nadan.gregoric@kclj.si (N.G.); katja.zaletel@kclj.si (K.Z.); adrijana.oblak@kclj.si (A.O.); josko.osredkar@kclj.si (J.O.)
8 Faculty of Medicine, University of Ljubljana, Vrazov trg 2, SI-1000 Ljubljana, Slovenia
9 Faculty of Pharmacy, University of Ljubljana, Aškerčeva cesta 7, SI-1000 Ljubljana, Slovenia
* Correspondence: igor.pravst@nutris.org; Tel.: +386-590-68871; Fax: +386-310-07981

Citation: Pravst, I.; Lavriša, Ž.; Hribar, M.; Hristov, H.; Kvarantan, N.; Seljak, B.K.; Gregorič, M.; Blaznik, U.; Gregorič, N.; Zaletel, K.; et al. Dietary Intake of Folate and Assessment of the Folate Deficiency Prevalence in Slovenia Using Serum Biomarkers. *Nutrients* **2021**, *13*, 3860. https://doi.org/10.3390/nu13113860

Academic Editor: Ulrika Ericson

Received: 29 September 2021
Accepted: 25 October 2021
Published: 28 October 2021

Abstract: Folate deficiency is associated with various health issues, including anemia, cardiovascular disease, and birth defects. Low folate intake and suboptimal folate status were found in several countries; however, this topic has not yet been investigated in Slovenia. Dietary folate intake and serum folate status were investigated through the nationally representative food consumption study SI.Menu/Nutrihealth. Folate intake was estimated using a sample of $N = 1248$ subjects aged 10–74 years, stratified in three age groups (adolescents, adults, elderly population), through two 24 h-dietary recalls and food propensity questionnaire. Data on serum folate and homocysteine was available for 280 participants. Very low folate intake (<300 µg/day) was observed in 59% of adolescents, 58% of adults and 68% of elderlies, and only about 12% achieved the WHO recommended level of 400 µg/day. Major dietary contributors were vegetables and fruit, and cereal products. Living environment, education, employment status and BMI were linked with low folate intake in adults; BMI, and sex in adolescents; and sex in elderlies. Considering low serum folate (<7 nmol/L) and high serum homocysteine (>15 nmol/L), folate deficiency was found in 7.6 and 10.5% in adults and elderlies, respectively. Additional public health strategies should be employed to promote the consumption of folate-rich foods. With current folate intakes, supplementation with folic acid is relevant especially in specific vulnerable populations, particularly in women planning and during pregnancy.

Keywords: folate; folic acid; folate intake; folate deficiency; homocysteine; Slovenia

1. Introduction

Folate is one of the water-soluble B-vitamins, which cannot be synthesized in the human body. Folate is an essential micronutrient required for the synthesis of both RNA and DNA, for cell division, growth and development [1]. These processes are regulated through many metabolic processes, which can be affected by the lack of folate. Insufficient dietary intake of folate and subsequent deficiency has been associated with various health issues, such as megaloblastic anaemia and cardiovascular disease [2]. Important role of folate in pregnancy is also well established [3]; as adequate folate intake during periconceptional period and the first trimester of pregnancy helps to prevent neural tube defects and other adverse birth outcomes [4].

Diets low in fresh fruits, leafy green vegetables, unrefined grains, and legumes have been linked with folate deficiency [5]. This can occur especially in populations where dietary intake relies more extensively on processed foods, while intake of fresh fruits, vegetables and legumes is often insufficient. On the other hand, regular consumption of such folate-rich foods has been shown as a protective measure [6]. Besides nutrition, lactation and alcoholism are among major factors, contributing to folate deficiency [5,6]. Folate status can be also affected by methylenetetrahydrofolate reductase gene mutation [7] and various diseases [8]. It should be noted that in foods folate is commonly bound to protein or carbohydrate food matrices [9,10], which limit its bioavailability. Losses of folate from foods can occur during cooking, particularly during heating and oxidation [11]. On the other hand, grinding of foods can help release more folates [12] and ascorbic acid improves folate stability [13]. Foods can be also enriched with folate during manufacturing. In the European Union, folic acid (pteroylmonoglutamic acid) and calcium-L-methylfolate are allowed sources of folate for voluntarily enrichment of foods [14], while (6S)-5-methyltetrahydrofolic acid derivatives are also allowed in food supplements [15]. Bioavailability of folate added to such products is generally higher [16] than in foods, where folate is bound to cell structures.

Considering that folate status is affected by such a wide range of parameters, dietary folate intake alone cannot be used to identify those at risk for folate deficiency. Different biological markers are therefore used for this purpose. Most commonly used is serum folate level [17]. Considering that serum folate levels increase up to 2 h after ingestion of folate (followed by rapid decline), it is recommended this biomarker is measured under fasted conditions [18]. With consideration of macrocytic anaemia as a haematological indicator, serum folate cut-off level of 14 nmol/L (6 ng/mL) is typically used as indicator of possible folate deficiency, while levels below 7 nmol/L (3 ng/mL) are interpreted as folate deficiency [19]. Additionally, elevated serum homocysteine is considered as crucial metabolic indicator of folate deficiency [17], with upper reference range limit between 10–19 µmol/L [17], most typically at 15 µmol/L [20,21]. Independently, hyperhomocysteinemia is also presenting an important risk factor of atherosclerosis and other cardiovascular disease, including arterial endothelial dysfunction and thromboembolism [22–24]. Besides serum folate and homocysteine, there are several other biomarkers of folate deficiency. Very valuable biomarker is folate concentration in red cells [25], which is not affected by short-term dietary interferences. Examples of other possible biomarkers are 5-methyltetrahydrofolate in cerebrospinal fluid, urinary folate, etc. [17].

In this paper term "folate" is used for all active forms of this vitamin, while folic acid is used specifically for pteroylmonoglutamic acid. The latter one is also one of the authorised forms of folate for use in enriched foods, food supplements and medicines. Speaking of requirements, these are set for folate equivalents and refer to all the natural and synthetic forms of folate. According to US Institute of Medicine (IOM) [26] and World Health Organisation (WHO) [27], estimated average requirements (EARs) of folate for adolescents (10–18 years) and adults/elderly (above 19 years) is 330 and 320 µg, respectively, while its recommended nutrient intakes (RNI) is 400 µg. European Food Safety Authority (EFSA) set population reference intake at 330 µg/day and average requirement (AR) at 250 µg/day [28]. According to D-A-CH recommendations (D, A and CH assign Germany, Austria and Switzerland, where these recommendations were prepared), which are also implemented in Slovenia [29], in year 2013 the reference daily intake for folate equivalents in adults was lowered from 400 µg [30] to 300 µg [29,31]. While there is no nationally representative data available on the actual folate intake in the general adult population, nor on the prevalence of folate deficiency, low folate intake/status was reported in specific vulnerable groups [32,33]. On the other hand, Rippin et. all reviewed data available from other European countries and reported that only a few populations met recommended folate intakes [34]. Pooled mean folate intake was 268 µg and 318 µg in women and men, respectively. Due to a public health concern of folate deficiency, some countries introduced the mandatory fortification of certain foods with folate [35], which resulted in increased folate intakes and improved folate status [5,36]. With consideration of possible adverse outcomes of excess vitamin intakes, changes in fortification policies should be planned very carefully [37]. To assure adequate folate intake, in Slovenia supplementation with folic acid is currently recommended to pregnant women and those planning pregnancy [38], however, there are no requirements or recommendations regarding the general population or mandatory food fortification. According to research, folate is also one of the micronutrients which can be critical in the elderly [39], as its insufficiency can cause various health problems, including cognitive performance, so supplementation could be also considered [40]. On the other hand, for example, in athletes, higher plasma folate combined with training could help to keep homocysteine levels at optimum [41]. There are general EU rules enabling voluntarily enrichment of foods with vitamins [42], but foods are quite rarely enriched with folate. Reported prevalence of labelled folate content on prepacked foods in Slovenia is 3% [43], but higher prevalence was observed on processed breakfast cereals (34%) and functional drinks (27%).

The objective of the present study was to estimate dietary folate intake and to determine the prevalence of folate deficiency in different Slovenian population groups. We exploited the data collected in the nationally representative food consumption study on adolescents, adults and elderly population (SI.Menu study), upgraded with collection of biological biomarkers on sub-sample of adults and elderlies (Nutrihealth study). Secondary objectives were to determine main sources of folate in peoples' diets, and to investigate determinants affecting low folate intake and folate deficiency.

2. Material and Methods

2.1. Study Design and Subjects

Data for estimation of folate intake was obtained from a cross-sectional national food consumption study (SI.Menu), which was conducted using cross-sectional approach in period from March 2017 to April 2018. Study was conducted in line with the EFSA's Guidance on European Union Menu Methodology [44]. More details on the methodology of the SI.Menu study was published previously [45]. Altogether, $N = 2280$ participants were randomly selected from Central Register of Population of Slovenia, with separate quotas for adolescent (10–17 years), adult (18–64 years), and elderly population (65–74 years), with 62% response rate. Exclusion of subjects with missing data and under/over-reporters ($N = 97$) is previously described [45,46]. Final SI.Menu study sample included $N = 1248$ subjects (468 adolescents, 364 adults and 416 elderlies).

Data for estimation of folate deficiency was obtained within Nutrihealth study, which was detailly described elsewhere [47]. Nutrihealth was conducted as an extension of the SI.Menu study. In short—after finished participation in the above described SI.Menu study, a sub-sample of adults and elderlies from SI.Menu study was invited to participate with blood and spot urine sample. A final sample of participants who provided samples for biomarker analysis included $N = 280$ participants; for these serum folate and homocysteine was determined (125 and 155 adults and elderlies, respectively).

2.2. SI.Menu Study Data Collection and Analyses

In the SI.Menu study, food consumption data and corresponding metadata was collected using General Questionnaire for socio-demographic and socio-economic data, and Food Propensity Questionnaire (FPQ) for collecting consumption frequency of key foods from selected food categories [48]. Both questionnaires were completed during personal visit, based on participants' answers. At same visit, anthropometric data, body height (m) and body mass (kg) were also collected. For adults body mass index (BMI) (kg/m^2) was interpreted using overweight cut-off point at 25 kg/m^2, while sex/age adjusted cut-off points (>1 standard deviation (SD)) were applied for adolescents [49,50]. Participant's physical activity was estimated using International Physical Activity Questionnaire (IPAQ) score [51]. General Questionnaire also contained questions which were used to identify participants' smoking status (non-smoker; smoker/occasional smoker/ex-smoker), medical (no diet, medical/weight loss diet) or other diets (no diet; vegetarian/vegan). Demographic characteristics was also collected and used for analyses coding: residential area (rural, intermediate, urban), education level (university degree, no university degree), self-reported financial status (below/above average, using monthly income cut-off at 1300€), employment status (employed/unemployed, student, retired).

Additionally, participant's dietary habits were investigated with two non-consecutive 24-h dietary recalls, executed 1–3 weeks apart (71% and 29% of the recalls were performed on workdays and weekends, respectively). Nationally developed picture book was used to support estimation of portion sizes. This tool was developed specifically for the conduction of the national food consumption study, and was composed of 46 pictures of foods, each presented in six portions [52].

2.3. Assessment of Dietary Folate Intake

Available food consumption data (24-h recalls) was analysed using the Open Platform for Clinical Nutrition (OPEN) [53], which contain Slovenian food composition database. This database consists of nutritional composition data for both generic and branded foods and provides traditional recipes of foods which are frequently consumed in Slovenia. Altogether, N = 2377 different foods were extracted from the SI.Menu study consumption dataset. For foods without data about folate content in the OPEN tool, missing information was taken from other sources, primarily from the FINELI [54] and USDA [55] food composition databases. Manual food-matching was done by trained nutritionist. Altogether, 89.5% of the reported foods were determined to be source of dietary folate. To enable estimation of daily folate intake with the use of Multiple Source Method (see Section 2.4), each reported food was assigned into one of pre-defined food categories, with consideration of FPQ. For the reporting purpose, all foods were additionally assigned into food categories, adapted by Global Food Monitoring Initiative [56]. Such categorisation was applied to describe the relative contribution of food categories in population usual daily dietary folate intake.

2.4. Serum Folate and Homocysteine Concentration

Folate and homocysteine concentrations were measured in serum of the participants of the Nutrihealth study at the University Medical Center (Ljubljana, Slovenia). Folate was analysed at the Department of Nuclear Medicine, while homocysteine was analysed at the Institute of clinical chemistry and biochemistry.

Folate was measured in serum with the chemiluminescence immunoassay determined on an Immulite 2000 XPi analyzer (Siemens Healthineers, Gwynedd, UK). Performance characteristics for the assay are as follows: Limit of detection is 1.8 nmol/L, linearity of the assay is in the range from 2.3 to 54 nmol/L with recovery range 96% to 106%. The intra-assay and inter-assay coefficients of variation range from 4.2% to 5.0% and from 4.6% to 5.5%, respectively. Cross-reactivity of the assay showed 0.9% cross-reactivity with 100 µg/L of Methotrexate. A cut-off of 7 nmol/L was used to identify subjects with low serum folate level. Additionally, serum concentration lower than 10 nmol/L was considered as marginally low folate concentration.

Homocysteine was measured using chemiluminescence immunoassay method on an IDS-iSYS analyzer (Immunodiagnostic Systems, Boldon, UK). Literature data showed good correlation (r = 0.95–0.99) between immunoassay methods and HPLC method [57]. Criteria for elevated serum homocysteine concentration was set at 15 µmol/L [17,20]. Additionally, more than 10 µmol/L was considered as marginally elevated serum homocysteine concentration.

As recommended [17], a combination of both—low serum folate concentration (<7 nmol/L) and elevated serum homocysteine concentration (>15 µmol/L) was used as criteria for identification of subjects with high risk for folate deficiency. Study protocol did not enable analyses of other parameters, such as red cell folate.

2.5. Data Analyses

Data collected in two 24-h recalls and the FPQs was used to estimate usual daily folate intakes, separately for all three age cohorts. Folate intake distributions per age group were adjusted for individual day-to-day variation using Multiple Source Method (MSM) [58]. A similar approach was used in our previous study, investigating vitamin D intake [59]. Age, sex, and BMI were considered as covariates. With the use of MSM we calculated individual daily usual folate intakes [60]. Year 2017 census data was used for population-weighting by iterative proportional fitting [61]. All population-weighted descriptive characteristics are reported separately for all three age groups, males and females. Population weighted mean folate intake was calculated (using SI.Menu study data) in µg/day, and in µg per daily intake of 1000 kcal. Proportion of the population meeting recommended daily folate intake was calculated two thresholds: nationally implemented D-A-CH recommendation [31] 300 µg and IOM/WHO recommendation 400 µg [26,27]. Population-weighted serum folate and homocysteine concentrations was also calculated (using Nutrihealth study data) separately for all three age groups. Prevalence of low serum folate and high serum homocysteine concentration was calculated using two cut-off values (<7 and <10 nmol/L for folate, and >10 and 15 µmol/L for homocysteine). We also calculated population-weighted prevalence of folate deficiency, using criteria of low serum folate (<7 nmol/L) and high serum homocysteine level (>15 µmol/L).

A series of regression analyses were conducted. Linear regression was employed to calculate energy-adjusted mean daily folate intakes, separately for all three cohorts, with consideration of sex, residential area, education level and financial status (only for adults and elderlies); BMI; IPAQ scores; employment status (only for adults); smoking status; and diet type. These parameters were also used in the logistic regression analyses to calculate odd ratios (OR) for meeting daily recommended intake of 300 µg folate. For models with serum folate concentrations, regression analysis was done on combined study sample of adults and elderlies ($N = 271$; 9 subjects were excluded due to missing dietary data). Odd ratios for low serum folate concentration (<7 nmol/L) were therefore calculated also with parameter of age group. Additional parameters in the analyses were sex, residential area, education level, financial status; smoking status, BMI; IPAQ; diet; supplement use, daily folate and energy intake, and serum homocysteine concentration. Similar approach was also used for OR for folate deficiency (serum folate <7 nmol/L and homocysteine >15 µmol/L), with following parameters in the model: age group, sex, residential area, education level, financial status; smoking status, BMI; IPAQ and daily folate energy intake. Again, model was adjusted for daily energy intake.

STATA (version 17.0; StataCorp LLC, Coledge Station, TX, USA) was used for statistical analyses. Statistical significance was set at $p < 0.05$. Means are reported with either a standard deviation (SD) or standard error (SE).

3. Results

In Table 1, the demographic characteristics of the SI.Menu sample are described ($N = 1248$). Only few participants ($N = 9$) explicitly reported supplementation with folate, while occasional use of multivitamin food supplements was reported by about a quarter of participants. About one third of adult and elderly SI.Menu participants (34% of adults and 37% of elderlies) were also included to Nutrihealth study, where biological samples (serum) were also collected.

Table 1. Demographic characteristics of the SI.Menu study sample for all three age cohorts (adolescents: 14–17 years; adults: 18–64 years; elderly: 65–74 years).

Variable		Adolescents N (%)	Adults N (%)	Elderly N (%)
Overall (SI.Menu study)		468 (100)	364 (100)	416 (100)
Age (mean ± SD)		13.4 (2.4)	43.6 (13.8)	68.7 (2.7)
Residential area	rural	270 (57.7)	202 (55.5)	229 (55.1)
	intermediate	76 (16.2)	56 (15.4)	71 (17.1)
	urban	122 (26.1)	106 (29.1)	116 (27.9)
Sex	male	238 (50.9)	173 (47.5)	213 (51.2)
	female	230 (49.1)	191 (52.5)	203 (48.8)
Education	no university degree	n.a.	249 (68.4)	342 (82.2)
	university degree	n.a.	115 (31.6)	74 (17.8)
Financial status	below average	n.a.	118 (38.4)	269 (71.5)
	above average	n.a.	189 (61.6)	107 (28.5)
Employment	employed	n.a.	226 (62.1)	n.a.
	unemployed	n.a.	42 (11.5)	n.a.
	student	n.a.	32 (8.8)	n.a.
	retired	n.a.	64 (17.6)	n.a.
BMI (mean ± SD)		21.0 (4.2)	26.7 (5.2)	28.4 (5.0)
BMI	normal	301 (64.6)	148 (40.7)	108 (26.0)
	overweight and obese	167 (35.7)	216 (59.3)	308 (74.0)
Smoking status	current, occasional, ex-smoker	30 (6.4)	165 (45.3)	185 (44.5)
	non smoker	438 (93.6)	199 (54.7)	231 (55.5)
IPAQ	low level	108 (23.3)	127 (35.3)	137 (33.4)
	moderate level	141 (30.5)	108 (30.0)	133 (32.4)
	high level	214 (46.2)	125 (34.7)	140 (34.2)
Supplement use	folate	1 (0.2)	7 (1.9)	1 (0.2)
	multivitamins	128 (27.5)	133 (36.5)	94 (22.6)
	does not use	339 (72.3)	224 (61.6)	321 (77.2)
Behavioural diet	vegetarian/vegan	12 (2.6)	8 (2.2)	3 (0.7)
	no diet	456 (97.4)	356 (97.8)	413 (99.3)
Medical diet	medical and/or weight loss	13 (2.8)	32 (8.8)	51 (12.3)
	no special diet	455 (97.2)	332 (91.2)	465 (87.7)
Subsample of the Nutrihealth study *			125 (34.3)	155 (37.3)

Notes: n.a. = not applicable. SD = standard deviation; BMI = body mass index; For adults and elderlies normal BMI was considered below 25 kg/m^2, while sex/age adjusted cut-off points were used for adolescents [49,50]; IPAQ = Physical activity according to International Physical Activity Questionnaire; * Serum folate and homocysteine levels available for sub-group participating in the SI.Menu study (Nutrihealth study sample).

Population-weighted descriptive statistics are shown Table 2, while distribution of estimated usual daily intake of folate is presented in histograms in Figure 1. In all age groups mean daily folate intake was below the recommended 300 µg/day, however somewhat higher mean was observed in specific subgroups, e.g., male adolescents and elderly females. The usual mean daily folate intake was quite similar in elderly (295.5 µg/day) and adult population (294.6 µg/day), and somewhat lower in adolescents (289.8 µg/day) (Table 2). The mean daily folate intake was generally lower for women, except in elderly, opposite trend was observed after consideration of daily energy intake. Folate intake calculated per 1000 kcal/day was higher for women in all three population groups. Same trend was also observed in a regression analyses model, when higher adjusted mean daily folate intake was observed in women in all population groups (Supplementary Table S1). Analyses showed sex a significant predictor of mean daily folate intake in elderlies and adolescents. In adolescents and adults, body mass index was also found a significant parameter, with lower folate levels observed in those with overweight/obesity. Additionally, adults with higher education and with employment had higher daily folate intakes. Population-weighted intake data show that about one third of the population met daily D-A-CH recommended folate intake of 300 µg/day; index was lower in the elderly (32.2%) than in adults (41.9%) and adolescents (41.2%). About 12% of each population group met IOM/WHO target value of 400 µg/day. Predictors associate with daily folate intake level (cut-off 300 µg) were determined using logistic regression analyses, with separate model for each population group (Table 3). In both, adolescents and adults, females compared to males had significantly lower odds for meeting recommended daily folate intake level (OR 0.63 and 0.44, respectively). Other factors significantly associate with daily dietary folate intake level were BMI in adolescents (OR 0.65 for overweight/obese) and education in adults (OR 1.93 for higher education). Employment status and residential area were notably associated with the daily dietary folate intake level ($p = 0.06$ and 0.07, respectively), with lowest odds observed for unemployed and those living in rural areas. Similar trends were observed in elderly.

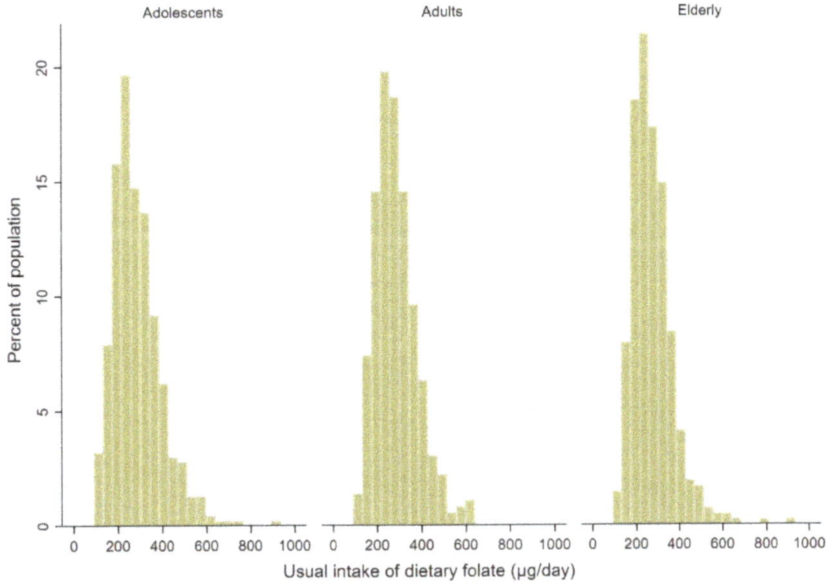

Figure 1. Histograms of estimated usual daily intake of folate for all three age cohorts (adolescents: 10–17 years; Adults: 18–64 years; Elderly: 65–74 years).

Table 2. Estimated population-weighted usual daily folate intake and indicators of folate deficiency (95% CI).

	Adolescents N (%)			Adults N (%)			Elderly N (%)		
	All	**Male**	**Female**	**All**	**Male**	**Female**	**All**	**Male**	**Female**
Si.Menu study; N(%)	468 (100)	238 (50.85)	230 (49.15)	364 (100)	173 (47.53)	191(52.47)	416 (100)	213 (51.20)	203 (48.80)
Daily folate intake									
Mean (95%CI) (μg /day)	289.8 (277.7–301.9)	308.1 (291.9–324.2)	270.1 (255.0–285.3)	294.6 (283.4–305.8)	311.4 (293.8–329.0)	277.5 (264.2–290.9)	295.5 (263.0–327.9)	278.6 (263.7–293.4)	311.0 (252.7–369.3)
Std. Err.	6.16	8.21	7.71	5.70	8.95	6.78	16.5	7.56	29.64
Median (μg /day)	271.5	301.4	247.5	281.2	297.3	267.2	274.1	280.1	270.9
Mean (95% CI) (μg/per 1000 kcal/day) *	127.7 (122.5–132.9)	119.8 (113.0–126.6)	136.3 (129.1–143.6)	138.4 (133.4–143.4)	132.3 (124.7–140.0)	144.6 (138.4–150.8)	140.7 (128.1–153.2)	125.8 (120.9–130.7)	154.3 (133.2–175.3)
Proportion of population with insufficient daily folate intake **									
<300 μg/day	58.7 (51.0–66.1)	49.9 (39.1–60.7)	68.3 (60.1–75.4)	58.1 (52.2–63.8)	52.9 (44.4–61.1)	63.5 (55.3–70.9)	67.8 (58.7–75.7)	73.5 (62.6–82.1)	62.6 (49.9–73.8)
<400 μg/day	87.9 (84.0–90.9)	84.3 (77.8–89.122.2)	91.8 (86.9–95.0)	87.8 (83.5–91.1)	83.5 (76.4–88.7)	92.1 (86.8–95.4)	87.6 (76.9–93.7)	93.3 (88.3–96.2)	82.4 (63.8–92.5)
Nutrihealth study; N (%)				125 (100)	52 (41.6)	73 (58.4)	155 (100)	76 (49.0)	79 (51.0)
Serum folate level									
Mean (95%CI) (nmol/L)				10.6 (9.6–11.7)	10.5 (8.9–12.2)	10.8 (9.5–12.1)	11.4 (10.0–12.9)	11.0 (8.6–13.5)	11.8 (10.3–13.3)
Std. Err.				0.54	0.84	0.66	0.72	1.24	0.78
Median (nmol/L)				10.0	9.0	10.0	10.0	9.0	10.0
Prevalence of low serum folate (%) (95% CI)									
<7 nmol/L				16.6 (10.8–24.6)	16.5 (8.6–29.5)	16.6 (9.5–27.6)	18.5 (13.1–25.4)	22.4 (14.3–33.2)	15.0 (8.7–24.7)
<10 nmol/L				49.0 (39.4–58.6)	50.4 (35.7–65.0)	47.5 (35.8–59.4)	48.6 (40.8–56.5)	51.3 (40.1–62.4)	46.3 (35.6–57.3)
Serum homocysteine level									
Mean (95%CI) (μmol/L)				12.6 (11.9–13.3)	13.6 (12.6–14.6)	14.6 (13.9–15.2)	14.6 (13.9–15.2)	16.1 (15.2–17.0)	13.2 (12.5–13.9)
Std. Err.				0.35	0.50	0.42	0.32	0.47	0.36
Median (μmol/L)				12.1	12.6	15.7	14.2	10.9	13.0
Prevalence of high homocysteine level (%) (95% CI)									
>10 μmol/L				75.3 (66.4–82.4)	88.7 (76.5–95.0)	61.0 (48.6–72.2)	88.9 (82.7–93.0)	96.0 (88.3–98.7)	82.5 (72.5–89.4)
>15 μmol/L				20.5 (13.9–29.1)	26.4 (15.8–40.8)	14.2 (8.0–23.9)	39.9 (32.4–47.8	56.6 (45.2–67.3)	25.0 (16.7–35.7)
Prevalence (%) of folate deficiency using criteria of low serum folate (<7 nmol/L) and high serum homocysteine (15 μmol/L)									
				6.9 (3.5–13.0)	9.7 (4.3–20.4)	3.9 (1.8–12.0)	10.1 (6.3–16.0)	14.5 (8.1–24.4)	5.3 (2.6–14.3)

Notes: Estimated daily folate intake with consideration of regular foods (excluding use of food supplements). * conversion factor for μg/MJ is 0.239; ** Cut-off values for daily folate intake according to national/D-A-CH [29,31] (300 μg) and IOM/WHO (400 μg) [27] recommendations; CI: confidence interval.

Table 3. Association between daily intake of dietary folate level (>300 μg/day) and sex, residential area, education, income, employment, smoking status, BMI, IPAQ, dietary pattern.

Variable		Adolescents (10–17 Years Old)		Adults (18–64 Years Old)		Elderly (65–74 Years Old)	
		(>300 μg/day) *n* (%)	Odds Ratio *	(>300 μg/day) *n* (%)	Odds Ratio *	(>300 μg/day) *n* (%)	Odds Ratio *
All		181 (38.7)		140 (38.5)		139 (33.4)	
Sex	Male	102 (42.9)	1	78 (45.1)	1	74 (34.7)	1
	Female	79 (34.4)	0.63 (0.43–0.93)	62 (32.5)	0.44 (0.26–0.75)	65 (32.0)	0.87 (0.54–1.42)
Residential area	Rural	104 (38.5)	1	74 (32.3)	1	20 (8.7)	1
	Intermediate	35 (46.1)	1.37 (0.81–2.32)	23 (32.4)	2.25 (1.79–10.18)	6 (8.5)	1.04 (0.57–1.90)
	Urban	42 (34.4)	0.79 (0.50–1.426)	42 (36.2)	1.51 (1.07–4.24)	9 (7.8)	1.23 (0.73–2.07)
Education	No university degree	n.a.	n.a.	83 (33.3)	1	115 (33.6)	1
	University degree			57 (49.6)	1.93 (1.07–3.47)	24 (32.4)	0.93 (0.50–1.73)
Financial status	Below average	n.a.	n.a.	38 (32.2)	1	89 (33.1)	1
	Above average			82 (43.4)	1.11 (0.62–2.00)	38 (35.5)	1.09 (0.65–1.82)
BMI	Normal	126 (41.9)	1	60 (40.5)	1	34 (31.5)	1
	Overweight/obese	55 (32.9)	0.65 (0.43–0.98)	80 (37.0)	0.71 (0.41–1.21)	105 (34.1)	0.98 (0.58–1.64)
IPAQ	Low intensity	40 (37.0)	1	53 (41.7)	1	42 (30.7)	1
	Moderate	62 (44.0)	1.55 (0.91–2.64)	43 (39.8)	0.83 (0.45–1.53)	47 (35.3)	1.19 (0.70–2.03)
	High intensity	79 (36.9)	1.01 (0.62–1.64)	42 (33.6)	0.56 (0.30–1.03)	50 (35.7)	1.12 (0.66–1.90)
Employment status	Employed	n.a.	n.a.	100 (44.3)	1	n.a.	n.a.
	Unemployed			10 (23.8)	0.29 (0.10–0.76)		
	Student			11 (34.4)	0.73 (0.26–2.06)		
	Retired			19 (29.7)	0.60 (0.28–1.28)		
Smoking status	Non smoker	172 (39.3)	1	76 (38.2)	1	75 (32.5)	1
	Current, occasional, ex-smoker	9 (30.0)	0.57 (0.25–1.30)	64 (38.8)	1.34 (0.79–2.28)	64 (34.6)	1.02 (0.63–1.66)
Medical diet	No diet	176 (38.7)	1	129 (38.9)	1	123 (33.7)	1
	Medical and/or weight loss	5 (38.5)	1.17 (0.37–3.74)	11 (34.4)	0.75 (0.29–1.96)	16 (31.4)	0.95 (0.49–1.84)
Behavioural diet	No diet	180 (39.5)	n.a	136 (38.2)	1	138 (33.4)	1
	Veget./vegan	1 (8.33)		4 (50.0)	1.03 (0.22–4.74)	1 (33.3)	0.92 (0.07–11.40)

Note: n.a. = not applicable. BMI = Body mass index; For adults and elderlies normal BMI was considered bellow 25 kg/m^2, while sex/age adjusted cut-off points were used for adolescents [49,50]; IPAQ = Physical activity according to International Physical Activity Questionnaire; Logistic regression analysis conducted on samples with excluded missing values (Family net income: $N = 57$ (adults) and 40 (elderly); IPAQ: $n = 5$ (adolescents), 4 (adults), 6 (elderly)); * Cut-off odds ratio for intake of over 300 μg of folate per day; Association was significant with following variables: sex, $p < 0.05$ (adolescents), BMI, $p < 0.05$; sex, $p < 0.005$ (adults), education, $p < 0.05$ (adults).

To provide further insights, we investigated relative contribution of different foods to daily folate intakes (Figure 2, Supplementary Table S2). Food categories with the greatest contribution to daily folate intake were fruit and vegetables, bread, bakery products and other cereal products. The latter includes various types of breakfast cereals, which were more important folate contributors in adolescents. Among bread and bakery products, brown bread had higher contribution of folate in elderly than in other population groups. Altogether, most notable difference between groups was observed in vegetables, which contributed 27% folate intake in adolescents, and about 40% in adults and elderlies. Milk, meat, and products of thereof were also shown as notable contributors to folate intake.

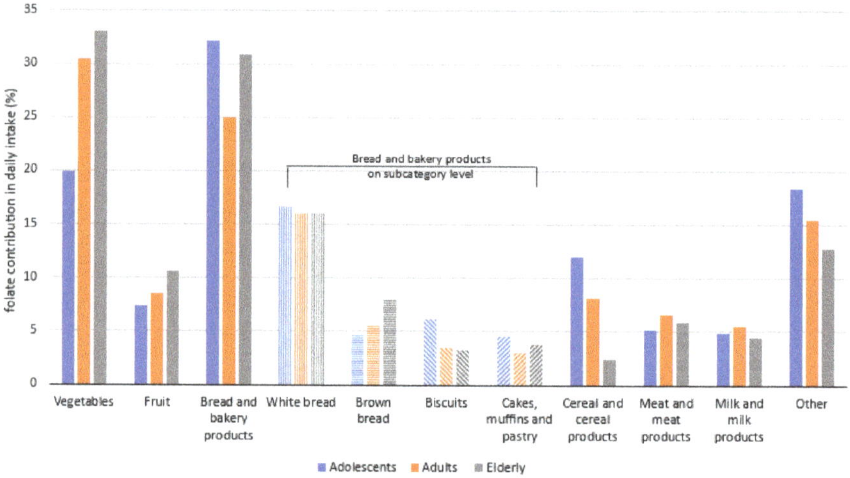

Figure 2. Relative contribution of selected food categories to daily folate intake among all three age cohorts (% of total folate intake; adolescents: 10–17 years; Adults: 18–64 years; Elderly: 65–74 years).

For adults and elderlies (but not for adolescents) we also investigated folate status with the use of biomarkers, i.e., serum folate and homocysteine concentration (Table 2). Mean population-weighted serum folate level in adults and elderlies was 10.6 and 11.4 nmol/L, respectively. Almost half of adult and elderly population had serum folate levels lower than 10 nmol/L, while very low folate concentrations (<7 mol/L) were observed in 17% adults and 18.5% elderlies (22% in males, and 15% in females). Logistic regression analyses model identified smoking (OR 2.18; 1.04, 4.43, $p = 0.04$) and serum homocysteine levels above 10 µmol/L (OR 5.0; 1.1, 22.5; $p = 0.04$) as predictors for folate deficiency (Figure 3). Education and diet were also found close to significant ($p = 0.07$ and 0.08, respectively), with lowest OR in those with higher education and on medical diet.

Interestingly, use of folate/multivitamin supplements and daily folate intake were not found significant predictors. However, we observed expected trend of lowest OR in supplement users and those with adequate daily folate intake.

Further, we investigated the prevalence of those with folate deficiency, with consideration of both low serum folate (<7 nmol/L) and high homocysteine (>15 µmol/L) concentration. Notably higher population-weighted folate deficiency prevalence was observed in elderlies (10.1%; 95%CI: 6.3–16.0), in comparison to other adults (6.9%; 95% CI: 3.5–13.0%). In both population groups, notably higher prevalence was observed in men (9.7% in adults and 14.5% in elderlies). Low number of subjects with identified folate deficiency limited our ability for identification of significant deficiency predictors, but the observed trends are comparable with the low folate levels. Notably higher prevalence of folate deficiency was observed in elderlies, males, smokers, living in rural areas, with lower education and lower income (Supplementary Table S3).

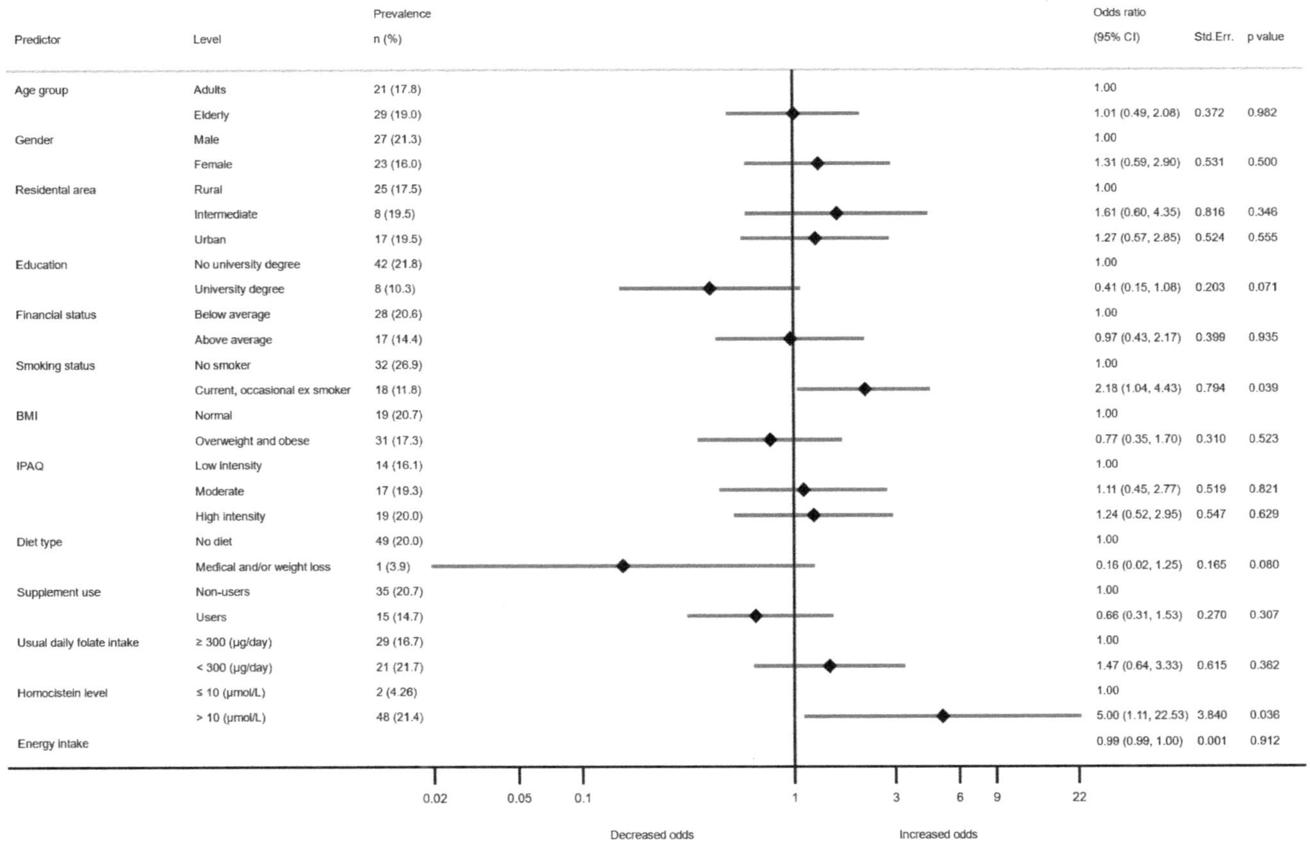

Figure 3. Association between serum folate concentration level (<7 nmol/L) and age, sex, residential area, education, family income, BMI (Body mass index), IPAQ (International Physical Activity Questionnaire) score, diet type, supplementation with multivitamins/folate, daily dietary folate intakes, and homocysteine status. Model adjusted for daily energy intake ($N = 271$).

4. Discussion

The present study is the first one evaluating folate intake and status in the Slovenian general population. With consideration of both—daily folate intakes, and biomarkers of folate status, we identified risks that are comparable with several other countries within and outside Europe [5]. Looking into folate intakes, the observed current situation is comparable to situation in the USA before the introduction of food fortification in the year 1998, where folate intake in the population was below the recommended requirement in 49% of the population [36,62]. Our data show a somewhat higher risk for low folate intake in elderlies, and a similar trend was also observed with folate biomarkers. Folate deficiency (low serum folate (<7 nmol/L) and high serum homocysteine (>15 μmol/L) concentration) was found in 7% of adults and 10% of elderly population. It should be noted that mean daily folate intake in elderlies can be deceiving, despite being the highest among population groups (Table 2), we observed higher inter-individual variability, in comparison to adults and adolescents, meaning that on one hand there are individuals with low folate intakes and on the other, those with high folate intakes. As seen from Figure 1, among elderly population there is a high proportion of those with usual folate intakes below 300 μg/day and on the other hand we also identified higher proportion of subjects with above average folate intakes, which affected estimation of mean daily folate intake. This is also evident from data on the prevalence of adequate folate intake. Elderlies were population group with the lowest proportion (32%) of individuals meeting D-A-CH recommendation for daily folate intake (300 μg). Folate related risks in the elderly population were also highlighted in other studies [63–66].

Available data show that dietary folate intake in European countries varies, with no clear gradient between countries' geographical regions. With the exception of some populations in Lithuania, Ireland, and Turkey, folate intake in EU countries is typically well below the WHO/NIH recommendation of 400 μg/day, and mostly even lower than 300 μg/day [34]. Particularly low folate intakes were reported in some Nordic countries [34,67]. In our study, except for the elderly population, males had significantly higher folate intakes, which was also observed in some other studies [68–70]. However, after adjustment for energy intake, in all three age cohorts, women had a higher folate intake (per 1000 kcal) than men (Table 2). A similar observation was noted in a study on a Finnish population, after folate intake was adjusted for daily energy intake [71]. Adjusting folate intake for daily energy intake provides very interesting insights. While men tend to consume a higher quantity of food compared to women, which can result in higher intakes of energy and certain (macro)nutrients, the nutritional density of such diet is clearly lower, at least for folate. Adjusting for energy intake can be an important factor, when studying sex differences in dietary intakes [72,73]. Because same daily recommended folate intake was used for both males and females, the regression analyses for meeting such intake (Table 3) was not adjusted for daily energy intakes. However, energy adjustment was used in the regression analyses on mean folate intakes (Supplementary Table S3).

Lower energy-adjusted mean dietary folate intakes were associated with certain sociodemographic and lifestyle factors, especially in adults. In the elderly population, besides sex, other investigated parameters did not significantly influence daily folate intake (Supplementary Table S1). In the adult population, significant determinants of daily folate intake were residential area, education, employment status and BMI, which was also significant predictor in adolescents. Adults living in rural areas had significantly lower folate intake compared to others, and similar was evident for unemployed, retired, as well as for those less educated. The influence of some of these factors was reported elsewhere as well, with those living in rural areas [74], lower level of education [75] and higher BMI [76] being at higher risk for poorer nutrition, including lower dietary folate intake. Besides, lower economic status is often associated with higher BMI [77–79], which even strengthens the link between factors leading to nutritionally inadequate diets. This implies that such populations are more vulnerable in terms of nutrition, which was seen also in Slovenia [79]. They tend to consume more processed foods and less nutritiously rich foods, such as fruit, vegetables, unrefined grains, which can result in inadequate nutrient intake. Those with lower incomes are also considerably more affected by food prices [80]; as various processed foods are often available for a lower price than whole fresh foods, this can pose an additional threat to such vulnerable populations [81]. Our observations support calls that public health policies should also consider economic barriers in these fragile population groups because an abundance of cheaper processed foods of poor nutritional density is making nutritious health diets more and more challenging [82].

We observed that in all investigated population groups the consumption of "Bread and bakery products" has high impact on folate intake, similar to "Vegetables". This can be explained by the fact, that in Slovenia bread is consumed on daily basis, sometimes even more than once per day. On the other hand, some vegetables are very rich in folate, so compared to bread, the higher folate intake can be achieved with smaller amounts of such foods consumed. This explains high folate intake from vegetables, although according to national food consumption data, the daily intake of vegetables intake in the population is generally still below recommendations [79]. Our study indicates that more regular consumption of folate-rich foods such as fruits, vegetables and wholegrains would be helpful to improve folate intake in the population. Vegetables, fruit, bread, and other cereal products were also seen as the major dietary folate contributors in other studies [71,76,83]. The folate contribution from daily diet is the greatest from fruits and vegetables in the elderly, where these food categories present more than 40% of total daily dietary folate intake. Additionally, we should also note that number of participants following vegan or vegetarian diet was too low for reliable insights. Interestingly, in adolescents, the contribution of folate from fruits and vegetables was below 30%, but a higher contribution was observed in bread and bakery products, particularly in white bread. They also get more folate from other cereal products, including (fortified)breakfast cereals, which are more popular in adolescents than in other population groups. These results are in line with the observations that consumption of such processed/refined foods is more common in the younger population. Milk, dairy, and meat products were also identified as notable sources of folate. The strategies to encourage adolescents into consuming more unprocessed foods, such as fruit, vegetables, and wholemeal foods (including bread) could help them to achieve higher folate intake, as currently their usual daily folate intake is the lowest among populations and is often achieved from foods which should not be consumed in abundance (white bread, biscuits, etc.). This is also supported with the results for folate density in diets of adolescents, which is notably lower in adolescents (128 μg/1000 kcal), than in adults and elderlies (138 and 141 μg/1000 kcal, respectively).

We also investigated biomarkers for folate deficiency. Interestingly, dietary intake of folate was not found among the strongest predictors of low serum folate levels. This phenomenon has been previously reported [84], and could be explained also with notable differences in bioavailability of folate from different food matrixes [85,86]. Significantly increased odds for low serum folate were observed in smokers, which is comparable with literature data [87,88]. Furthermore, the trends, although not significant, but corresponding to the literature data [89,90] also show a higher likelihood for low folate or folate deficiency in elderlies, males, with lower education or financial status. Interestingly, similar trends were observed to impact dietary folate intakes. As expected, higher serum homocysteine concentration was found a significant predictor of lower folate status. However, homocysteine levels can be affected by several different lifestyle factors, as well as personal characteristics, including age and sex [91,92]. The link between serum folate and homocysteine should be interpreted with caution, as not every increase in serum homocysteine is caused by low folate status [93]. However, low serum folate combined with elevated serum homocysteine is considered a better indicator for folate deficiency than serum folate alone [17].

With consideration of folate deficiency rates observed in our study, particular attention is required in the elderlies, although we should be aware that the deficiency, to some extent, is also present in a general adult population. Nutrihealth study was designed in a way, to also provide deeper insights into the vulnerable population of elderlies. In elderlies, low folate status is associated with adverse health effects, such as psychiatric and cognitive problems [63,66,94]. Timely intervention with encouragement of consuming folate-rich foods and possible folic acid supplementation should be considered in order to prevent negative outcomes, related to low folate intake and status in this vulnerable population.

However, women of childbearing age are also considered a very important population group, where sufficient folate status is crucial [95,96], but they were not investigated as a separate sample. It was recognized that the risk for neural tube defects is halved if the standard diet is supplemented with 400 µg of folic acid [96]. Therefore, women are recommended to take folic acid at least one month before to at least three months after conception. In many countries, less than 50% of women follow these recommendations [97,98]. Furthermore, pregnancies are not always planned, and some researchers note that fortification of foods with folic acid could substantially lower risks for neural tube defects in Europe [99]. In our study, adult women were the population with the lowest prevalence of folate deficiency, but we should note that the prevalence of vitamin supplements use was also the highest in this group. Altogether, the diet was supplemented with folate/multivitamins by 51% of adult females and 26% of males, but the study design did not enable further insights into the use of food supplements (i.e., to investigate typical daily dosages etc.). Although mean folate intake (270 µg/day), suffices the folate AR set by EFSA (250 µg/day), only 32% of this population met recommended daily folate intake of 300 µg (while only 8% met WHO/NIH recommendation of 400 µg folate), supplementation with folic acid is extremely important for women in childbearing age to reduce risks for neural tube defects in offsprings. Suboptimal folate intakes of women of childbearing age were reported elsewhere as well [100–102]. It should be also mentioned that the incidence of neural tube defects in Slovenia did not improve after voluntarily supplementation was advised to women who plan pregnancy [103], meaning that additional efforts are needed to achieve sufficient folate intake in this vulnerable population. A possible approach would be the introduction of reimbursed prescription of folate medicines to all pregnant women and those planning pregnancy, but such approach would also be notably affected by compliance rates and would not provide for those with unplanned pregnancies.

This study highlighted that non-optimal folate intake could be addressed by promoting the consumption of folate-rich foods, particularly fruit, vegetables, unrefined grains, and legumes. It has been previously shown that by a 10% increase in folate intake from foods, a 6% increase in serum folate could be achieved [104]. Such a shift in people's diets would also provide other benefits, for example, higher intake of dietary fibre, which is also not sufficient in the Slovenian population [105]. As adolescents tend to consume more unrefined grain products, processed food and less fruit and vegetables, innovative educational approaches and careful meal planning in schools could help to improve the situation. There are some inequalities observed considering demographic factors. Public health strategies should therefore focus on promoting and assuring the availability of healthy food for everyone as well as offering knowledge on healthy nutrition, with particular focus to vulnerable groups of the population, such as those less educated, unemployed, living in rural places, elderly, and those with higher BMI. These fragile population groups were also identified to be at risk for other nutrition and health-related factors [79].

Another option for increasing folate intakes in fortification of certain foods with folate, as practiced in some countries, such as USA [106], Canada [107] and Chile [108]. Such practice resulted not only in improvement of the folate status but also in notably lower prevalence of neural tube defects [5]. The latter was also the key reason for the decision to introduce mandatory fortification of non-wholemeal wheat flour in the United Kingdom very recently [109]; they expect the intervention will help to prevent neural tube defects for about 20% (around 200 per year). Interestingly, folate food fortification in the USA led to a higher mean increase in folate intake than initially predicted [36]. After fortification, the mean dietary folate intake in the population increased from 275 to 351 µg/day, while the mean serum folate concentration increased from 11.4 to 26.9 nmol/L [110]. To achieve maximal reductions in serum homocysteine concentrations by folate supplementation, daily doses of more than 800 μg folate is typically required [93]. Recently, a collaborative study performed by the European Commission Joint Research Centre (JCR) and the European network of population-based registries for the epidemiological surveillance of congenital anomalies (EUROCAT) highlighted that folic acid fortification of grain products could contribute to prevention of at least 1000 birth anomalies in the EU every year, despite the fact that folic acid supplementation is already advised, but obviously not always efficiently implemented [111]. However, in many countries, including in the EU, we have no practice of mandatory fortification of foods with folate. This is partly because of the limited evidence on the expected additional health benefits in clinical trials, due to feared health risks and because of the issue of freedom of choice [5], and also because only some specific populations could benefit from high(er) folate intake [37]. Requirements for folate intake are complex and affected by various factors. In foods, folate can be found in various forms, which have different properties, functionalities and bioavailability, while some individual (including genetic) factors should be also appraised [112]. This makes the decision for mandatory fortification even more challenging, however, any other approach might be very limited in assuring sufficient folate intake in future mothers, at least in cases of unplanned pregnancies.

Major strengths of this study are that folate intakes were estimated using food consumption data, collected in a nationally representative population of adolescents, adults, and elderlies, and that for sub-sample of adults and elderlies we were also able to investigate serum biomarkers of folate deficiency. Study strength is also that SI.Menu food consumption study was done with internationally harmonised EU Menu methodology, employing both 24 h recalls and an FPQ. However, some study limitations also need to be mentioned. Considering that folate content was not included in some foods in the OPEN food composition dataset, international food composition databases were also used. Also, the cooking and processing of foods can result in folate losses, which was not considered in the calculation of the intakes. Different forms of folate with different bioavailability were also not considered. We should also mention that this study estimated folate intake with foods, and not with dietary supplements or medicines. While in the SI.Menu study we tracked the use of multivitamin and folate supplements, the collected data did not allow us to estimate folate intakes from these sources. Folate was very rarely supplemented alone; reported use of multivitamin supplements was highest in adults (37%), particularly in women (51%). Considering that supplementation can present an important additional source of folate, this topic should be addressed in further studies. This is even more relevant in the context of the COVID-19 pandemic, which affected people's lifestyle and behaviours [82], and the use of dietary supplements [83]. We should also mention that while serum folate and homocysteine concentrations are considered as key biomarkers for the identification of folate deficiency [17–19], red blood cell folate would provide a better indication for long-term folate status [25,113]. Also, while we used a standard medical diagnostic method for measuring folate in human serum, the chemiluminescence immunoassay is sensitive to very high biotin concentrations. Considering that in serum with biotin levels up to 1500 μg/L such a measurement error is lower than 10%, we did not use control for serum biotin concentration. Another limitation is, that while quota-sampling in the Nutrihealth study provided a good study sample for the vulnerable elderly population, this study was not designed in a way to detailly address another very important population group—pregnant women and those planning pregnancy. Our sample of adults provided interesting insight about this group, but the study sample size limited our ability for deeper analyses of sub-groups. Furthermore, the Nutrihealth study did not address folate deficiency in adolescents and children. It would make sense to further investigate the abovementioned groups, particularly in the context of different supplementation practices.

5. Conclusions

For the first time, daily folate intakes with foods were estimated for the Slovenian general population, with a focus on adolescents, adults, and elderlies. In the latter two groups, the prevalence of folate deficiency was also investigated using serum folate and homocysteine as biomarkers. Prevalence of sub-optimal folate intake was observed in all studied populations; however, the prevalence of folate deficiency was noted to a smaller extent. Folate intake (<300 μg/day) was observed in 59% of adolescents, 58% of adults and 68% of elderlies, while about 12% achieved the WHO recommended level of 400 μg/day. Residential area, education, employment status and BMI were linked with low folate intake in adults; BMI, and sex in adolescents; and sex in elderlies. Major dietary contributors were vegetables and fruit, and cereal products (particularly bread). Vegetables were a much more notable contributor in adults and elderlies, in comparison to adolescents. Contrary, folate contribution was higher from processed foods in adolescents, compared to the elderly population. Considering low serum folate (<7 nmol/L) and high serum homocysteine (>15 nmol/L), folate deficiency was found in 7.6 and 10.5% in adults and elderlies, respectively. We observed a trend of higher proportions of folate deficiency in elderlies, males, those living in rural residential areas, with no university degree, below average family net income, higher body mass index and in smokers. Our study indicated that public health strategies should focus on promoting the consumption of fruit, vegetables, legumes, and whole grain foods, particularly in vulnerable population groups. This would be also beneficial, to assure adequate intake of various other micronutrients. Another option for increasing folate intake in the general population is also the introduction of food fortification, but such decisions need to be taken very carefully, and with consideration of the wider European context, because a considerable proportion of the processed foods in the national food supply is originating from other countries. With current folate intakes, supplementation with folic acid should be additionally encouraged in pregnant women and women who plan pregnancy. Standardized prescription of folate medicines (with costs covered from health insurance) for pregnant women and those planning pregnancy would be very beneficial. Additionally, folic acid supplementation should be also considered in some other vulnerable populations, particularly in elderlies.

Supplementary Materials: The following are available online at https://www.mdpi.com/article/10.3390/nu13113860/s1, Supplementary Table S1. Mean (SD) and model adjusted mean (95% CI) usual daily intake of folate by sociodemographic and lifestyle parameters for the three age cohorts. Models adjusted for daily energy intakes (kcal/day), dietary fruit/vegetables intakes (g/day), dietary bread intakes (g/day). Supplementary Table S2. Relative contribution of selected food categories to usual daily dietary folate intake in different age groups (% of total dietary folate intake). Supplementary Table S3. Sample prevalence of folate deficiency and prevalence adjusted odds ratios (95% CI)-by socio-demographic and lifestyle parameters (*N* = 271).

Author Contributions: I.P. was responsible for assuring the Nutrihealth study set-up of the analyses and participated in the data analyses; J.O. was responsible for the data collection in Nutrihealth study. K.Z. and A.O. were responsible for the sample collection and laboratory analyses in Nutrihealth study. Ž.L. wrote first draft of the manuscript and prepared the submission; H.H. performed data analyses; N.G. participated in data interpretation and discussion; A.K. and K.Ž. participated in study conduction and data interpretation. M.H. and N.K. conducted food-matching to estimate nutritional composition of foods. M.G. and U.B. were responsible for SI.Menu study design and food consumption data; B.K.S. was responsible for information technology; K.Ž., A.K. made a revision of the final draft. All authors have read and agreed to the published version of the manuscript.

Funding: The SI.Menu 2017/18 national food consumption survey was funded by the European Food Safety Authority (EFSA contract No. OC/EFSA/DATA/2014/02-LOT2-CT03), Ministry of Health of the Republic of Slovenia, and Slovenian Research Agency. The Nutrihealth study was part of the research project L3-8213 "Children's and adults' nutrition as a protective or health-risk factor", funded by the Slovenian Research Agency and Ministry of Health of Republic of Slovenia. Study was conducted within research programs P3-0395 "Nutrition and Public Health" and P2-0098 "Computer Systems and Structures", funded by the Slovenian Research Agency. We also acknowledge that contribution of Ž.P. has been supported by the Slovenian Research Agency, which has been funding post-doctoral research project "Dietary challenges in the population of nursing homes' residents Z3-3213".

Institutional Review Board Statement: The study was conducted according to the guidelines of the Declaration of Helsinki. SI.Menu study was approved by the National Medical Ethics Committee, Ljubljana, Slovenia (KME 53/07/16; approval No. 0120-337/2016 issued on 19 July 2016). Nutrihealth study was conducted as an extension of the SI.Menu study. Study protocol was also approved by the Slovenian National Medical Ethics Committee, Ljubljana, Slovenia (KME 72/07/16; approval No. 0120-337/2016-4 issued on 7 July 2017). Study was registered at ClinicalTrials.gov (ID: NCT03284840).

Informed Consent Statement: Informed consent was obtained from all subjects involved in the study.

Data Availability Statement: The data presented in this study are available on request from the corresponding author.

Acknowledgments: We acknowledge support of all SI.Menu project partners and their researchers, particularly Nataša Delfar, Metka Zaletel, Darja Lavtar, Ada Hočevar Grom, and Ivan Eržen (National Institute of Public Health, Ljubljana, Slovenia); Petra Golja and Katja Zdešar Kotnik (Group of Anthropology at Biotechnical Faculty of University of Ljubljana, Slovenia); Stojan Kostanjevec (University of Ljubljana, Faculty of Education, Ljubljana, Slovenia; Majda Pajnkihar (University of Maribor, Faculty of Health Sciences, Maribor, Slovenia); Tamara Poklar Vatovec (University of Primorska, Faculty of Health Sciences, Koper, Slovenia); and Evgen Benedik (University Medical Centre Ljubljana, University Children' s Hospital, Ljubljana, Slovenia). We acknowledge support of Veronika Belec and her team and interviewers at GfK Research Institute (Slovenia) for conducting the survey field work. We also acknowledge Irena Rogelj (Biotechnical Faculty, University of Ljubljana) for support in project management and funding acquisition, and the regional community healthcare centers and hospitals for collection of biological samples in the Nutrihealth study. We are grateful to the staff of UKC Ljubljana (Troha Poljančič Vera, Gaberšček Simona) and all medical laboratory teams across Slovenia led by Spomenka Lajtner (Community Health Center Ljubljana); Jana Žnidaršič (Community Health Center Cerknica), Katarina Furar (Community Health Center Ivančna Gorica); Tomaž Vargazon (Community Health Center Kamnik); Ljubica Štefanič (Community Health Center Kočevje); Irena Sluga (Community Health Center Litija); Jolanda Prah (Community Health Center Maribor); Jelica Štanta (Community Health Center Gornja Radgona); Ana Gruškovnjak (Community Health Center Ljutomer); Edita Ružič (Community Health Center Murska Sobota); Mojca Završnik (Hospital Ptuj); Tanja Lađić (General Hospital Slovenj Gradec); Tina Kochler (Community Health Center Slovenska Bistrica); Mojca Kozmelj (Community Health Center Celje); Maja Škerjanc (Community Health Center Velenje); Branka Svetic (Community Health Center Kranj); Sabina Nabernik (Community Health Center Jesenice); Marko Rudež (Community Health Center Škofja Loka); Vanja Pahor (General Hospital Izola); Damjana Bogataj (Community Health Center Ajdovščina); Andreja Seljak (Community Health Center Idrija); Kristina Gruden (Community Health Center Nova Gorica); Valentina Buda (Community Health Center Postojna); Polonca Stopar (Community Health Center Sežana); Zalka Murovec (Community Health Center Tolmin); Danijela Furlan (General Hospital Novo mesto); and Andreja Kerin (General hospital Brežice). We also acknowledge support of Sanja Krušič and Nina Zupanič (Nutrition Institute, Ljubljana, Slovenia) for support in literature review.

Conflicts of Interest: The authors declare no conflict of interests. The funders had no role in the design of the study; in the collection, analyses, or interpretation of data; in the writing of the manuscript; or in the decision to publish the results. I.P. has led and participated in various other research projects in the area of nutrition, public health, and food technology, which were (co)funded by the Slovenian Research Agency, Ministry of Health of the Republic of Slovenia, the Ministry of Agriculture, Forestry and Food of the Republic of Slovenia, and in case of specific applied research projects, also by food businesses. I.P., M.G., and U.B. are members of a workgroup for food reformulation at Ministry of Health of Republic of Slovenia.

References

1. Duthie, S.J.; Narayanan, S.; Brand, G.M.; Pirie, L.; Grant, G. Impact of Folate Deficiency on DNA Stability. *J. Nutr.* **2002**, *132*, 2444S–2449S. [CrossRef]
2. Mahmood Al-Sammak, N.; Ibrahim, H. Trends of under-five mortality in Nineveh (2004–2013): A time series analysis. *J. Health Res. Rev.* **2014**, *1*, 5–9. [CrossRef]

3. Greenberg, J.A.; Bell, S.J.; Guan, Y.; Yu, Y.H. Folic Acid supplementation and pregnancy: More than just neural tube defect prevention. *Rev. Obstet. Gynecol.* **2011**, *4*, 52–59.
4. Molloy, A.M.; Kirke, P.N.; Brody, L.C.; Scott, J.M.; Mills, J.L. Effects of folate and vitamin B12 deficiencies during pregnancy on fetal, infant, and child development. *Food Nutr. Bull.* **2008**, *29*, S101–S111. [CrossRef]
5. Eichholzer, M.; Tönz, O.; Zimmermann, R. Folic acid: A public-health challenge. *Lancet* **2006**, *367*, 1352–1361. [CrossRef]
6. Allen, L.H. Causes of vitamin B12 and folate deficiency. *Food Nutr. Bull.* **2008**, *29*, S20–S34. [CrossRef]
7. Crider, K.S.; Zhu, J.-H.; Hao, L.; Yang, Q.-H.; Yang, T.P.; Gindler, J.; Maneval, D.R.; Quinlivan, E.P.; Li, Z.; Bailey, L.B.; et al. MTHFR 677C→T genotype is associated with folate and homocysteine concentrations in a large, population-based, double-blind trial of folic acid supplementation. *Am. J. Clin. Nutr.* **2011**, *93*, 1365–1372. [CrossRef]
8. Calvert, H. Folate status and the safety profile of antifolates. *Semin. Oncol.* **2002**, *29*, 3–7. [CrossRef]
9. Sauberlich, H.E.; Kretsch, M.J.; Skala, J.H.; Johnson, H.L.; Taylor, P.C. Folate requirement and metabolism in nonpregnant women. *Am. J. Clin. Nutr.* **1987**, *46*, 1016–1028. [CrossRef]
10. Tamura, T. Determination of food folate. *J. Nutr. Biochem.* **1998**, *9*, 285–293. [CrossRef]
11. McKillop, D.J.; Pentieva, K.; Daly, D.; McPartlin, J.M.; Hughes, J.; Strain, J.; Scott, J.M.; McNulty, H. The effect of different cooking methods on folate retention in various foods that are amongst the major contributors to folate intake in the UK diet. *Br. J. Nutr.* **2002**, *88*, 681–688. [CrossRef]
12. Castenmiller, J.J.; van de Poll, C.J.; West, C.E.; Brouwer, I.A.; Thomas, C.M.; van Dusseldorp, M. Bioavailability of folate from processed spinach in humans. Effect of food matrix and interaction with carotenoids. *Ann. Nutr. Metab.* **2000**, *44*, 163–169. [CrossRef]
13. Malin, J. Total folate activity in Brussels sprouts: The effects of storage, processing, cooking and ascorbic acid content. *Int. J. Food Sci. Technol.* **1977**, *12*, 623–632. [CrossRef]
14. EC. Community Register on the Addition of Vitamins and Minerals and of Certain Other Substances to Foods. Available online: https://ec.europa.eu/food/system/files/2021-01/labelling_nutrition-vitamins_minerals-comm_reg_en.pdf (accessed on 31 August 2021).
15. EC. Directive 2002/46/EC on the Approximation of the Laws of the Member States Relating to Food Supplements. Available online: https://eur-lex.europa.eu/legal-content/EN/TXT/?uri=CELEX:32002L0046 (accessed on 15 February 2021).
16. Gregory, J.F., III. Case Study: Folate Bioavailability. *J. Nutr.* **2001**, *131*, 1376S–1382S. [CrossRef]
17. Bailey, L.B.; Stover, P.J.; McNulty, H.; Fenech, M.F.; Gregory, J.F., III.; Mills, J.L.; Pfeiffer, C.M.; Fazili, Z.; Zhang, M.; Ueland, P.M.; et al. Biomarkers of Nutrition for Development—Folate Review. *J. Nutr.* **2015**, *145*, 1636S–1680S. [CrossRef]
18. Sobczyńska-Malefora, A.; Harrington, D.J. Laboratory assessment of folate (vitamin B9) status. *J. Clin. Pathol.* **2018**, *71*, 949. [CrossRef]
19. WHO. Serum and Red Blood Cell Folate Concentrations for Assessing Folate Status in Populations. Available online: http://apps.who.int/iris/bitstream/handle/10665/75584/WHO_NMH_NHD_EPG_12.1_eng.pdf (accessed on 31 August 2021).
20. Ueland, P.M.; Refsum, H.; Stabler, S.P.; Malinow, M.R.; Andersson, A.; Allen, R.H. Total homocysteine in plasma or serum: Methods and clinical applications. *Clin. Chem.* **1993**, *39*, 1764–1779. [CrossRef]
21. Green, R. Indicators for assessing folate and vitamin B-12 status and for monitoring the efficacy of intervention strategies. *Am. J. Clin. Nutr.* **2011**, *94*, 666s–672s. [CrossRef] [PubMed]
22. Refsum, H.; Ueland, P.; Nygård, O.; Vollset, S. Homocysteine and cardiovascular disease. *Annu. Rev. Med.* **1998**, *49*, 31–62. [CrossRef]
23. Woo, K.S.; Chook, P.; Lolin, Y.I.; Cheung, A.S.; Chan, L.T.; Sun, Y.Y.; Sanderson, J.E.; Metreweli, C.; Celermajer, D.S. Hyperhomocyst(e)inemia is a risk factor for arterial endothelial dysfunction in humans. *Circulation* **1997**, *96*, 2542–2544. [CrossRef]
24. Welch, G.N.; Loscalzo, J. Homocysteine and atherothrombosis. *N. Engl. J. Med.* **1998**, *338*, 1042–1050. [CrossRef] [PubMed]
25. Leeming, R.J.; Pollock, A.; Melville, L.J.; Hamon, C.G.B. Measurement of 5-methyltetrahydrofolic acid in man by high-performance liquid chromatography. *Metabolism* **1990**, *39*, 902–904. [CrossRef]
26. IOM Dietary Reference Intakes for Thiamin, Riboflavin, Niacin, Vitamin B6, Folate, Vitamin B12, Pantothenic Acid, Biotin, and Choline. Available online: https://www.ncbi.nlm.nih.gov/books/NBK114318/ (accessed on 31 August 2021).
27. WHO. Vitamin and Mineral Requirements in Human Nutrition. Available online: https://apps.who.int/iris/bitstream/handle/10665/42716/9241546123.pdf (accessed on 31 August 2021).
28. EFSA Panel on Dietetic Products, N. Allergies. Scientific opinion on dietary reference values for folate. *EFSA J.* **2014**, *12*, 3893.
29. NIJZ. Referenčne Vrednosti za Energijski Vnos ter Vnos Hranil. Available online: https://www.nijz.si/sites/www.nijz.si/files/uploaded/referencne_vrednosti_za_energijski_vnos_ter_vnos_hranil_obl.pdf (accessed on 31 August 2021).
30. Krawinkel, M.B.; Strohm, D.; Weissenborn, A.; Watzl, B.; Eichholzer, M.; Bärlocher, K.; Elmadfa, I.; Leschik-Bonnet, E.; Heseker, H. Revised D-A-CH intake recommendations for folate: How much is needed? *Eur. J. Clin. Nutr.* **2014**, *68*, 719–723. [CrossRef]
31. D-A-CH. Reference Values DACH. Available online: https://www.sge-ssn.ch/grundlagen/lebensmittel-und-naehrstoffe/naehrstoffempfehlungen/dachreferenzwerte/ (accessed on 31 August 2021).
32. Fidler Mis, N.; Kobe, H.; Stimec, M. Dietary Intake of Macro- and Micronutrients in Slovenian Adolescents: Comparison with Reference Values. *Ann. Nutr. Metab.* **2012**, *61*, 305–313. [CrossRef]
33. Mijatov, M.A.; Mičetić-Turk, D. Dietary Intake In Adult Female Coeliac Disease Patients In Slovenia: Prehranski Vnos Odraslih Bolnic S Celiakijo V Sloveniji. *Zdr. Varst.* **2016**, *55*, 86–93.

34. Rippin, H.L.; Hutchinson, J.; Jewell, J.; Breda, J.J.; Cade, J.E. Adult nutrient intakes from current national dietary surveys of European populations. *Nutrients* **2017**, *9*, 1288. [CrossRef]
35. Crider, K.S.; Bailey, L.B.; Berry, R.J. Folic acid food fortification-its history, effect, concerns, and future directions. *Nutrients* **2011**, *3*, 370. [CrossRef]
36. Choumenkovitch, S.F.; Selhub, J.; Wilson, P.W.F.; Rader, J.I.; Rosenberg, I.H.; Jacques, P.F. Folic Acid Intake from Fortification in United States Exceeds Predictions. *J. Nutr.* **2002**, *132*, 2792–2798. [CrossRef]
37. Smith, A.D.; Kim, Y.-I.; Refsum, H. Is folic acid good for everyone? *Am. J. Clin. Nutr.* **2008**, *87*, 517–533. [CrossRef]
38. Žmitek, K.; Krušič, S.; Pravst, I. An Approach to Investigate Content-Related Quality of Nutraceuticals Used by Slovenian Consumers: A Case Study with Folate and Vitamin D Supplements. *Foods* **2021**, *10*, 845. [CrossRef]
39. Ubbink, J.B. Should all elderly people receive folate supplements? *Drugs Aging.* **1998**, *13*, 415–420. [CrossRef]
40. Hooshmand, B.; Solomon, A.; Kåreholt, I.; Rusanen, M.; Hänninen, T.; Leiviskä, J.; Winblad, B.; Laatikainen, T.; Soininen, H.; Kivipelto, M. Associations between serum homocysteine, holotranscobalamin, folate and cognition in the elderly: A longitudinal study. *J. Intern. Med.* **2012**, *271*, 204–212. [CrossRef] [PubMed]
41. König, D.; Bissé, E.; Deibert, P.; Müller, H.M.; Wieland, H.; Berg, A. Influence of Training Volume and Acute Physical Exercise on the Homocysteine Levels in Endurance-Trained Men: Interactions with Plasma Folate and Vitamin B_{12}. *Ann. Nutr. Metab.* **2003**, *47*, 114–118. [CrossRef] [PubMed]
42. EC. Regulation (EC) No 1925/2006 of the European Parliament and of the Council of 20 December 2006 on the Addition of Vitamins and Minerals and of Certain Other Substances to Foods. Available online: https://eur-lex.europa.eu/legal-content/EN/AUTO/?uri=CELEX:02006R1925-20210408 (accessed on 31 August 2021).
43. Pravst, I.; Kušar, A. Consumers' exposure to nutrition and health claims on pre-packed foods: Use of sales weighting for assessing the food supply in Slovenia. *Nutrients* **2015**, *7*, 9353–9368. [CrossRef]
44. Authority, E.F.S. Guidance on the EU Menu methodology. *EFSA J.* **2014**, *12*, 39–44.
45. Zupanič, N.; Hribar, M.; Hristov, H.; Lavriša, Ž.; Kušar, A.; Gregorič, M.; Blaznik, U.; Koroušić Seljak, B.; Golja, P.; Vidrih, R.; et al. Dietary Intake of trans Fatty Acids in the Slovenian Population. *Nutrients* **2021**, *13*, 207. [CrossRef]
46. Zupanič, N.; Hristov, H.; Gregorič, M.; Blaznik, U.; Delfar, N.; Koroušić Seljak, B.; Ding, E.L.; Fidler Mis, N.; Pravst, I. Total and Free Sugars Consumption in a Slovenian Population Representative Sample. *Nutrients* **2020**, *12*, 17–29. [CrossRef]
47. Hribar, M.; Hristov, H.; Gregorič, M.; Blaznik, U.; Zaletel, K.; Oblak, A.; Osredkar, J.; Kušar, A.; Žmitek, K.; Rogelj, I.; et al. Nutrihealth Study: Seasonal Variation in Vitamin D Status Among the Slovenian Adult and Elderly Population. *Nutrients* **2020**, *12*, 1838. [CrossRef]
48. Gregorič, M.; Blaznik, U.; Delfar, N.; Zaletel, M.; Lavtar, D.; Seljak, B.K.; Golja, P.; Kotnik, K.Z.; Pravst, I.; Mis, N.F. Slovenian national food consumption survey in adolescents, adults and elderly. *EFSA Supporting Publ.* **2019**, *16*. [CrossRef]
49. De Onis, M.; Onyango, A.; Borghi, E.; Siyam, A.; Blössner, M.; Lutter, C. Worldwide implementation of the WHO child growth standards. *Public Health Nutr.* **2012**, *15*, 1603–1610. [CrossRef] [PubMed]
50. Onis, M.d.; Onyango, A.W.; Borghi, E.; Siyam, A.; Nishida, C.; Siekmann, J. Development of a WHO growth reference for school-aged children and adolescents. *Bull. World Health Organ.* **2007**, *85*, 660–667. [CrossRef]
51. Lee, P.H.; Macfarlane, D.J.; Lam, T.H.; Stewart, S.M. Validity of the international physical activity questionnaire short form (IPAQ-SF): A systematic review. *Int. J. Behav. Nutr. Phys. Act.* **2011**, *8*, 1–11. [CrossRef]
52. Vede, T. Izdelava in validacija slikovnega gradiva za določanje vnosa živil. In *Design and Validation of Food Picture Book*; University of Ljubljana, Biotechnical Faculty: Ljubljana, Slovenia, 2016.
53. Korošec, M.; Golob, T.; Bertoncelj, J.; Stibilj, V.; Seljak, B.K. The Slovenian food composition database. *Food Chem.* **2013**, *140*, 495–499. [CrossRef]
54. Fineli. Finnish Food Composition Database. Available online: https://fineli.fi/fineli/en/index (accessed on 31 August 2021).
55. Bodner-Montville, J.; Ahuja, J.K.; Ingwersen, L.A.; Haggerty, E.S.; Enns, C.W.; Perloff, B.P. USDA food and nutrient database for dietary studies: Released on the web. *J. Food Compos. Anal.* **2006**, *19*, S100–S107. [CrossRef]
56. Dunford, E.; Webster, J.; Metzler, A.B.; Czernichow, S.; Ni Mhurchu, C.; Wolmarans, P.; Snowdon, W.; L'Abbe, M.; Li, N.; Maulik, P.K.; et al. International collaborative project to compare and monitor the nutritional composition of processed foods. *Eur. J. Prev. Cardiol.* **2012**, *19*, 1326–1332. [CrossRef] [PubMed]
57. La'ulu, S.L.; Rawlins, M.L.; Pfeiffer, C.M.; Zhang, M.; Roberts, W.L. Performance Characteristics of Six Homocysteine Assays. *Am. J. Clin. Pathol.* **2008**, *130*, 969–975. [CrossRef] [PubMed]
58. Harttig, U.; Haubrock, J.; Knüppel, S.; Boeing, H. The MSM program: Web-based statistics package for estimating usual dietary intake using the Multiple Source Method. *Eur. J. Clin. Nutr.* **2011**, *65* (Suppl. 1), 87–91. [CrossRef] [PubMed]
59. Hribar, M.; Hristov, H.; Lavriša, Ž.; Koroušić Seljak, B.; Gregorič, M.; Blaznik, U.; Žmitek, K.; Pravst, I. Vitamin D Intake in Slovenian Adolescents, Adults, and the Elderly Population. *Nutrients* **2021**, *13*, 3528. [CrossRef] [PubMed]
60. Haubrock, J.; Nöthlings, U.; Volatier, J.-L.; Dekkers, A.; Ocké, M.; Harttig, U.; Illner, A.-K.; Knüppel, S.; Andersen, L.F.; Boeing, H.; et al. Estimating Usual Food Intake Distributions by Using the Multiple Source Method in the EPIC-Potsdam Calibration Study. *J. Nutr.* **2011**, *141*, 914–920. [CrossRef]
61. Kolenikov, S. Calibrating Survey Data using Iterative Proportional Fitting (Raking). *Stata J.* **2014**, *14*, 22–59. [CrossRef]
62. Quinlivan, E.P.; Gregory, J.F., III. Effect of food fortification on folic acid intake in the United States. *Am. J. Clin. Nutr.* **2003**, *77*, 221–225. [CrossRef] [PubMed]

63. Rampersaud, G.C.; Kauwell, G.P.; Bailey, L.B. Folate: A key to optimizing health and reducing disease risk in the elderly. *J. Am. Coll. Nutr.* **2003**, *22*, 1–8. [CrossRef] [PubMed]
64. Luchsinger, J.A.; Tang, M.-X.; Miller, J.; Green, R.; Mayeux, R. Relation of higher folate intake to lower risk of Alzheimer disease in the elderly. *Arch. Neurol.* **2007**, *64*, 86–92. [CrossRef]
65. Smith, R.L. Evaluation of vitamin B12 and folate status in the nursing home. *J. Am. Med Dir. Assoc.* **2001**, *2*, 230–238. [CrossRef]
66. Fioravanti, M.; Ferrario, E.; Massaia, M.; Cappa, G.; Rivolta, G.; Grossi, E.; Buckley, A. Low folate levels in the cognitive decline of elderly patients and the efficacy of folate as a treatment for improving memory deficits. *Arch. Gerontol. Geriatr.* **1997**, *26*, 1–13. [CrossRef]
67. Dhonukshe-Rutten, R.; De Vries, J.; De Bree, A.; Van Der Put, N.; Van Staveren, W.; De Groot, L. Dietary intake and status of folate and vitamin B12 and their association with homocysteine and cardiovascular disease in European populations. *Eur. J. Clin. Nutr.* **2009**, *63*, 18–30. [CrossRef]
68. de Bree, A.; Verschuren, W.M.; Blom, H.J.; Kromhout, D. Association between B vitamin intake and plasma homocysteine concentration in the general Dutch population aged 20–65 y. *Am. J. Clin. Nutr.* **2001**, *73*, 1027–1033. [CrossRef] [PubMed]
69. O'Brien, M.M.; Kiely, M.; Harrington, K.E.; Robson, P.J.; Strain, J.J.; Flynn, A. The North/South Ireland Food Consumption Survey: Vitamin intakes in 18–64-year-old adults. *Public Health Nutr.* **2008**, *4*, 1069–1079. [CrossRef] [PubMed]
70. Waśkiewicz, A.; Sygnowska, E.; Broda, G. Dietary intake of vitamins B 6, B 12 and folate in relation to homocysteine serum concentration in the adult Polish population-WOBASZ Project. *Kardiol. Pol. (Pol. Heart J.)* **2010**, *68*, 275–282.
71. Alfthan, G.; Laurinen, M.S.; Valsta, L.M.; Pastinen, T.; Aro, A. Folate intake, plasma folate and homocysteine status in a random Finnish population. *Eur. J. Clin. Nutr.* **2003**, *57*, 81–88. [CrossRef] [PubMed]
72. Schröder, H.; Marrugat, J.; Covas, M.; Elosua, R.; Pena, A.; Weinbrenner, T.; Fito, M.; Vidal, M.; Masia, R. Population dietary habits and physical activity modification with age. *Eur. J. Clin. Nutr.* **2004**, *58*, 302–311. [CrossRef]
73. Tucker, K.L.; Selhub, J.; Wilson, P.W.; Rosenberg, I.H. Dietary intake pattern relates to plasma folate and homocysteine concentrations in the Framingham Heart Study. *J. Nutr.* **1996**, *126*, 3025–3031. [CrossRef] [PubMed]
74. Aranceta, J.; Serra-Majem, L.; Pérez-Rodrigo, C.; Llopis, J.; Mataix, J.; Ribas, L.; Tojo, R.; Tur, J.A. Vitamins in Spanish food patterns: The eVe Study. *Public Health Nutr.* **2001**, *4*, 1317–1323. [CrossRef] [PubMed]
75. Freisling, H.; Elmadfa, I.; Gall, I. The effect of socioeconomic status on dietary intake, physical activity and Body Mass Index in Austrian pregnant women. *J. Hum. Nutr. Diet.* **2006**, *19*, 437–445. [CrossRef] [PubMed]
76. Pounis, G.; Di Castelnuovo, A.F.; de Lorgeril, M.; Krogh, V.; Siani, A.; Arnout, J.; Cappuccio, F.P.; van Dongen, M.; Zappacosta, B.; Donati, M.B. Folate intake and folate serum levels in men and women from two European populations: The IMMIDIET project. *Nutrition* **2014**, *30*, 822–830. [CrossRef]
77. Goodman, E.; Adler, N.E.; Daniels, S.R.; Morrison, J.A.; Slap, G.B.; Dolan, L.M. Impact of objective and subjective social status on obesity in a biracial cohort of adolescents. *Obes. Res.* **2003**, *11*, 1018–1026. [CrossRef]
78. Reidpath, D.D.; Burns, C.; Garrard, J.; Mahoney, M.; Townsend, M. An ecological study of the relationship between social and environmental determinants of obesity. *Health Place* **2002**, *8*, 141–145. [CrossRef]
79. Gregorič, M.; Blaznik, U.; Turk, V. Različni Vidiki Prehranjevanja Prebivalcev Slovenije (V Starosti Od 3 Mesecev Do 74 Let). Available online: https://www.nijz.si/sites/www.nijz.si/files/publikacije-datoteke/razlicni_vidiki_prehranjevanja_prebivalcev_slovenije.pdf (accessed on 31 August 2021).
80. Steenhuis, I.H.M.; Waterlander, W.E.; de Mul, A. Consumer food choices: The role of price and pricing strategies. *Public Health Nutr.* **2011**, *14*, 2220–2226. [CrossRef]
81. Waterlander, W.E.; de Haas, W.E.; van Amstel, I.; Schuit, A.J.; Twisk, J.W.R.; Visser, M.; Seidell, J.C.; Steenhuis, I.H.M. Energy density, energy costs and income—How are they related? *Public Health Nutr.* **2010**, *13*, 1599–1608. [CrossRef]
82. Cassady, D.; Jetter, K.M.; Culp, J. Is price a barrier to eating more fruits and vegetables for low-income families? *J. Am. Diet. Assoc.* **2007**, *107*, 1909–1915. [CrossRef]
83. Konings, E.J.; Roomans, H.H.; Dorant, E.; Goldbohm, R.A.; Saris, W.H.; van den Brandt, P.A. Folate intake of the Dutch population according to newly established liquid chromatography data for foods. *Am. J. Clin. Nutr.* **2001**, *73*, 765–776. [CrossRef] [PubMed]
84. McLean, E.; de Benoist, B.; Allen, L.H. Review of the magnitude of folate and vitamin B12 deficiencies worldwide. *Food Nutr. Bull.* **2008**, *29*, S38–S51. [CrossRef] [PubMed]
85. Yates, A.A. Bioavailability of nutrients and other bioactive components from dietary supplements. *J. Nutr.* **2001**, *131*, 1331S–1334S.
86. Saini, R.K.; Nile, S.H.; Keum, Y.S. Folates: Chemistry, analysis, occurrence, biofortification and bioavailability. *Food Res. Int.* **2016**, *89*, 1–13. [CrossRef] [PubMed]
87. Vardavas, C.I.; Linardakis, M.K.; Hatzis, C.M.; Malliaraki, N.; Saris, W.H.M.; Kafatos, A.G. Smoking status in relation to serum folate and dietary vitamin intake. *Tob. Induc. Dis.* **2008**, *4*, 8. [CrossRef] [PubMed]
88. Piyathilake, C.J.; Macaluso, M.; Hine, R.J.; Richards, E.W.; Krumdieck, C.L. Local and systemic effects of cigarette smoking on folate and vitamin B−12. *Am. J. Clin. Nutr.* **1994**, *60*, 559–566. [CrossRef]
89. Selhub, J.; Jacques, P.F.; Wilson, P.W.; Rush, D.; Rosenberg, I.H. Vitamin status and intake as primary determinants of homocysteinemia in an elderly population. *JAMA* **1993**, *270*, 2693–2698. [CrossRef] [PubMed]
90. Ganji, V.; Kafai, M.R. Demographic, Lifestyle, and Health Characteristics and Serum B Vitamin Status Are Determinants of Plasma Total Homocysteine Concentration in the Post-Folic Acid Fortification Period, 1999–2004. *J. Nutr.* **2008**, *139*, 345–352. [CrossRef]

91. BRATTSTRÖM, L.; LINDGREN, A.; ISRAELSSON, B.; ANDERSSON, A.; HULTBERG, B. Homocysteine and cysteine: Determinants of plasma levels in middle-aged and elderly subjects. *J. Intern. Med.* **1994**, *236*, 633–641. [CrossRef] [PubMed]
92. Clarke, R.; Refsum, H.; Birks, J.; Evans, J.G.; Johnston, C.; Sherliker, P.; Ueland, P.M.; Schneede, J.; McPartlin, J.; Nexo, E.; et al. Screening for vitamin B-12 and folate deficiency in older persons. *Am. J. Clin. Nutr.* **2003**, *77*, 1241–1247. [CrossRef] [PubMed]
93. Collaboration, H.L.T. Dose-dependent effects of folic acid on blood concentrations of homocysteine: A meta-analysis of the randomized trials. *Am. J. Clin. Nutr.* **2005**, *82*, 806–812.
94. D'Anci, K.E.; Rosenberg, I.H. Folate and brain function in the elderly. *Curr. Opin. Clin. Nutr. Metab. Care* **2004**, *7*, 659–664. [CrossRef] [PubMed]
95. Forges, T.; Monnier-Barbarino, P.; Alberto, J.M.; Gueant-Rodriguez, R.M.; Daval, J.L.; Gueant, J.L. Impact of folate and homocysteine metabolism on human reproductive health. *Hum. Reprod. Update* **2007**, *13*, 225–238. [CrossRef]
96. EFSA. Scientific Opinion on the substantiation of a health claim related to increasing maternal folate status by supplemental folate intake and reduced risk of neural tube defects pursuant to Article 14 of Regulation (EC) No 1924/2006. *EFSA J.* **2013**, *11*, 3328.
97. Ray, J.G.; Singh, G.; Burrows, R.F. Evidence for suboptimal use of periconceptional folic acid supplements globally. *BJOG: Int. J. Obstet. Gynaecol.* **2004**, *111*, 399–408. [CrossRef]
98. Force, U.P.S.T. Folic Acid Supplementation for the Prevention of Neural Tube Defects: US Preventive Services Task Force Recommendation Statement. *JAMA* **2017**, *317*, 183–189.
99. Morris, J.K.; Addor, M.-C.; Ballardini, E.; Barisic, I.; Barrachina-Bonet, L.; Braz, P.; Cavero-Carbonell, C.; Den Hond, E.; Garne, E.; Gatt, M.; et al. Prevention of Neural Tube Defects in Europe: A Public Health Failure. *Front. Pediatr.* **2021**, *9*, 513. [CrossRef]
100. Zekovic, M.; Djekic-Ivankovic, M.; Nikolic, M.; Gurinovic, M.; Krajnovic, D.; Glibetic, M. Validity of the Food Frequency Questionnaire Assessing the Folate Intake in Women of Reproductive Age Living in a Country without Food Fortification: Application of the Method of Triads. *Nutrients* **2017**, *9*, 128. [CrossRef] [PubMed]
101. Mojtabai, R. Body mass index and serum folate in childbearing age women. *Eur. J. Epidemiol.* **2004**, *19*, 1029–1036. [CrossRef]
102. Rogers, L.M.; Cordero, A.M.; Pfeiffer, C.M.; Hausman, D.B.; Tsang, B.L.; De-Regil, L.M.; Rosenthal, J.; Razzaghi, H.; Wong, E.C.; Weakland, A.P.; et al. Global folate status in women of reproductive age: A systematic review with emphasis on methodological issues. *Ann. N. Y. Acad. Sci.* **2018**, *1431*, 35–57. [CrossRef] [PubMed]
103. Kokalj, T.S.; Rejc, B.; Gersak, K. Incidence and prevention of neural tube defects in Slovenia. *Eur. J. Obstet. Gynecol. Reprod. Biol.* **2011**, *156*, 119–120. [CrossRef] [PubMed]
104. Marchetta, C.M.; Devine, O.J.; Crider, K.S.; Tsang, B.L.; Cordero, A.M.; Qi, Y.P.; Guo, J.; Berry, R.J.; Rosenthal, J.; Mulinare, J.; et al. Assessing the Association between Natural Food Folate Intake and Blood Folate Concentrations: A Systematic Review and Bayesian Meta-Analysis of Trials and Observational Studies. *Nutrients* **2015**, *7*, 2663–2686. [CrossRef] [PubMed]
105. Koch, V.; Pokorn, D. Comparison of nutritional habits among various adult age groups in Slovenia. *Nutr. Res.* **1999**, *19*, 1153–1164. [CrossRef]
106. Food Administration, D. Food standards: Amendment of standards of identity for enriched grain products to require addition of folic acid; final rule (21 CFR Parts 136, 137, and 139). *Fed. Regist.* **1996**, *61*, 8781–8797.
107. Canada, H. Regulations amending the Food and Drug Regulations (1066). *Can Gaz Part 1* **1997**, *131*, 3702–3737.
108. Hertrampf, E.; Cortés, F.; Erickson, J.D.; Cayazzo, M.; Freire, W.; Bailey, L.B.; Howson, C.; Kauwell, G.P.; Pfeiffer, C. Consumption of folic acid–fortified bread improves folate status in women of reproductive age in Chile. *J. Nutr.* **2003**, *133*, 3166–3169. [CrossRef]
109. DHSC. Folic Acid Added to Flour to Prevent Spinal Conditions in Babies. Available online: https://www.gov.uk/government/news/folic-acid-added-to-flour-to-prevent-spinal-conditions-in-babies (accessed on 21 September 2021).
110. Dietrich, M.; Brown, C.J.; Block, G. The effect of folate fortification of cereal-grain products on blood folate status, dietary folate intake, and dietary folate sources among adult non-supplement users in the United States. *J. Am. Coll. Nutr.* **2005**, *24*, 266–274. [CrossRef]
111. EC. Fortification with Folic Acid Could help Prevent Severe Birth Defects in at Least 1000 Pregnancies per Year. Available online: https://ec.europa.eu/jrc/en/news/fortification-folic-acid-could-help-prevent-severe-birth-defects-least-1000-pregnancies-year (accessed on 21 September 2021).
112. West, A.A.; Caudill, M.A. Genetic variation: Impact on folate (and choline) bioefficacy. *Int. J. Vitam. Nutr. Res.* **2010**, *80*, 319. [CrossRef]
113. Raiten, D.J.; Namasté, S.; Brabin, B.; Combs, G., Jr.; L'Abbe, M.R.; Wasantwisut, E.; Darnton-Hill, I. Executive summary:Biomarkers of Nutrition for Development: Building a Consensus. *Am. J. Clin. Nutr.* **2011**, *94*, 633s–650s. [CrossRef]

nutrients

MDPI

Article

Inadequate Intake of Dietary Fibre in Adolescents, Adults, and Elderlies: Results of Slovenian Representative SI. Menu Study

Barbara Koroušić Seljak [1,*], Eva Valenčič [1,2,3], Hristo Hristov [4], Maša Hribar [4], Živa Lavriša [4], Anita Kušar [4], Katja Žmitek [4,5], Sanja Krušič [4], Matej Gregorič [6], Urška Blaznik [6], Blaž Ferjančič [7], Jasna Bertoncelj [7], Mojca Korošec [7] and Igor Pravst [4,5,7]

1 Computer Systems Department, Jožef Stefan Institute, SI-1000 Ljubljana, Slovenia; eva.valencic@ijs.si
2 Jožef Stefan International Postgraduate School, SI-1000 Ljubljana, Slovenia
3 Priority Research Centre in Physical Activity and Nutrition, Faculty of Health and Medicine, School of Health Sciences, The University of Newcastle, Callaghan, NSW 2308, Australia
4 Nutrition Institute, Tržaška Cesta 40, SI-1000 Ljubljana, Slovenia; hristo.hristov@nutris.org (H.H.); masa.hribar@nutris.org (M.H.); ziva.lavrisa@nutris.org (Ž.L.); anita.kusar@nutris.org (A.K.); katja.zmitek@vist.si (K.Ž.); sanja.krusic@nutris.org (S.K.); igor.pravst@nutris.org (I.P.)
5 Higher School of Applied Sciences (VIST), Gerbičeva Cesta 51A, SI-1000 Ljubljana, Slovenia
6 National Institute of Public Health, Trubarjeva 2, SI-1000 Ljubljana, Slovenia; matej.gregoric@nijz.si (M.G.); urska.blaznik@nijz.si (U.B.)
7 Biotechnical Faculty, University of Ljubljana, Jamnikarjeva 101, SI-1000 Ljubljana, Slovenia; blaz.ferjancic@bf.uni-lj.si (B.F.); jasna.bertoncelj@bf.uni-lj.si (J.B.); mojca.korosec@bf.uni-lj.si (M.K.)
* Correspondence: barbara.korousic@ijs.si; Tel.: +386-1-4773-363

Citation: Seljak, B.K.; Valenčič, E.; Hristov, H.; Hribar, M.; Lavriša, Ž.; Kušar, A.; Žmitek, K.; Krušič, S.; Gregorič, M.; Blaznik, U.; et al. Inadequate Intake of Dietary Fibre in Adolescents, Adults, and Elderlies: Results of Slovenian Representative SI. Menu Study. *Nutrients* **2021**, *13*, 3826. https://doi.org/10.3390/nu13113826

Academic Editor: Iain A. Brownlee

Received: 21 September 2021
Accepted: 24 October 2021
Published: 27 October 2021

Abstract: Dietary fibre has proven to promote healthy body mass and reduce the risk of non-communicable diseases. To date, in Slovenia, there were only a few outdated studies of dietary fibre intake; therefore, we explored the dietary fibre intake using food consumption data collected in the SI.Menu project. Following the EU Menu methodology, data were collected on representative samples of adolescents, adults, and elderlies using a general questionnaire, a food propensity questionnaire, and two 24 h recalls. The results indicate that the intake of dietary fibre in Slovenia is lower than recommended. The proportion of the population with inadequate fibre intakes (<30 g/day) was 90.6% in adolescents, 89.6% in adults, and 83.9% in elderlies, while mean daily fibre intakes were 19.5, 20.9, and 22.4 g, respectively. Significant determinants for inadequate dietary fibre intake were sex in adolescents and adults, and body mass index in adults. The main food groups contributing to dietary fibre intake were bread and other grain products, vegetables and fruits, with significant differences between population groups. Contribution of fruits and vegetables to mean daily dietary fibre intake was highest in elderlies (11.6 g), followed by adults (10.6 g) and adolescents (8.5 g). Public health strategies, such as food reformulation, promoting whole-meal alternatives, consuming whole foods of plant origin, and careful planning of school meals could beneficially contribute to the overall dietary fibre intake in the population.

Keywords: dietary fibre; dietary fibre intake; 24 h recall; EU Menu; Slovenian population

1. Introduction

Though the importance of dietary fibre in human health is well established, knowledge of dietary fibre intake in some countries is still lacking. In Slovenia, fairly recently, a national food consumption survey was carried out which enabled estimation of dietary fibre intake among all the populations. Food consumption data collected in such a way are valuable; however, their matching with compositional data on dietary fibre is a challenging task because dietary fibre refers to a number of compounds with different chemical structures for which compositional data for a complete set of foods are still missing.

The Codex Alimentarius [1] and the European Food Safety Authority (EFSA) [2] define dietary fibre as a compound consisting of at least 10 monomeric units which are not

hydrolysed in the human small intestine and are passing to the large intestine. Oligosaccharides with 3 to 9 monomeric units are also considered as dietary fibre [2]. Some types of dietary fibre can be metabolised by the gut microbiota and can form energy-yielding substances [3]. Therefore, the energy value of 2 kcal per gram of dietary fibre is considered in the calculation of the energy value of foods [4].

Dietary fibre has been recognised as a very important part of healthy diets. By increasing stool frequency and bulk volume, dietary fibre helps with defecation and might protect us against heart diseases, hypertension, and certain types of cancer [5,6]. Several recent sources also reveal the importance of dietary fibre intake in local colon microbiota [7–9]. Colonic bacteria produce enzymes that can partly degrade dietary fibre molecules through the process of fermentation. By-products of the dietary fibre fermentation (i.e., short-chain fatty acids and gases) are very important as they influence the composition of the gut microbiota, leading to a healthier gut microbiome which helps to regulate the immune system [10].

The terms soluble and insoluble dietary fibre are sometimes used to describe the types of dietary fibre in our diet. In the review of Stephen et al. [11], a difference in lowering risk factors for cardiovascular diseases CVD based on the solubility of dietary fibre was reported, in which soluble dietary fibre shows a greater effect. A similar result was confirmed for the reduction in appetite in weight management—soluble dietary fibre shows better appetite suppression. Despite some exceptions, the division of dietary fibre based on its solubility is the closest approximation for predicting the fermentability of dietary fibre. The majority of soluble dietary fibre is fermentable, and the fermentation of dietary fibre in the colon is an important process for colon health influencing the human organism as a whole [12]. However, very few food composition databases (FCDBs) include information about the content of (in)soluble dietary fibre. National FCDBs also lack complete data on other types of dietary fibre, such as non-starch polysaccharides (NPS; cellulose, hemicelluloses, pectins, hydrocolloids (i.e., gums, mucilages, β-glucans), and fructans); resistant (short-chain) oligosaccharides (fructooligosaccharides (FOS), galactooligosaccharides (GOS), and other resistant oligosaccharides); resistant starch (physically enclosed starch, some types of raw starch granules, retrograded amylose, and chemically and/or physically modified starches); lignin (aromatic polymer associated with the dietary fibre polysaccharides). These types of dietary fibre have various physicochemical properties, which are of key importance for their physiological effects.

Recommendations about adequate dietary fibre intake are not harmonised. Based on the available evidence on bowel function, the EFSA's NDA Panel concluded that dietary fibre intake of 25 g per day is adequate for normal laxation in adults, while in children, from the age of one year, they recommended 2 g of dietary fibre per MJ [2]. Considering the evidence of benefit to health associated with consumption of diets rich in dietary fibre-containing foods, the NDA Panel also noted that dietary fibre intake of greater than 25 g per day can reduce risks of coronary heart disease and type 2 diabetes, and improve body mass maintenance. In Slovenia, D-A-CH recommendations (D-A-CH present country identification for the countries Germany (D), Austria (A), and Switzerland (CH), for which these guidelines were developed) have been adopted by the Slovenian Ministry of Health since 2004 [13]. This guidance generally recommends an intake of 30 g of dietary fibre per day [14].

Daily dietary fibre requirements can be met by consuming plenty of fresh fruits and vegetables, potatoes, whole-grain cereals, legumes, and pulses, nuts, and seeds. The EFSA's NDA Panel estimated that the average dietary fibre intakes in Europe vary from 10 to 20 g per day in young children (of age less than 10 to 12 years), from 15 to 30 g per day in adolescents, and from 16 to 29 g per day in adults [2]. However, no recent data on the intake of dietary fibre are available for Slovenia. The only nationally representative study on adults was conducted back in 1995/96, indicating a mean dietary fibre daily intake of 20.1 g (20.3 in men and 19.9 in women) [15,16]. Additionally, almost two decades ago, Fidler Mis et al. conducted a nationally representative study among adolescents and reported a dietary fibre intake of 28 $\pm$ 9 g among boys and 31 $\pm$ 11 g among girls [17].

The main objective of the study presented in this paper was to estimate intake of dietary fibre exploiting the most recent data, collected within the national food consumption survey (SI.Menu 2017/2018). Our focus was on adolescents, adults, and elderly populations. Due to the lack of studies on the estimation of dietary fibre intake that differentiate between different types of fibre, which have different physiological roles in the human body, we also aimed to evaluate a proportion of the (in)soluble fibre in people's diets. Secondary objectives were to investigate the prevalence of inadequate dietary fibre intakes and associations with different socio-demographic and lifestyle determinants and to identify major sources of dietary fibre in different age groups.

2. Material and Methods

2.1. Study Design and Subjects

The study was conducted as part of a national cross-sectional food consumption survey, named SI.Menu, which was aimed at collecting food consumption data in Slovenia in the period from March 2017 to April 2018. The survey was designed considering the EFSA Guidance on European Union Menu Methodology [18]. Details about the methodology and sample characteristics are published elsewhere [19,20]. We should note that Slovenia has a population of only 2.06 million people (2007 census data). The participants were stratified into three age groups: adolescents (10–17 years old), adults (18–64 years old), and the elderly (65–74 years old). Sampling was carried out for individuals, which could not be substituted with another household member. A total of 2280 subjects were selected using the Central Register of Population (CRP) of Slovenia according to age, sex, and place of residency (to cover all NUTS-3 statistical regions). In Slovenia, CPR is administered by the Ministry of the Interior of the Republic of Slovenia and represents the central repository and processing of data concerning residents of the Republic of Slovenia; data were provided from the Statistical Office of the Republic of Slovenia. Individuals living abroad, and institutionalised, ill, or people with disabilities were excluded from the study. The selected residents were visited by survey interviewers, who checked the eligibility of the respondents and collected required information by interviews. The interviewers underwent a course in nutrition surveys and nutritional surveillance in practice. In case of any questions or problems during an interview, they had a possibility to contact the project leaders. The survey was completed by a total of 62% of the invited participants (n = 1319); the lowest participation rate was observed in adults (57%), while the rate was higher in adolescents (69%) and elderlies (65%). In these two groups, participation rates were very similar between males and females, while in adults, the participation rate of males was notably lower (50 vs. 61%). The study protocol was approved by the National Medical Ethics Committee (KME 53/07/16; approval No. 0120-337/2016 issued on 19.7.2016). Prior to inclusion in the study, all subjects were informed about the study and signed an informed consent form. In the case of adolescents, informed consent was also obtained from the parent or legal guardian.

2.2. Food Consumption Data

In the survey SI.Menu, two questionnaires were prepared to enable the collection of food consumption data and corresponding metadata. Additionally, food consumption data were collected with two 24 h dietary recalls:

1. General Questionnaire (GQ) enabled the collection of data for accessing general socio-demographic, socio-economic, and lifestyle determinants, such as place of living, number of household members, marital status, level of education, monthly net income of the household, dietary and consumer habits, as well as usual frequency and duration of physical activity.
2. Food Propensity Questionnaire (FPQ) was used to record the usual frequency of consumption of specific foods in the last 12 months: Altogether, 78 food items were allocated into 9 food groups. For example, within cereals and cereal products, FPQ included separate questions for white bread, whole-grain bread, etc. The FPQ consid-

ered the following frequency response options: *never, 1–3 times per month or less, once per week, 2–3 times per week, 4–6 times per week, 1–2 times per day or more*. More details about the SI.Menu FPQ can be found elsewhere [19].

3. Information about the participant's dietary habits was collected by two non-consecutive 24 h dietary recalls which were carried out up to three weeks apart (71% of the recalls were performed on workdays and 29% on weekends). The majority (87%) of the second 24 h recalls was collected within 7 days; the rest was completed within the next two weeks. The first 24 h recalls occurred at the participant's home and were completed by the participant's interviewer. The second 24 h recall was performed over the telephone or at participants' homes (in cases when they could not be reached over the telephone). The latter was only performed for a small sample of participants. Portion sizes were estimated with the help of a nationally adjusted picture book, containing 46 pictures of different food products or simple recipes presented in 6 different portion sizes [19]. For home-cooked (mixed) dishes, participants were asked to provide recipes. When this was not available, and for all outdoor dining, standard recipes from the Open Platform for Clinical Nutrition (OPEN) [21] were used.

At the first interviewer's visit, the general questionnaire and the FPQ were completed by the interviewer. The interviewer also collected anthropometric data, measured the participant's body height (m) and body mass (kg) using a portable, calibrated scale. These data were used to determine the body mass index (BMI; kg/m^2) considering the overweightness cut-off point at 25 kg/m^2, except for adolescents, where sex/age-adjusted cut-off points (>1SD) were applied [22,23]. Additionally, using the provided information about the usual frequency and duration of physical activity, the participant's International Physical Activity Questionnaire (IPAQ) score [24] was calculated. In the GQ, participants were asked to select a type of residential area (city/town; suburbs; village type areas), and this information was used to assign them into urban, intermediate, and rural areas, respectively.

Food consumption data collected by the 24 h recalls were analysed using a food composition database created as part of the Open Platform for Clinical Nutrition (OPEN) [21]. The OPEN database includes compositional data on generic foods and some branded foods and provides a list of ingredients for traditional and other recipes frequently consumed in Slovenia. For foods not covered in national compositional data, the OPEN database uses European (EuroFIR) and United States Department of Agriculture (USDA) food composition databases [25].

2.3. *Assessment of Dietary Fibre Content*

Data on dietary fibre intake were extracted from the SI.Menu food consumption dataset. The SI.Menu dataset contains both simple foods and complex foods and recipes. In some cases, the ingredient information and recipe were provided by the subjects, while in other cases, traditional or commonly used recipes (pre-collected in OPEN) were used. For these complex foods and recipes, a disaggregation method was applied considering both the yield and retention factors [26], to calculate their dietary fibre contents. Foods extracted from the 24 h recalls were inserted into OPEN and checked by a nutrition expert. In case of missing information on the amount of dietary fibre in specific food products, missing data were supplemented with the data from the Fineli database [27]. This is the Finish National Food Composition Database, maintained by the National Institute for Health and Welfare, which also contains information about soluble and insoluble dietary fibre. If a product was not found in this database, the Danish [28] and German [29] databases were also used. Altogether, 67.9% of food items were determined to be sources of total dietary fibre, and for 60.0% of those food items, we were able to estimate the amount of (in)soluble dietary fibre, corresponding to 77.3%, 79.3%, and 81.7% in the total fibres intake for adolescents, adults, and elderly populations, respectively. For the remaining part of the dietary fibre, we were unable to distinguish between soluble and insoluble fractions. Missing data on the content of (in)soluble dietary fibre were observed in spices, nuts, and seeds, dried vegetables and fruits, processed meat, processed fish, ready-to-eat foods, baby foods, cakes, muffins and pastry, sauces, side dishes, food supplements, and meal replacements. Each food was allotted into one of the 101 food categories, of which 78 were included in FPQ.

For the purpose of identifying major sources of dietary fibre, foods were additionally categorised into categories developed within the Global Food Monitoring Initiative (GFMI) [30], adapted for use in Slovenia [20], as in previously published SI.Menu studies [20]. The following major food categories were included: fruit and vegetables; bread and bakery products; cereal and cereal products; convenience foods; snack foods; other.

2.4. Study Sample

Details about the exclusion criteria are previously explained [20]. Under- and overreporting were assessed using the cut-off points method initially described by Goldberg et al. [31] and further adapted by Black et al. [32]. The method is based on the ratio of reported daily energy intake and basic metabolic rate (BMR). BMR was calculated based on sex, age, body height, and body weight using the method described by Harris et al. [33] and adapted by Roza and Shizgal [34]. The calculated cut-off points for 24 h recalls for under- and overreporting were 0.41 and 2.46, respectively. Participants reporting energy intakes of less than 500 kcal were also excluded from the analyses. After exclusion of 97 subjects (incomplete or missing anthropometric data: $n = 12$; missing one of the 24 h recall data: $n = 36$; under/overreporting: $n = 49$), a study sample included 1248 subjects—468 adolescents, 364 adults, and 416 elderlies. Exclusion of subjects with missing anthropometric data was conducted because body height and weight were also needed to determine BMR, while subjects with only one 24 h recall were excluded because dietary fibre intakes were estimated with the use of the multiple source method [35], which only works with at least two 24 h recalls.

2.5. Data Analyses

Usual intake distributions per age group adjusted for within individual day-to-day variation were modelled with the multiple source method (MSM) [35], using both 24 h recalls and PFQ data. The method is characterised by a two-part shrinkage technique applied to residuals of two regression models—one for the daily food intake data and one for consumption frequency (FPQ). The shrunken residuals are back-transformed to their original scale, and the individual usual intake is obtained by multiplication of the frequency and amounts result. The MSM was used to correct dietary data for intra- and inter-personal variabilities. Assessment of dietary fibre intake was performed only for food categories which contained food items as sources of fibre. Age group, sex, and body mass index (BMI) were considered as covariates. After the MSM was applied, the usual daily intake of dietary fibre was calculated for each individual. The same approach was also applied separately for soluble and insoluble dietary fibre. The share of insoluble dietary fibre in the total dietary fibre content (calculated as the sum of soluble and insoluble dietary fibre) was further calculated for each individual. Similarly, daily energy intake was calculated for each subject and used for the calculation of total dietary fibre in grams per daily energy intake of 1000 Kcal.

Descriptive characteristics (mean, median, and proportions) are presented for age cohorts. Prevalence for inadequate daily intake of total dietary fibre was calculated using two cut-off values: 30 g (nationally adapted D-A-CH recommendation [13,14]) and 25 g (EFSA's guidance [2]). To account for the differences in participation rates in different population groups, population weighing was used in the presentation of these epidemiological results. Weighting was performed using iterative proportional fitting [36] for deviances in age and sex, with consideration of 2017 census data.

Logistic regression analysis was used to investigate associations between the prevalence of inadequate dietary fibre intakes with different socio-demographic and lifestyle determinants, separately for each population group. For this analysis, inadequate dietary fibre intake was determined using a cut-off value of 30 g/day, using individual usual dietary fibre intakes, estimated by MSM [35]. The following parameters were used in the models: sex, place of living, BMI, and IPAQ levels for all age groups, education, income for adults and elderly, and employment status for adults only. Logistic regression model parameters were estimated by the maximum likelihood method; odds ratios (ORs) with 95% confidence intervals (CIs) were calculated. All regression analyses were conducted on samples with excluded missing values (family net income: $n = 57$ (adults) and 40 (elderly); IPAQ: $n = 5$ (adolescents), 4 (adults), 6 (elderly)).

The relative contribution of different food categories to usual daily dietary fibre intake was calculated using previously described GFMI food categories. Separately for each age group, we summed contributions to daily fibre intake in all food groups and calculated their percentage (%) in the total dietary fibre intake in the age group. Further, we calculated mean usual dietary fibre intakes for three food categories, which were identified as major contributors to fibre intake (fruits and vegetables, bread and bakery products, cereal and cereal products). For each subject, food category contribution to daily fibre intake (in grams per day) was used for the calculation of age group mean and SD. To evaluate the differences in contributions of these food categories in the daily fibre intakes between age groups, we used one-way ANOVA, followed by Levene's test for homogeneity of variance. Multiple-comparison post hock test with Bonferroni correction was used in cases with equal variances; alternatively, Games–Howell adjustment was used.

All statistical analyses were performed using STATA version 15.1 (StataCorp LLC, College Station, TX, USA), while individual dietary intakes were estimated using the online tool MSM V1.0.1 (https://msm.dife.de/ (assessed on 26 October 2021); the Department of Epidemiology of the German Institute of Human Nutrition Potsdam-Rehbrücke, Germany). Differences were considered as significant at $p < 0.05$.

3. Results

Characteristics of the study sample are presented in Table 1. Sex ratios in male and female adolescents, adults, and the elderly were close to 1:1. Most subjects in all age groups lived in rural places of living (approx. 55%), while subjects living in urban places were in minority (slightly more than 25%). The remaining participants lived in intermediate places of living. Almost 70% of adults and more than 80% of the elderly had no university degree. The largest proportion of adults was employed (62.1%), 11.5% of subjects were unemployed, 8.8% were students, and 17.6% were retired. Body mass index (BMI) was estimated as normal in the majority of adolescents (64.3%), while in adults and the elderly, the percentages of overweight and obese subjects were dominant (59.3% of adults and 74% of the elderly).

Population-weighting was used to account for different participation rates in different population groups. Population weighted mean intake of total dietary fibre was 19.5, 20.9, and 22.4 g/day for adolescents, adults, and elderly populations, respectively (Table 2). The distribution of the mean usual intake of total dietary fibre in the age cohorts is presented in Supplementary Figure S1. With consideration of energy intake, women had a higher mean intake of dietary fibre in all three age groups (11.9, 13.7, and 14.4 g per 1000 kcal, respectively), while their actual mean daily dietary fibre intake was somewhat lower than in men.

Next, we examined a proportion of the population with insufficient dietary fibre intake, with consideration of nationally adapted D-A-CH recommended intake of at least 30 g of total dietary fibre daily (Table 2). While 90.6% of adolescents and 89.6% of adults had insufficient dietary fibre intake, the percentage of the elderly who daily consume less than 30 g/day of total dietary fibre was 83.9%. The trend was similar if the cut-off point was set to 25 g dietary fibre per day, which is an amount recommended by the EFSA to prevent constipation. The proportion of the population consuming less than 25 g/day of total dietary fibre was again the lowest among the elderly (70.8%), while among adults and adolescents, these proportions accounted for 75.5% and 83.0%. We also estimated the proportion of insoluble dietary fibre in people's diets. This proportion was estimated with consideration of all foods, for which data about the content of (in)soluble dietary fibre was available. Insoluble dietary fibres prevailed in all three age groups with small differences in the mean share of insoluble dietary fibre (63.9, 64.9, and 65.2, respectively; Table 2).

Multivariable logistic regression analyses were used to investigate associations of the prevalence of inadequate dietary fibre intakes (<30 g per day) with different socio-demographic and lifestyle determinants. Due to the study sampling approach, which provided representative samples separately for adolescents, adults, and elderlies, regression analysis was carried out separately for each population group. Most of the socio-economic and lifestyle variables were used in all the modes, except for education and income (only used in adults and elderlies) and employment (only used for adults). Results are presented in Table 3. It should be noted that in contrast to results in Table 2, herein, reported prevalence of insufficient dietary fibre intake is provided for the study sample, without population weighting. Sex was identified as a notable predictor of dietary fibre intake. In women, odd ratios for inadequate dietary fibre were significantly higher in adolescents (OR 2.00; 95%CI 1.05, 3.81) and adults (OR 2.78; 95%CI 1.21, 6.38). A contrary trend was observed in elderlies (OR 0.68, 95%CI: 0.36–1.29) but without a statistically significant difference (p = 0.237). In adults, overweight/obesity was found strong and significant predictor for lower OR (2.86; 95%CI 1.24, 6.59) for meeting recommended dietary fibre intake. Without statistical difference, a similar trend was observed in adolescents but not in elderlies. Surprisingly, higher family net income was found close to significant ($p < 0.1$) predictor for insufficient dietary fibre intake (OR 2.07, 95%CI: 0.85, 5.05) in adults but not in elderlies.

Table 1. Demographic and lifestyle characteristics of the study sample.

		Age Cohorts		
		Adolescents (10–17 Years)	**Adults (18–64 Years)**	**Elderly (65–74 Years)**
		n = 468	n = 364	n = 416
Age; years—mean (SD)		13.4 (2.37)	43.6 (13.81)	68.7 (2.7)
Place of living—n (%)	Rural	270 (57.7)	202 (55.5)	229 (55.1)
	Intermediate	76 (16.2)	56 (15.4)	71 (17.1)
	Urban	122 (26.1)	106 (29.1)	116 (27.9)
Sex—n (%)	Male	238 (50.9)	173 (47.5)	213 (51.2)
	Female	230 (49.1)	191 (52.5)	203 (48.8)
Education—n (%)	No university degree	n.a.	249 (68.4)	342 (82.2)
	University degree	n.a.	115 (31.6)	74 (17.8)
Family monthly net income—n (%)	Below average	n.a.	118 (38.4)	269 (71.5)
	Above average	n.a.	189 (61.6)	107 (28.5)
BMI—mean (SD)		21.0 (4.2)	26.7 (5.2)	28.4 (5.0)
n (%)	Normal	301 (64.6)	148 (40.7)	108 (26.0)
	Overweight and obese	167 (35.7)	216 (59.3)	308 (74.0)
IPAQ—n (%)	Low intensity	108 (23.3)	127 (35.3)	137 (33.4)
	Moderate	141 (30.5)	108 (30.0)	133 (32.4)
	High intensity	214 (46.2)	125 (34.7)	140 (34.2)
Employment status—n (%)	Employed	n.a.	226 (62.1)	n.a.
	Unemployed	n.a.	42 (11.5)	n.a.
	Student	n.a.	32 (8.8)	n.a.
	Retired	n.a.	64 (17.6)	n.a.

Notes: Body mass index (BMI) was considered to be normal when it was below 25 kg/m^2, except for adolescents, where sex/age-adjusted cut-off points [22,23] were used; International Physical Activity Questionnaire (IPAQ); standard deviation (SD); not applicable (n.a). Table adapted from [20].

Table 2. Population-weighted usual daily dietary fibre intake and proportion of the population with inadequate daily fibre intakes.

	Adolescents (10–17)			Adults (18–64)			Elderly (65–74)		
	All	**Male**	**Female**	**All**	**Male**	**Female**	**All**	**Male**	**Female**
Weighted * N (%)	150,674 (78.2)	75,580 (50.2)	73,094 (49.8)	1,302,132 (78.2)	670,464 (51.5)	631,668 (48.5)	212,793 (12.8)	100,247.5 (47.1)	112,545.5 (52.9)
Sample N (%)	468 (100)	238 (50.9)	230 (49.2)	364 (100)	173 (47.5)	191(52.5)	416 (100)	213 (51.2)	203 (48.8)
Intake of total dietary fibre									
Mean (95%CI) [g/day]	19.5 (18.8–20.2)	20.5 (19.6–21.5)	18.3 (17.3–19.3)	20.9 (20.1–21.7)	21.1 (19.9–22.3)	20.7 (19.6–21.8)	22.4 (20.5–24.3)	20.9 (19.6–22.1)	23.9 (20.7–27.0)
Median [g/day]	18.8	19.6	17.5	19.7	20.1	19.2	20.6	18.8	21.8
Mean (95%CI) [g per 1000 Kcal/day] **	11.2 (10.8–11.7)	10.6 (10.0–11.9)	11.9 (11.3–12.5)	12.2 (11.7–12.7)	10.7 (10.1–11.3)	13.7 (13.0–14.3)	13.2 (12.3–14.0)	11.8 (10.8–12.8)	14.4 (13.5–15.3)
Prevalence for inadequate daily intake of total dietary fibre ***									
<25 g/day	83.0 (78.4–86.7)	79.1 (71.6–85.0)	87.2 (81.6–91.3	75.5 (69.9–80.3)	74.2 (66.1–80.9)	76.8 (68.7–83.4)	70.8 (61.5–78.7)	77.6 (67.6–85.2)	64.6 (51.2–76.1)
<30 g/day	90.6 (87.1–93.1)	88.1 (82.4–92.2)	93.2 (88.9–95.9)	89.6 (85.6–92.6)	88.6 (82.5–92.7)	90.7 (84.7–94.6)	83.9 (74.0–90.5)	91.0 (85.1–94.7)	77.4 (61.0–88.2)
Share of insoluble dietary fibre intake as % of total daily fibre intake ****									
Mean (95%CI)	63.9 (63.3–64.5)	63.9 (63.0–64.7)	63.9 (63.1–64.6)	64.9 (64.4–65.4)	64.9 (64.1–65.7)	64.9 (64.2–65.5)	65.2 (63.9–66.4)	64.6 (62.3–66.9)	65.7 (65.0–66.4)
Median	63.9	63.7	64.2	65.3	65.4	65.3	65.3	64.6	65.6

Notes: * Number of citizens and respective share in the population in terms of age and sex cohorts (census data in 2017). ** Conversion factor into g/MJ is 0.239; *** Prevalence for inadequate daily intake of total dietary fibre was calculated using two cut-off vales: 30 g (nationally adapted D-A-CH recommendation) [13,14], and 25 g (EFSA's guidance [2]); **** Based on the available data for content of (in)soluble dietary fibre in foods (corresponding to 77.3%, 79.3% and 81.7% in the total fibres' intake for adolescents, adults, and elderly population, respectively).

Table 3. Association between prevalence of inadequate daily intake of dietary fibre (<30 g/day) and sex, place of living, education, family net income, BMI, IPAQ, employment for diferent age groups.

Variable		Adolescents (10–17 Years old)			Adults (18–64 Years old)			Elderly (65–74 Years old)		
		Prevalence (%)	Crude OR	Adjusted OR	Prevalence (%)	Crude OR	Adjusted OR	Prevalence (%)	Crude OR	Adjusted OR
Overall		422 (90.2)			329 (90.4)			367 (88.2)		
Sex	Male	208 (87.4)	1	1	152 (87.9)	1	1	191 (89.7)	1	1
	Female	214 (93.0)	1.93 (0.98–3.90)	2.00 (1.05–3.81)	177 (92.7)	1.75 (0.81–3.85)	2.78 (1.21–6.38)	176 (86.7)	0.75 (0.39–1.43)	0.68 (0.36–1.29)
Place of living	Village	240 (88.9)	1	1	184 (91.1)	1	1	200 (87.3)	1	1
	Town	70 (92.1)	1.46 (0.57–4.46)	1.45 (0.57–3.67)	51 (91.1)	1.00 (0.34–3.61)	0.68 (0.22–2.09)	65 (91.6)	1.57 (0.60–4.83)	1.89 (0.74–4.89)
	City	112 (91.8)	1.40 (0.64–3.23)	1.35 (0.63–2.89)	94 (88.7)	0.77 (0.33–1.82)	0.80 (0.33–1.90)	102 (87.9)	1.06 (0.51–2.26)	1.34 (0.64–2.82)
Education	No university degree		n.a.	n.a.	228 (91.6)	1	1	306 (89.5)	1	1
	University degree				101 (87.8)	0.66 (0.31–1.48)	0.76 (0.31–1.74)	61 (82.4)	0.55 (0.27–1.21)	0.48 (0.22–1.04)
Family net income	Below average		n.a.	n.a.	105 (89.0)	1	1	237 (88.1)	1	1
	Above average				170 (90.0)	1.10 (0.48–2.48)	2.07 (0.85–5.05)	91 (85.1)	0.77 (0.39–1.58)	0.87 (0.43–1.76)
BMI	Normal	267 (88.7)	1	1	130 (87.8)	1	1	99 (91.7)	1	1
	Overweight and obese	155 (92.8)	1.64 (0.80–3.59)	1.75 (0.87–3.51)	199 (92.1)	1.62 (0.76–3.48)	2.86 (1.24–6.59)	268 (87.0)	0.61 (0.25–1.33)	0.61 (0.27–1.35)
IPAQ	Low intensity	95 (88.0)	1	1	113 (89.0)	1	1	118 (86.1)	1	1
	Moderate	127 (90.1)	1.24 (0.51–2.99)	1.14 (0.50–2.58)	96 (88.9)	0.99 (0.40–2.47)	0.96 (0.38–2.45)	119 (89.5)	1.36 (0.62–3.10)	1.53 (0.72–3.25)
	High intensity	195 (91.1)	1.40 (0.61–3.14)	1.39 (0.65–2.97)	117 (93.6)	1.81 (0.68–5.18)	2.00 (0.72–5.58)	124 (88.6)	1.25 (0.58–2.73)	1.33 (0.63–2.79)
Employment	Employed		n.a.	n.a.	198 (87.6)	1	1		n.a.	n.a.
	Unemployed				39 (92.9)	1.84 (0.53–9.89)	3.49 (0.71–17.10)			
	Student				30 (93.8)	2.12 (0.49–19.25)	1.81 (0.37–8.94)			
	Retired				62 (98.9)	4.38 (1.05–38.89)	5.52 (1.13–27.07)			

Notes: Confidence interval (CI); Body mass index (BMI) was considered as normal below 25 kg/m^2, except for adolescents, where gender/age adjusted cut-off points [22,23] were used. Logistic regression analysis conducted on samples with excluded missing values (family net income: n = 57 (adults) and 40 (elderly); IPAQ (International Physical Activity Questionnaire): n = 5 (adolescents), 4 (adults), 6 (elderly)). Cut-off odds ratios calculated with threshold of 30 g dietary fibre daily; Following parameters were found significant: $p < 0.05$ sex (adolescents); $p < 0.05$ sex (adults), $p < 0.1$ income (adults), $p < 0.05$ BMI (adults); $p < 0.1$ education (elderly).

Since dietary fibre includes heterogeneous components found in different foods (mostly of plant origin), we explored collected data to figure out which food categories mostly contributed to the dietary fibre intake among different age groups. First, we calculated the relative contribution of different food categories to the total consumption of dietary fibre, separately for all three age groups (Supplementary Table S1). Fruits and vegetables, bread and bakery products, and cereal products were found as the main contributors to the total dietary fibre intake, and therefore subject to further analyses, enabling statistical comparison between age groups. Mean usual dietary fibre intakes from these three food categories are presented in Figure 1. Significant age-dependent trend was observed in fruit and vegetable products, which contributed 8.5 g, 10.6 g, and 11.6 g of dietary fibre in adolescents, adults, and elderlies, respectively. Somewhat similar trends were observed for bread and bakery products, where the difference was only significant between adolescents and elderlies (6.2 g vs. 6.9 g). However, a contrary trend with significant differences was observed in cereals and cereal products, which contributed to 2.9 g, 2.5 g, and 1.8 g of dietary fibre in adolescents, adults, and elderlies, respectively. We should note that in all three age groups subjects consumed more dietary fibre from vegetables (mostly fresh, and including mushrooms, legumes, pulses, and sprouts) than from fruits. The diversity of foods in the category of vegetables was high and included all types of foods usually consumed in the Mediterranean diet and also traditional products for our region. For example, fermented vegetables (such as sauerkraut and fermented turnip) are regularly consumed in Slovenia, especially in wintertime.

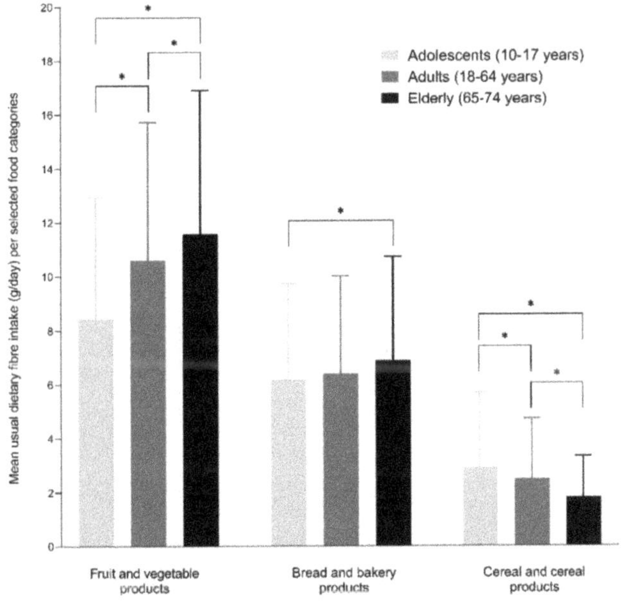

Figure 1. Mean usual dietary fibre intakes (g/day) from selected food categories among different age groups. The symbol * denotes a trend in the difference.

4. Discussion

Reported results indicate that the mean usual intake of total dietary fibre among the residents in Slovenia is notably lower than the national recommendations (30 g/day, adopted D-A-CH guidelines). Highest dietary fibre intake was observed in the elderly population (mean 22.4 g/day; 13.2 g/1000 kcal = 3.2 g/MJ), followed by adults (20.9 g; 12.2 g/1000 kcal = 2.9 g/MJ) and adolescents (19.5 g/day; 11.2 g/1000 kcal = 2.7 g/MJ).

In comparison, a previous study reported a slightly lower mean dietary fibre daily intake of 20.1 g in Slovenian adults in 1995/96 [15,16]. In adolescents, in the period 2003–2005, Fidler Mis et al. also reported a very similar intake of dietary fibre (2.6 and 2.8 g/MJ in boys and girls, respectively) [17]. However, it should be noted, that the latter study was conducted using a very different methodology (food frequency questionnaires without food recalls), and their reported intake of dietary fibre in grams per day is somewhat higher (28 and 31 g in boys and girls, respectively).

Regarding the proportion of insoluble dietary fibre, the results of the presented study showed that the largest proportion could be observed in the elderly (65.2%; 14.6 g/day), followed by adults (64.9%; 13.6 g/day) and adolescents (63.9%; 12.5 g). Although there is no dietary reference intake for soluble or insoluble fibre, many experts recommend about one-fourth (25%) of total dietary fibre intake—6 to 8 g per day—should come from soluble fibre, and the remaining 75% should come from insoluble fibre. Considering this recommendation, we can conclude that the consumption of soluble fibre (from foods such as oats, brussels sprouts, beans, peas, apples, oranges, nuts, and flax, and other seeds) is well satisfied by all the Slovenian populations (7.8 g, 7.3 g, and 7 g are daily consumed by the elderly, adults and adolescents, respectively). It seems that the more significant problem is with the consumption of insoluble fibre. No Slovenian population satisfies the daily dietary reference intake of 19 to 22 g of insoluble fibre (from foods such as wheat bran; vegetables such as green beans and dark leafy greens; root vegetables such as carrots, beets, and radish; fruit skins; and intact whole grains). However, with this interpretation, we need to emphasise that for about 20% of the total dietary fibre intake (23%, 21%, and 18% for adolescents, adults, and elderly population, respectively), data about the content of (in)soluble fibre were not available. This issue was addressed with a methodological approach—the proportion of insoluble fibre was not calculated as a percentage of total dietary fibre but as a percentage of the sum of soluble and insoluble fibre. This was possible under the assumption that the distribution of (in)soluble fibre in the diet from foods with missing data is comparable with the dietary contribution of foods, for which data were available.

Alarmingly, the proportion of participants with insufficient daily fibre intake (<30 g) was over 80% in all age groups. The results of our study are quite consistent with studies in other countries, which also reported inadequate dietary fibre intake [2,37–40]. The proportion of the population with insufficient dietary fibre intake was lowest in elderlies (especially females). This could be because they are more likely to suffer from chronic diseases than other populations and are, therefore, more concerned about their nutrition and some of them also follow specific diets [41]. This population also prepares more meals at home from raw ingredients, such as fruits, vegetables, and grains, which also contain more dietary fibre [41].

An interesting finding of our study is that in comparison with men, adolescent and adult women were more likely to have insufficient dietary fibre intake, while the contrary was observed in elderlies. Conversely, in the Irish elderly population [39], women consumed less dietary fibre than men, but their intake of dietary fibre per energy intake was higher than in men, which was also observed in our study. We should also mention a Dutch study on adults (19–69 years) [42], which only reported dietary fibre per energy intake and also observed higher fibre intakes in women. Nevertheless, the majority of the population is far from meeting the recommendations for daily dietary fibre intake, regardless of gender, which is in line with previous reports from Slovenia [43] and elsewhere [39,44].

As mentioned, according to the opinion of the EFSA's NDA Panel, daily intake of 25 g of dietary fibre is adequate for normal laxation [2], but this lower threshold value is also not met by 83.0% of adolescents, 75.5% of adults, and 70.8% of the elderly population. We should also mention that the European Society for Paediatric Gastroenterology Hepatology and Nutrition (ESPGHAN) also recommends that during adolescence the dietary fibre intake should be gradually increased to reach a dietary fibre intake of 30 g/day [45], which is far from the current status in Slovenia. Results of our study highlighted that adolescents should be encouraged to swipe white bread for wholemeal varieties, while comparison with other age groups (Figure 1) also indicates the high potential of vegetables and fruits to increase intake of dietary fibre. For example, in adolescents, the contribution of fruits and vegetables to daily dietary fibre intake is 3.1 g lower than in elderlies. At the same time, adolescents consume approximately twice more amount of fibre from white bread than from brown bread. On the other hand, other cereal products were more important sources of dietary fibre for adolescents than for adults and elderlies. Other studies also highlighted that adolescents more commonly consume snacks (usually high energy foods), and lack main meals (most commonly breakfast) or have irregular meals, and have low consumption of fruits and vegetables [46]. To improve intake of dietary fibre, consuming more of other wholesome, fibre-rich foods should be supported in adolescents, as well as in other populations.

In a UK study [40], an investigation of the associations between dietary fibre intakes from different food sources with measures of body composition was performed. Comparison of our results with this study showed that in both countries, the percentage of the adult population which meets the recommended intake for total dietary fibre of more than 30 g per day is about 10% (9% in the UK and 10.4% in Slovenia). In both studies, the main total dietary fibre contributions come from grains, fruits, and vegetables. Vegetables and fruits are good sources of non-starch polysaccharides (NSP) (hemicelluloses and pectin), resistant oligosaccharides, and resistant starch. On the other hand, nuts and seeds—sources of cellulose, hemicelluloses, and also resistant starch [47]—were also notable sources of dietary fibre in Slovenian adults (4.1%), while this was not the case in the UK study (0.7%). This could be explained by the fact that nuts and seeds are popular snacks in Slovenia. However, such snacks are often high in added fat and salt; therefore, healthy alternatives should be encouraged. In both studies, it appeared that the main sources of total dietary fibre from the category of grains are white bread and other grain sources, which are not necessarily whole grain. The adolescents could benefit from the reformulation of breakfast cereals to contain more dietary fibre, as this food group is more popular in this population group than in adults or elderlies. The UK study observed that higher whole-grain and non-whole-grain cereal dietary fibre intakes are associated with lower BMI. However, dietary fibre from whole-grain sources (but not from non-whole-grain sources) was associated with lower waist circumference and percentage of body fat, suggesting that some benefits of dietary fibre from whole grains might also be attributable to other nutrients and non-nutrient components (e.g., phytochemicals) in whole grains. Several other studies also associated insufficient dietary fibre intake with obesity, but results are not always conclusive [38,48,49]. Whole grains include bran as a source of NSPs, and they provide lignin, resistant oligosaccharides, and resistant starch. The SI.Menu data showed that Slovenian residents consume a very wide variety of different whole-grain cereals, including wheat, corn, buckwheat, millet, etc.

In 2017, Stephen et al. published a comprehensive review of dietary fibre intake in European countries, considering data from nearly 140,000 individuals covering a broad age range from early childhood to the elderly [10]. Overall, daily dietary fibre intake for adults living in European countries was estimated to be 18–24 g for men and 16–20 g for women, which is comparable with our results (21.1 g for men and 20.7 g for women). This review highlighted bread and grain products as the largest source of total dietary fibre. Additionally, in our study, a combination of bread, bakery products, and other cereal products were a major source of dietary fibre (45.0%, 40.0%, and 39.0% for adolescents, adults, and elderly, respectively), followed by vegetables and fruits (Supplementary Table S1).

In order to improve the situation in Slovenia, promoting whole-meal options and encouraging food reformulations to increase dietary fibre content could result in higher overall dietary fibre intake in all population groups. Large meta-analyses covering over 1.7 million subjects [50] showed that with every 10 g increase in dietary fibre intake per day, the risk of premature death was lowered by 11%. Considering this, the target recommendation of daily intake of 30 g dietary fibre is very relevant, but we are still far from this target. Our study showed that the dietary fibre intake is very comparable with the situation two decades ago, despite the fact that we have observed notable changes in people's behaviours [51], and also in the food supply. These changes could result in a lower intake of dietary fibre. Nowadays, food stores are packed with tens of thousands of highly processed foods [52], which became major food sources in different population groups. It is becoming clear that this trend will not be turned around, making reformulation policies even more important. Furthermore, in adolescents, dietary fibre intake could be also improved by careful planning of school meals to include more whole foods of plant origin and refined cereal alternatives, such as whole-meal bread and cereals.

The strength of the study lies in that it was built on the comprehensive nationally representative dataset, collected using internationally harmonised European Union Menu Methodology for food consumption surveys. This cross-sectional study included randomised selection of residents aged 10–74 years. To enable the provision of insights into more vulnerable population groups, sampling was performed separately for adolescents, adults, and the elderly population. Another strength is that data were collected over 12 months to cover all calendar seasons, with quota sampling for each quarter of the year. This means that nationally representative sampling was also achieved during seasons. Furthermore, estimated usual intake was estimated in both 24 h recalls and food propensity questionnaire data. Some limitations should be also mentioned. While we did our best to increase the participation rate, 38% of participants did not respond to our invitation. However, the observed participation rate is common in national dietary studies with randomised sampling, because participation in the survey is voluntary. We also did not use any financial compensations, which could, on one the one hand, increase participation rate, but on the other hand, such an approach would have different effects on different economic groups. However, after a successful face-to-face visit, subjects received a small gift (glass water bottle, bib, headphones, umbrella, or ice bags), as a gratuity gesture and to stimulate participation in a second 24 h dietary recall. Subjects were able to select between these incentives, which were very well accepted. Participation was also affected by season. A lower participation rate was observed during summer when people are come commonly out of home (holidays), and it was highest during the winter season. We should, however, mention that interviewer had to achieve five contacts with the selected subjects. If none were successful, the subject was coded as a non-respondent. We did not use within a household or other substitutions, which would increase the participation rate. In the case of soft refusals, the subject was contacted one more time by a different interviewer. An important limitation of the study is also that national food composition tables did not enable us to estimate fibre content in all of the reported foods. Additional food composition databases were, therefore, used. As already mentioned, the estimation of the proportion of the insoluble dietary fibre (as % of total fibre) was carried out only with consideration of foods for which we were able to estimate the amount of soluble and insoluble dietary fibre. Another limitation is that we did not include the consumption of food supplements with dietary fibre (e.g., *Plantago psyllium*, some prebiotics, mushrooms, algae), which can also present an important source of dietary fibre. We should mention that the penetration of the consumption of such supplements in Slovenia is low. About 10% of Slovenian adults reported at least monthly consumption of probiotic and prebiotic formulations, and 7% are using herbal formulations and plant extracts, while the use of algae and mushrooms formulations is about 2% [41].

5. Conclusions

In Slovenia, as in most other European countries, the mean daily intake of the total dietary fibre is much lower than recommended. Sex and BMI were found as the strongest determinants for insufficient daily dietary fibre intake in specific population groups. The main food groups contributing to dietary fibre intake in all population groups were fruits and vegetables, bread, and cereal products. While for the elderly population, there are no previous data for comparison, dietary fibre intake in adolescents and the adult population seems comparable with results of studies, conducted about two decades ago. However, major methodological differences between the studies limit these comparisons. The results of this study indicate that more efficient approaches are needed to increase dietary fibre intake and to meet recommendations, which should exceed food reformulation activities. Considerably higher intake of dietary fibre could be achieved by encouraging consumers for whole-meal options in bread and cereal products and with increased overall consumption of plant-based foods.

Supplementary Materials: The following are available online at https://www.mdpi.com/article/10.3390/nu13113826/s1, Supplementary Figure S1: Histogram of distribution of daily dietary fibre intakes for different age groups (adolescents: age 10–17 years; adults: age 18–64 years; elderly: age 65–74 years), Supplementary Table S1: Relative contribution of selected food categories to usual daily dietary fibre intake among different age groups (% of total dietary fibre intake).

Author Contributions: B.K.S. wrote the first draft of the manuscript and was responsible for information technology in the SI.Menu study, and H.H. for the preparation of the database and for data analyses. B.F., M.H. and S.K. conducted food-matching to estimate the content of dietary fibre. M.G. and U.B. were responsible for SI.Menu study design and food consumption data. I.P. was responsible for assuring the set-up of the analyses, collaborated in the data analyses, and reviewed the manuscript. Ž.L. contributed to the preparation of the final draft and categorisation of foods, B.F., J.B. and M.K. contributed a critical assessment of dietary fibre data from different FCDBs and a revision of the final draft. K.Ž., A.K. and E.V. made a revision of the final draft and prepared the submission. All authors have read and agreed to the published version of the manuscript.

Funding: This study was conducted within research programs P3-0395 'Nutrition and Public Health' and P2-0098 'Computer Systems and Structures', funded by the Slovenian Research Agency. The original SI.Menu 2017/18 national food consumption survey was funded by the European Food Safety Authority (EFSA Contract No. OC/EFSA/DATA/2014/02-LOT2-CT03), Ministry of Health of the Republic of Slovenia, and Slovenian Research Agency.

Institutional Review Board Statement: The study was conducted according to the guidelines of the Declaration of Helsinki and approved by the National Medical Ethics Committee, Ljubljana, Slovenia (KME 53/07/16; Approval No. 0120-337/2016 issued on 19 July 2016).

Informed Consent Statement: Informed consent was obtained from all subjects involved in the study.

Data Availability Statement: The data presented in this study are available on request from the corresponding author.

Acknowledgments: We acknowledge the support of all SI.Menu project partners and their researchers, particularly Nataša Delfar, Metka Zaletel, Darja Lavtar, Ada Hočevar Grom, and Ivan Eržen (National Institute of Public Health, Ljubljana, Slovenia); Petra Golja (Group of Anthropology at Biotechnical Faculty of University of Ljubljana, Slovenia); Stojan Kostanjevec (the University of Ljubljana, Faculty of Education, Ljubljana, Slovenia; Majda Pajnkihar (the University of Maribor, Faculty of Health Sciences, Maribor, Slovenia); Tamara Poklar Vatovec (the University of Primorska, Faculty of Health Sciences, Koper, Slovenia); and Evgen Benedik (University Medical Centre Ljubljana, University Children's Hospital, Ljubljana, Slovenia). We also acknowledge the support of Veronika Belec and her team and interviewers at GfK Research Institute (Slovenia) for conducting the survey fieldwork.

Conflicts of Interest: The authors declare no conflict of interest. The funders had no role in the design of the study; in the collection, analyses, or interpretation of data; in the writing of the manuscript; or in the decision to publish the results. I.P. has led and participated in various other research projects in the area of nutrition, public health, and food technology, which were (co)funded by the Slovenian Research Agency, Ministry of Health of the Republic of Slovenia, the Ministry of Agriculture, Forestry and Food of the Republic of Slovenia, and in the case of specific applied research projects, also by food businesses. I.P., M.G., and U.B. are members of a workgroup for food reformulation at the Ministry of Health of the Republic of Slovenia.

References

1. Codex Alimentarius Commission. Guidelines on nutrition labelling CAC/GL 2–1985 as last amended 2017. In *Joint FAO/WHO Food Standards Programme*; Secretariat of the Codex Alimentarius Commission, FAO: Rome, Italy; p. 10.
2. EFSA Panel on Dietetic Products. Scientific Opinion on Dietary Reference Values for carbohydrates and dietary fibre. *EFSA J.* **2010**, *8*. [CrossRef]
3. Hervik, A.K.; Svihus, B. The Role of Fibre in Energy Balance. *J. Nutr. Metab.* **2019**, *2019*. [CrossRef] [PubMed]
4. Regulation (EU) No 1169/2011 of the European Parliament and of the Council of 25 October 2011 on the Provision of Food Information to Consumers. Available online: https://eur-lex.europa.eu/legal-content/EN/AUTO/?uri=CELEX:02011R1169-20180101&qid=1547797296601 (accessed on 28 December 2020).
5. Wan, Y.; Wang, F.; Yuan, J.; Li, J.; Jiang, D.; Zhang, J.; Li, H.; Wang, R.; Tang, J.; Huang, T.; et al. Effects of dietary fat on gut microbiota and faecal metabolites, and their relationship with cardiometabolic risk factors: A 6-month randomised controlled-feeding trial. *Gut* **2019**, *68*, 1417–1429. [CrossRef] [PubMed]
6. Linus Pauling Institute, Micronutrient Information Center. Fiber. Available online: https://lpi.oregonstate.edu/mic/other-nutrients/fiber (accessed on 26 October 2021).
7. Cani, P.D. Human gut microbiome: Hopes, threats and promises. *Gut* **2018**, *67*, 1716–1725. [CrossRef]
8. Cronin, P.; Joyce, S.; O'Toole, P.; O'Connor, E. Dietary Fibre Modulates the Gut Microbiota. *Nutrients* **2021**, *13*, 1655. [CrossRef]
9. Barber, T.M.; Kabisch, S.; Pfeiffer, A.F.H.; Weickert, M.O. The Health Benefits of Dietary Fibre. *Nutrients* **2020**, *12*, 3209. [CrossRef]
10. Zhao, L.; Zhang, F.; Ding, X.; Wu, G.; Lam, Y.Y.; Wang, X.; Fu, H.; Xue, X.; Lu, C.; Ma, J.; et al. Gut bacteria selectively promoted by dietary fibers alleviate type 2 diabetes. *Science* **2018**, *359*, 1151–1156. [CrossRef]
11. Stephen, A.M.; Champ, M.M.-J.; Cloran, S.J.; Fleith, M.; Van Lieshout, L.; Mejborn, H.; Burley, V.J. Dietary fibre in Europe: Current state of knowledge on definitions, sources, recommendations, intakes and relationships to health. *Nutr. Res. Rev.* **2017**, *30*, 149–190. [CrossRef]
12. Mehta, N.; Ahlawat, S.S.; Sharma, D.P.; Dabur, R. Novel trends in development of dietary fiber rich meat products—A critical review. *J. Food Sci. Technol.* **2013**, *52*, 633–647. [CrossRef]
13. Doets, E.L.; de Wit, L.S.; Dhonukshe-Rutten, R.A.M.; Cavelaars, A.E.J.M.; Raats, M.M.; Timotijevic, L.; Brzozowska, A.; Wijnhoven, T.M.A.; Pavlovic, M.; Totland, T.H.; et al. Current micronutrient recommendations in Europe: Towards understanding their differences and similarities. *Eur. J. Nutr.* **2008**, *47*, 17–40. [CrossRef]
14. NIJZ. Referenčne Vrednosti za Energijski vnos ter vnos Hranil. Tabelarična Priporočila za Otroke (od 1. leta Starosti Naprej), Mladostnike, Odrasle, Starejše Odrasle, Nosečnice ter Doječe Matere. Dopolnjena Izdaja. 2020. Available online: https://www.nijz.si/sl/referencne-vrednosti-za-energijski-vnos-ter-vnos-hranil (accessed on 26 October 2021).
15. Koch, V. Prehrambene Navade Odraslih Slovencev z Vidika Varovanja Zdravja. Doktorska Disertacija, Biotehnična fakulteta, Oddelek za živilstvo, Ljubljana, Slovenia. (In Slovene)
16. Verena Koch, V.; Pokorn, P. Comparison of nutritional habits among various adult age groups in Slovenia. *Nutr. Res.* **1999**, *19*, 1153–1164. [CrossRef]
17. Mis, N.F.; Kobe, H.; Štimec, M.; Mis, N.F.; Timec, M. Dietary Intake of Macro- and Micronutrients in Slovenian Adolescents: Comparison with Reference Values. *Ann. Nutr. Metab.* **2012**, *61*, 305–313. [CrossRef]
18. EFSA. Guidance on the EU Menu methodology. *EFSA J.* **2014**, *12*, 3944.
19. National Institute of Public Health (NIJZ); Gregorič, M.; Blaznik, U.; Delfar, N.; Zaletel, M.; Lavtar, D.; Seljak, B.K.; Golja, P.; Kotnik, K.Z.; Pravst, I.; et al. Slovenian national food consumption survey in adolescents, adults and elderly. *EFSA Support. Publ.* **2019**, *16*. [CrossRef]
20. Zupanič, N.; Hristov, H.; Gregorič, M.; Blaznik, U.; Delfar, N.; Koroušić Seljak, B.; Ding, E.L.; Fidler Mis, N.; Pravst, I. Total and Free Sugars Consumption in a Slovenian Population Representative Sample. *Nutrients* **2020**, *12*, 1729. [CrossRef]
21. Korošec, M.; Golob, T.; Bertoncelj, J.; Stibilj, V.; Seljak, B.K. The Slovenian food composition database. *Food Chem.* **2013**, *140*, 495–499. [CrossRef]
22. de Onis, M.; Onyango, A.; Borghi, E.; Siyam, A.; Blössner, M.; Lutter, C. Worldwide implementation of the WHO Child Growth Standards. *Public Health Nutr.* **2012**, *15*, 1603–1610. [CrossRef]
23. De Onis, M.; Onyango, A.W.; Borghi, E.; Siyam, A.; Nishida, C.; Siekmann, J. Development of a WHO growth reference for school-aged children and adolescents. *Bull. World Health Organ.* **2007**, *85*, 660–667. [CrossRef]
24. Lee, P.H.; Macfarlane, D.J.; Lam, T.; Stewart, S.M. Validity of the international physical activity questionnaire short form (IPAQ-SF): A systematic review. *Int. J. Behav. Nutr. Phys. Act.* **2011**, *8*, 115. [CrossRef]
25. Bodner-Montville, J.; Ahuja, J.K.; Ingwersen, L.A.; Haggerty, E.S.; Enns, C.W.; Perloff, B.P. USDA Food and Nutrient Database for Dietary Studies: Released on the web. *J. Food Compos. Anal.* **2006**, *19*, S100–S107. [CrossRef]
26. Vásquez-Caicedo, A.L.; Bell, S.; Hartmann, B. Report on Collection of Rules on Use of Recipe Calculation Procedures Including the Use of Yield and Retention Factors for Imputing Nutrient Values for Composite Foods. EuroFIR Technical Report. Available online: https://www.eurofir.org/wp-content/uploads/2014/05/6.-Report-on-collection-of-rules-on-use-of-recipe-calculation-procedures-including-the-use-of-yield-and-retention-factors-for-imputing-nutrient-values-for-composite-foods.pdf (accessed on 26 October 2021).
27. Fineli Food Composition Database. Available online: https://fineli.fi/fineli/en/avoin-data (accessed on 26 October 2021).
28. Danish Food Composition Database. Available online: https://frida.fooddata.dk (accessed on 26 October 2021).

29. Souci, S.W.; Fachmann, W.; Kraut, H. *Food Composition and Nutrition Tables=Die Zusammensetzung der Lebensmittel Nährwert-Tabellen = La Composition des Aliments TABLEAUX des valeurs Nutritives*; 8th rev. and Completed ed; an Imprint of Wissenschaftliche Verlagsgesellschaft; Andersen, G., Ed.; MedPharmScientific Publishers: Stuttgart, Germany, 2016.
30. Dunford, E.; Neal, B.; Macgregor, G.; Czernichow, S.; Ni Mhurchu, C.; Food Monitoring Group. International collaborative project to compare and track the nutritional composition of fast foods. *BMC Public Health* **2012**, *12*, 559.
31. Goldberg, G.R.; Black, A.E.; Jebb, S.A.; Cole, T.J.; Murgatroyd, P.R.; Coward, W.A.; Prentice, A.M. Critical evaluation of energy intake data using fundamental principles of energy physiology: 1. Derivation of cut-off limits to identify under-recording. *Eur. J. Clin. Nutr.* **1991**, *45*, 569–581. [PubMed]
32. Black, A.E. Critical evaluation of energy intake using the Goldberg cut-off for energy intake: Basal metabolic rate. A practical guide to its calculation, use and limitations. *Int. J. Obes.* **2000**, *24*, 1119–1130. [CrossRef] [PubMed]
33. Harris, J.A.; Benedict, F.G. A Biometric Study of Human Basal Metabolism. *Proc. Natl. Acad. Sci. USA* **1918**, *4*, 370–373. [CrossRef]
34. Roza, A.M.; Shizgal, H.M. The Harris Benedict equation reevaluated: Resting energy requirements and the body cell mass. *Am. J. Clin. Nutr.* **1984**, *40*, 168–182. [CrossRef]
35. Harttig, U.; Haubrock, J.; Knüppel, S.; Boeing, H. The MSM program: Web-based statistics package for estimating usual die-tary intake using the Multiple Source Method. *Eur. J. Clin. Nutr.* **2011**, *65*, S87–S91. [CrossRef]
36. Kolenikov, S. Calibrating Survey Data using Iterative Proportional Fitting (Raking). *Stata J.* **2014**, *14*, 22–59. [CrossRef]
37. Fayet-Moore, F.; Cassettari, T.; Tuck, K.; McConnell, A.; Petocz, P. Dietary Fibre Intake in Australia. Paper I: Associations with Demographic, Socio-Economic, and Anthropometric Factors. *Nutrients* **2018**, *10*, 599. [CrossRef]
38. Gonzalez-Rodriguez, L.G.; Perea-Sanchez, J.M.; Aranceta-Bartrina, J.; Gil, A.; Gonzalez-Gross, M.; Serra-Majem, L.; Varela-Moreiras, G.; Ortega, R.M. Intake and Dietary Food Sources of Fibre in Spain: Differences with Regard to the Prevalence of Excess Body Weight and Abdominal Obesity in Adults of the ANIBES Study. *Nutrients* **2017**, *9*, 326. [CrossRef]
39. Kehoe, L.; Walton, J.; McNulty, B.; Nugent, A.; Flynn, A. Energy, Macronutrients, Dietary Fibre and Salt Intakes in Older Adults in Ireland: Key Sources and Compliance with Recommendations. *Nutrients* **2021**, *13*, 876. [CrossRef]
40. Gibson, R.; Eriksen, R.; Chambers, E.; Gao, H.; Aresu, M.; Heard, A.; Chan, Q.; Elliott, P.; Frost, G. Intakes and Food Sources of Dietary Fibre and Their Associations with Measures of Body Composition and Inflammation in UK Adults: Cross-Sectional Analysis of the Airwave Health Monitoring Study. *Nutrients* **2019**, *11*, 1839. [CrossRef]
41. Gregorič, M.; Blaznik, U.; Turk, V.F.; Hočevar-Grom, A.; Delfar, N.; Korošec, A.; Lavtar, D.; Zaletel, M.; Koroušić-Seljak, B.; Golja, P.; et al. *Različni Vidiki Prehranjevanja Prebivalcev Slovenije: V Starosti od 3 Mesecev do 74 let*; Nacionalni Inštitut za Javno Zdravje: Ljubljana, Slovenia, 2019; ISBN 978-961-6945-01-1. (In Slovene)
42. Dinnissen, C.; Ocké, M.; Buurma-Rethans, E.; van Rossum, C. Dietary Changes among Adults in The Netherlands in the Period 2007–2010 and 2012–2016. Results from Two Cross-Sectional National Food Consumption Surveys. *Nutrients* **2021**, *13*, 1520. [CrossRef]
43. Gregorič, M.; Kotnik, K.Z.; Pigac, I.; Blenkuš, M.G. A Web-Based 24-H Dietary Recall Could Be a Valid Tool for the Indicative Assessment of Dietary Intake in Older Adults Living in Slovenia. *Nutrients* **2019**, *11*, 2234. [CrossRef]
44. King, D.E.; Mainous, A.G.; Lambourne, C.A. Trends in dietary fibre intake in the United States, 1999–2008. *J. Acad. Nutr. Diet.* **2012**, *112*, 642–648. [CrossRef]
45. Aggett, P.J.; Agostoni, C.; Axelsson, I.; Edwards, C.A.; Goulet, O.; Hernell, O.; Koletzko, B.; Lafeber, H.N.; Micheli, J.L.; Michaelsen, K.F.; et al. Nondigestible Carbohydrates in the Diets of Infants and Young Children: A Commentary by the ES-PGHAN Committee on Nutrition. *J. Pediatric Gastroenterol. Nutr.* **2003**, *36*, 329–337. [CrossRef]
46. Gherasim, A.; Arhire, L.I.; Niță, O.; Popa, A.D.; Graur, M.; Mihalache, L. The relationship between lifestyle components and dietary patterns. *Proc. Nutr. Soc.* **2020**, *79*, 311–323. [CrossRef]
47. Fuentes-Zaragoza, E.; Sánchez-Zapata, E.; Sendra, E.; Sayas-Barberá, E.; Navarro, C.; Fernández-López, J.; Pérez-Alvarez, J.A. Resistant starch as prebiotic: A review. *Starch-Stärke* **2011**, *63*, 406–415. [CrossRef]
48. Bozzetto, L.; Costabile, G.; Della Pepa, G.; Ciciola, P.; Vetrani, C.; Vitale, M.; Rivellese, A.A.; Annuzzi, G. Dietary Fibre as a Unifying Remedy for the Whole Spectrum of Obesity-Associated Cardiovascular Risk. *Nutrients* **2018**, *10*, 943. [CrossRef]
49. Kendall, C.W.; Esfahani, A.; Jenkins, D.J. The link between dietary fibre and human health. *Food Hydrocoll.* **2010**, *24*, 42–48. [CrossRef]
50. Liu, L.; Wang, S.; Liu, J. Fiber consumption and all-cause, cardiovascular, and cancer mortalities: A systematic review and meta-analysis of cohort studies. *Mol. Nutr. Food Res.* **2014**, *59*, 139–146. [CrossRef]
51. National Institute of Public Health. Prehrambene Navade Odraslih Prebivalcev Slovenije. Available online: https://www.nijz.si/sl/publikacije/prehrambene-navade-odraslih-prebivalcev-slovenije (accessed on 26 September 2021). (In Slovene)
52. Hafner, E.; Pravst, I. Evaluation of the Ability of Nutri-Score to Discriminate the Nutritional Quality of Prepacked Foods Using a Sale-Weighting Approach. *Foods* **2021**, *10*, 1689. [CrossRef] [PubMed]

nutrients

MDPI

Article

Vitamin D Intake in Slovenian Adolescents, Adults, and the Elderly Population

Maša Hribar [1,2], Hristo Hristov [1], Živa Lavriša [1], Barbara Koroušić Seljak [3], Matej Gregorič [4], Urška Blaznik [4], Katja Žmitek [1,5] and Igor Pravst [1,2,5,*]

1 Nutrition Institute, Tržaška Cesta 40, SI-1000 Ljubljana, Slovenia; masa.hribar@nutris.org (M.H.); hristo.hristov@nutris.org (H.H.); ziva.lavrisa@nutris.org (Ž.L.); katja.zmitek@vist.si (K.Ž.)
2 Biotechnical Faculty, University of Ljubljana, Jamnikarjeva 101, SI-1000 Ljubljana, Slovenia
3 Computer Systems Department, Jozef Stefan Institute, SI-1000 Ljubljana, Slovenia; barbara.korousic@ijs.si
4 National Institute of Public Health, Trubarjeva 2, SI-1000 Ljubljana, Slovenia; matej.gregoric@nijz.si (M.G.); urska.blaznik@nijz.si (U.B.)
5 VIST–Higher School of Applied Sciences, Gerbičeva cesta 51A, SI-1000 Ljubljana, Slovenia
* Correspondence: igor.pravst@nutris.org

Abstract: Vitamin D is involved in calcium and phosphorus metabolism, and is vital for numerous bodily functions. In the absence of sufficient UV-B light-induced skin biosynthesis, dietary intake becomes the most important source of vitamin D. In the absence of biosynthesis, the recommended dietary vitamin D intake is 10–20 μg/day. Major contributors to dietary vitamin D intake are the few foods naturally containing vitamin D (i.e., fish), enriched foods, and supplements. The present study aimed to estimate the vitamin D intake in Slovenia, to identify food groups that notably contribute to vitamin D intake, and to predict the effects of hypothetical mandatory milk fortification. This study was conducted using data collected by the national cross-sectional food consumption survey (SI.Menu) in adolescents (n = 468; 10–17 years), adults (n = 364; 18–64 years), and the elderly (n = 416; 65–74 years). Data collection was carried out between March 2017 and April 2018 using the EU Menu Methodology, which included two 24-h recalls, and a food propensity questionnaire. Very low vitamin D intakes were found; many did not even meet the threshold for very low vitamin D intake (2.5 μg/day). Mean daily vitamin D intake was 2.7, 2.9, and 2.5 μg in adolescents, adults, and the elderly, respectively. Daily energy intake was found to be a significant predictor of vitamin D intake in all population groups. In adolescents and adults, sex was also found to be a significant predictor, with higher vitamin D intake in males. The study results explained the previously reported high prevalence of vitamin D deficiency in Slovenia. An efficient policy approach is required to address the risk of vitamin D deficiency, particularly in vulnerable populations.

Keywords: vitamin D; Slovenia; dietary intake; EU Menu; food propensity questionnaire; 24 h recall

Citation: Hribar, M.; Hristov, H.; Lavriša, Ž.; Koroušić Seljak, B.; Gregorič, M.; Blaznik, U.; Žmitek, K.; Pravst, I. Vitamin D Intake in Slovenian Adolescents, Adults, and the Elderly Population. *Nutrients* **2021**, *13*, 3528. https://doi.org/10.3390/nu13103528

Academic Editor: Edgard Delvin

Received: 10 September 2021
Accepted: 5 October 2021
Published: 8 October 2021

1. Introduction

Vitamin D (VitD) deficiency and insufficiency are global health issues, posing a major public health risk [1,2]. Poor VitD status is connected with skeletal and non-skeletal health issues, including the functioning of the immune system [3–7]. Maintenance of an optimal VitD status is therefore of the utmost importance. The most common biomarker of VitD status is the serum concentration of 25-hydroxyvitamin D (25(OH)D). Optimal VitD status can be achieved by biosynthesis in the human skin when exposed to sufficient ultraviolet B light radiation (UVB) [8]. In the absence of or with insufficient UVB skin irradiation, VitD becomes an essential nutrient, and needs to be provided by dietary intake. Due to the changing solar zenith angle in European countries situated above the latitude of 35° N [9] (including Slovenia; 45–46° N [10]), the intensity of UVB light in wintertime is not sufficient to induce cutaneous synthesis of VitD [11]. In such cases, dietary intake becomes a major source of VitD [12,13]. The term *Vitamin D* usually covers two fat-soluble vitamers, namely ergocalciferol (D2) and cholecalciferol (D3). While D3 is obtained by endogenous synthesis via UVB exposure and diet (animal sources), the source of D2 is solely from the diet (from fungi). Overall, dietary VitD intake is obtained through food naturally rich in VitD, VitD-enriched food, or prescribed and over-the-counter medicines and supplements. The majority of foods are a naturally poor source of VitD, with the exception of oil-rich fish and eggs [12]. These VitD-rich foods tend to be consumed quite rarely or in smaller quantities [14], whereas poorer VitD sources (such as meat and meat products, milk products, and animal fats) are consumed regularly, and therefore constitute the majority of dietary VitD intake [15]. In addition to naturally occurring VitD, its content can be enhanced during production or processing; this is achieved through bioaddition and/or fortification. The term bioaddition is used to describe processes that are used during food production to enhance naturally occurring VitD, by feeding the animal a VitD-rich diet (used in meat and eggs) or by UVB irradiation of mushrooms or yeast [16]. The term fortification usually describes a process where either D2 or D3 is added near or at the end of food processing [16]. VitD fortification was initially used in cow's milk to prevent rickets in Northern America and Europe [17]. In addition to dairy products, other types of foods are now also used as a vehicle of fortification, such as orange juice, cereal-based foods, and infant formulas [15]. Fortification can be either voluntary or mandatory policy. While a few countries, such as the USA, Canada, Australia, and Finland, have implemented regulated food fortification to increase dietary VitD intake [18], most countries have voluntary VitD fortification (also called "enrichment") [18–22]. As in most European Union (EU) countries, this is also the case in Slovenia, where there are no recommendations regarding VitD fortification. The choice of fortification is in the hands of food manufacturers, and later is the choice of the end consumers. An important source of VitD intake is also supplementation with prescribed and over-the-counter medicines and food supplements. VitD is available both in multivitamin/multimineral and single-component products. These typically contain daily dosages of up to about 100 μg [16,23]. It should be noted that in Slovenia, VitD is routinely prescribed to children during the first year (10 μg), and further supplementation is recommended until the age of 18 years [24], while there is no official supplementation recommendation for the general adult population. However, in Slovenia during the COVID-19 pandemic, the typical non-medical prescribed VitD supplementation dosage in adults was 25 μg per day [23].

The low VitD status across the world is alarming [1,2,25,26], and this is also the case in Slovenia [11]. A previous study representative of Slovenia showed that during the extended winter period (November–April), 40.8% of adults in Slovenia had a serum 25(OH)D level below the critical level of 30 nmol/L, and 81.6% were below the recommended 50 nmol/L [11]; however, that study did not investigate dietary intake of VitD. It should be noted that thresholds for deficiency, insufficiency, and optimal status are still not fully harmonised across organisations [26–31], and there is a lack of consensus on the recommended daily intake of VitD [32]. The World Health Organisation (WHO) recommends 10 μg/day (400 International Units (IUs)) for those aged 51–65 years, and 15 μg/day (600 IU) for those aged over 65 years [27], whereas the daily recommended level by the European Food Safety Authority (EFSA) is 15 μg/day (600 IU) for all ages [30]. A level of 20 μg/day (800 IU) is recommended by D-A-CH (the nutrition societies of Germany, Austria, and Switzerland) [33]. These recommendations typically refer to VitD intake in the absence of UVB-induced endogenous synthesis [30,33]. Interestingly, for food labeling purposes, the nutrient reference value (NRV) for VitD is still set at 5 μg/day [34], and 50% of NRV (2.5 μg/day) has been used as a lower reference nutrient intake (LRNI) [35]. Several studies have reported VitD intakes well below the recommendations. In Europe, the reported VitD intake is generally between 3 and 5 μg per day [14,31,35–40], with higher intakes in Northern Europe (up to 11 μg/day) and lower in Southern Europe [14,26,37,38,41]. In Slovenia, only some specific population groups have been investigated (such as children [42,43], teenagers [44–46], and others [47–51]); nationally representative data for the healthy adult population are not available. Regarding the VitD intake, Lichthammer et al. [52] included populations (15–75 years) from four Central-Eastern European countries (including 81 subjects from Slovenia), and reported very low mean VitD intakes in all countries (the lowest in Austria with 2.2 μg daily, followed by Slovenia (2.6 μg), Poland (3.8 μg), and Hungary (4.1 μg)).

To address the challenge of the low VitD status in Slovenia, the *National expert working group on guidelines for sufficient vitamin D levels in the Slovenian population* was established by the National Institute of Public Health, at the request of the Ministry of Health of the Republic of Slovenia. The objective of the present study was to estimate the nationally representative VitD intake in the adolescent, adult, and elderly populations in Slovenia. We also aimed to identify food groups that notably contribute to VitD intake, and to estimate changes in dietary VitD intakes in the hypothetical scenario of mandatory milk fortification.

2. Materials and Methods

2.1. Study Design and Population

Data were collected within the scope of the cross-sectional Slovenian national food consumption survey (SI.Menu study). Data collection was carried out between March 2017 and April 2018 using the European Food Safety Agency (EFSA) Guidance on EU Menu Methodology [53]. The detailed study methodology is described in detail elsewhere [54]. In short, the participants were Slovenian residents, selected using the Central Register of the Population of Slovenia according to age, size and type of household, and place of residency. A total of 2280 individuals were allocated to three age groups: adolescents (10–17 years old), adults (18–64 years old), and the elderly (65–74 years old). Altogether, the response rate was 62.2% (n = 1319). The survey protocol was registered and accepted by the National Medical Ethics Committee (KME 0120-337/2016). All participants were informed about the details of the survey, and thereafter signed a written informed consent form. For the participants younger than 18 years of age, written consent was also obtained from the parent or legal guardian. Data were collected by skilled interviewers, during two interviews. The first interview included a general questionnaire, following by the food propensity questionnaire (FPQ), and the first 24 h dietary recall, while the second interview included second 24 h dietary recall.

2.2. Dietary Assessment Methods

2.2.1. General Questionnaire and Anthropometric Measurements

At the first face-to-face interview, participants completed a general questionnaire using computer-assisted personal interviewing. The questionnaire was adapted for adolescents and adults/the elderly. It included questions on socio-economic and socio-demographic determinants, such as marital status, place of living, level of education, employment status, and monthly income of the household. Participants also provided self-reported physical activity levels, which were converted to the International Physical Activity Questionnaire (IPAQ) score, as described by Craig et al. [55]. Participants' weight and height were measured by using calibrated instruments at the end of the first interview.

2.2.2. 24-h Dietary Recall

The interviewers performed two 24-h dietary recalls. The first recall was performed with a computer-assisted face-to-face interviewer, the second recall was repeated between 7 days to 3 weeks after the first one and was administered either by computer-assisted telephone interview or by face-to-face interview. During the recall, participants were asked to report their intake data for food and beverages consumed during the preceding day, following a daily meal timeline. Portion sizes were estimated using a nationally adjusted and validated picture book, developed in "Pilot study for the Assessment of Nutrient Intake and Food Consumption Among Kids in Europe" (PANCAKE), that contained 46 pictures of different food products or simple recipes, with each one photographed in six different portion sizes [54,56].

2.2.3. Food Propensity Questionnaire

As recommended by the EFSA, a FPQ was used to record the usual frequency of consumption of specific foods and food supplements in the last 12 months [53,57]. In total, 75 food items were allocated into nine food groups: cereals and cereal products; milk and milk products; fruit; vegetables; meat, fish, eggs, and meat products; fats and fatty food; sugar and sweeteners; beverages; and miscellaneous. The frequency response options for the food list were never, 1–3 times per month or less, once per week, 2–3 times per week, 4–6 times per week, and 1–2 times per day or more. A special field was dedicated to food supplement use, where examples were listed (e.g., multivitamins, vitamin D, proteins, omega 3 and omega 6 fatty acids, etc.), and there was the possibility to add more.

2.3. Assessment of Nutrient Intake

All foods and beverages reported during the 24-h recalls were assigned the appropriate energy and nutrient contents based on compositional data from the Open Platform for Clinical Nutrition (OPEN) [58]. The OPEN is a web-based application based on the national food composition database, which contains information and data for ingredients and recipes frequently used in Slovenia. To enable an accurate estimation of the nutritional composition of more complex foods and dishes, a disaggregation method was applied based on the recipes provided by the subjects, when applicable, or traditional recipes collected in the OPEN platform. To estimate the usual daily VitD intake in the population, we used the Multiple Source Method (MSM) analysis [59], in which reported foods were allotted into corresponding food categories, included in the FPQ [53].

All the extracted foods (n = 2377) from the SI.Menu consumption dataset were checked by a nutrition expert, and missing composition data were supplemented with VitD content. When the VitD content of the food was not found in OPEN, additional food composition databases were used (the National Food Composition Database in Finland (Fineli) [60], The Composition of Foods [61], or the United States Department of Agriculture Food Composition Database (USDA) [62]). Altogether, 63.4% of food items from the whole list were determined to be a source of dietary VitD. For assessment of the hypothetical VitD intake in the scenario of mandatory milk fortification, we assumed that all types of milk were enriched with an additional 2 μg of VitD per 100 mL [63].

2.4. Data Analysis

Exclusion criteria and assessment of under- and over-reporting subjects are explained in Zupanič et al. [64]. In short, subjects with incomplete anthropometric and/or 24-h recall data, and over-and under-reporting (based on the ratio of reported energy intake, with consideration of metabolic rate) were excluded. The final sample contained 1248 valid subjects: 468 adolescents, 364 adults, and 416 elderly subjects.

Usual VitD intake distributions per age group, adjusted for within individual day-to-day variation, were modelled with the MSM [59]. This method examines different food and nutrient distributions to estimate adjusted population distributions. It is characterised by a two-part shrinkage technique applied to residuals of two regression models, one for the positive daily intake data and one for the event of consumption. The shrunken residuals are back-transformed to their original scale, and the individual usual intake is obtained by multiplication of the frequency and amounts. The MSM was used to correct dietary intake data for intra- and inter-personal variability. Corrections were carried out only for food categories (28/101) identified as a source of VitD. Corrections for the reported frequency of food consumption were carried out for all categories which were included in the FPQ. We should note that while most (24/28) of the relevant food categories that are a source of VitD were included in the FPQ, this was not the case for a few products, including eggs. For such foods, the consumption was estimated only with consideration of 24 h recall data, without correction by FPQ data. Sex and body mass index (BMI) were included as covariates in the models. After the MSM was applied, individual usual daily intakes of VitD were calculated [65]. The same approach was used for the estimation of VitD intake in the hypothetical scenario of mandatory milk fortification, but different food composition data were used (details provided in Section 2.3).

Descriptive characteristics (mean, median, proportions) are presented for age cohorts and per different socio-demographic-, anthropometric-, and individual-based variables within each age group. Linear and logistic regression analyses were used to determine the significant differences between different sub-populations in terms of VitD intake. The adjusted means of VitD intake were determined by sex, place of living, BMI, and IPAQ levels for all age groups, while education and income were also used for adults and the elderly, and employment status was used only for adults. Some models were adjusted for energy intake. To report nationally representative epidemiological data, weighting was carried out with iterative proportional fitting [66], with consideration of age and sex, using census data from the 2017 reference population. The prevalence of very low VitD consumption was found using a previously defined LRNI threshold of 2.5 μg/day [35], separately for all age groups, with adjustments for socio-demographic, anthropometric, and lifestyle parameters. Model parameters were estimated by the maximum likelihood method. Odds ratios (ORs) with 95% confidence intervals (CIs) were used as a measure of relative risk for very low VitD intake (less than LRNI). Differences were considered significant at $p < 0.05$, except where it is stated otherwise.

The MSM online tool V1.0.1 (https://msm.dife.de/; accessed on 6 June 2021; the Department of Epidemiology of the German Institute of Human Nutrition Potsdam-Rehbrücke, Germany) was used for estimation of individual nutrient intakes, while statistical analyses were conducted using STATA V15.1 (StataCorp LLC, College Station, TX, USA).

3. Results

We analyzed data for three population groups in the SI.Menu study [54]; adolescents, adults, and the elderly. The study population was representative for sex and age (10–17/18–64/65–75 years). The most relevant demographic and lifestyle characteristics of the study sample are outlined in Table 1. VitD supplementation was explicitly reported by 3.6%, 6.0%, and 4.8% of adolescents, adults, and the elderly, respectively. When considering both the use of VitD and multivitamin products, the proportion of supplementation was 17.1%, 17.3%, and 7.0%, in adolescents, adults, and the elderly, respectively (Table 1). Our study design unfortunately did not provide enough detail about supplementation patterns to estimate intakes of VitD with medicines and/or food supplements, or to investigate seasonal differences in supplementation practices. VitD intakes in this study, therefore, corresponded to consumption of regular foods, without food supplements.

A comparison of the 24 h recalls and FPQ data showed notable differences in the ability of both methods to identify true consumers of specific food categories (Supplementary Table S1). For example, the proportion of true consumers of sea fish was up to 10% higher when FPQ data were considered. Therefore, the usual dietary VitD food intakes were estimated using the MSM method, with consideration of both 24 h recalls and FPQ data [53].

Population-weighted dietary VitD intake for all three study populations is presented in Table 2. The mean VitD intake was 2.73 μg/day (95% CI: 2.56–2.91), 2.85 μg/day (95% CI: 2.69–3.00), and 2.45 μg/day (95% CI: 2.34–2.57) for adolescents, adults, and the elderly, respectively. When adjusting for energy intakes, mean VitD intakes were 1.24, 1.34 and 1.20 μg per 1000 kcal per day, respectively. The highest prevalence of very low VitD intakes (below 2.5 μg/day) was observed in the elderly population (61.0%), followed by adolescents (55.0%) and adults (45.8%).

Table 1. Demographic and lifestyle characteristics of the study sample.

		Age Cohorts		
		Adolescents	**Adults**	**Elderly**
		(10–17 Years)	**(18–64 Years)**	**(65–74 Years)**
		***n* = 468**	***n* = 364**	***n* = 416**
Age; years—mean (SD)		13.4 (2.37)	43.6 (13.81)	68.7 (2.7)
Place of living—*n* (%)	Rural	270 (57.7)	202 (55.5)	229 (55.1)
	Intermediate	76 (16.2)	56 (15.4)	71 (17.1)
	Urban	122 (26.1)	106 (29.1)	116 (27.9)
Sex—*n* (%)	Male	238 (50.9)	173 (47.5)	213 (51.2)
	Female	230 (49.1)	191 (52.5)	203 (48.8)
Education—*n* (%)	No university degree	n.a.	249 (68.4)	342 (82.2)
	University degree	n.a.	115 (31.6)	74 (17.8)
Family monthly net income—*n* (%)	Below average	n.a.	118 (38.4)	269 (71.5)
	Above average	n.a.	189 (61.6)	107 (28.5)
BMI—mean (SD)		21.0 (4.2)	26.7 (5.2)	28.4 (5.0)
n (%)	Normal	301 (64.6)	148 (40.7)	108 (26.0)
	Overweight and obese	167 (35.7)	216 (59.3)	308 (74.0)
IPAQ—*n* (%)	Low intensity	108 (23.3)	127 (35.3)	137 (33.4)
	Moderate	141 (30.5)	108 (30.0)	133 (32.4)
	High intensity	214 (46.2)	125 (34.7)	140 (34.2)
Employment status—*n* (%)	Employed	n.a.	226 (62.1)	n.a.
	Unemployed	n.a.	42 (11.5)	n.a.
	Student	n.a.	32 (8.8)	n.a.
	Retired	n.a.	64 (17.6)	n.a.
Use of dietary supplements—*n* (%)	Vitamin D	17 (3.63)	22 (6.04)	20 (4.81)
	Multivitamin	72 (15.4)	52 (14.3)	11 (2.64)
	Vitamin D and/or multivitamin supplements	80 (17.1)	63 (17.3)	29 (6.97)

Notes: Body mass index (BMI) was considered to be normal when it was below 25 kg/m^2, except for adolescents, where sex/age adjusted cut-off points [22,23] were used; International Physical Activity Questionnaire (IPAQ); standard deviation (SD); not applicable (n.a).

Adjusted mean VitD intakes by sex, place of living, BMI, IPAQ score, education, income, and employment for different age groups are presented in Table 3, with separate models for all three study populations, and are additionally adjusted for energy intake. Linear regression analyses showed that energy intake was a significant predictor of VitD intake for all three population groups (higher VitD intake for higher energy intake), while sex, education level, and BMI were significant predictors only in some models. Sex determined the VitD intake in adolescents and adults; in both cases, higher intakes were observed in males. Body mass index (BMI) was found to be a significant predictor only in adolescents, with lower VitD intake in those who were overweight/obese, while education was found to be a significant determinant of VitD intake in the elderly population, with the highest VitD intake in those with a university degree. Similar trends were also observed in other models investigating the likelihood of a very low VitD intake (Figure 1). For all three population groups, the model used a threshold of lower reference nutrient intake (LRNI; 2.5 µg/day). Sex was again found to be a significant predictor in adolescents ($p < 0.001$), adults ($p < 0.001$), and the elderly ($p < 0.005$), with higher odds ratios in females. In comparison with adolescent males, adolescent females were 3.47 (95% CI: 2.3–5.2) times more likely to have intakes below the LRNI, while in adults, this difference was even more pronounced (OR 10.29; 95% CI: 5.8–18.4). BMI was also a significant ($p < 0.001$) predictor in adolescents, with an OR 2.88 (95% CI: 1.9–4.5) times higher for overweight/obese subjects.

Table 2. Population-weighted dietary vitamin D intake (μg/day), and prevalence of very low vitamin D intake.

	Adolescents (10–17)			**Adults (18–64)**			**Elderly (65–74)**		
	All	**Male**	**Female**	**All**	**Male**	**Female**	**All**	**Male**	**Female**
Sample Size, *n* (%)	**468 (100)**	238 (50.85)	230 (49.15)	**364 (100)**	173 (47.53)	191 (52.47)	**416 (100)**	213 (51.20)	203 (48.80)
				Vitamin D intake					
Mean [μg/day] (95% CI)	**2.73 (2.56–2.91)**	3.02 (2.83–3.22)	2.42 (2.14–2.70)	**2.85 (2.69–3.00)**	3.39 (3.17–3.62)	2.30 (2.14–2.44)	**2.45 (2.34–2.57)**	2.60 (2.42–2.78)	2.32 (2.16–2.48)
Std.Err.	**0.09**	0.10	0.14	**0.08**	0.12	0.08	**0.06**	0.09	0.08
Median [μg/day]	**2.33**	2.70	1.95	**2.66**	3.01	2.04	**2.21**	2.30	2.13
Mean [μg/1000 kcal per day] (95% CI)	**1.24 (1.15–1.33)**	1.23 (1.12–1.35)	1.25 (1.11–1.40)	**1.34 (1.27–1.41)**	1.47 (1.37–1.57)	1.21 (1.13–1.29)	**1.20 (1.13–1.27)**	1.20 (1.10–1.30)	1.20 (1.10–1.29)
				Prevalence of very low vitamin D intake (%)					
<2.5 [μg/day]	**55.0 (47.4–62.4)**	38.8 (30.0–48.5)	72.6 (64.6–79.3)	**45.8 (40.1–51.6)**	21.0 (15.2–28.2)	71.0 (63.5–77.5)	**61.0 (51.7–69.6)**	58.2 (43.6–71.5)	63.5 (50.6–74.7)

Notes: 95% CI: 95% confidence interval.

Table 3. Unadjusted and adjusted mean (95% CI) levels of vitamin D intake (μg/day) by sex, place of living, body mass index (BMI), International Physical Activity Questionnaire (IPAQ) score, education, income, employment, and energy intake (kcal/day) for different age groups. All three models were adjusted to account for differences in energy intake (kcal/day).

Variable		Adolescents (10–17 Years Old)		Adults (18–64 Years Old)		Elderly (65–74 Years Old)	
		Unadjusted	Adjusted	Unadjusted	Adjusted	Unadjusted	Adjusted
Sex	Male	3.01 (2.82–3.20)	2.95 (2.72–3.17)	3.35 (3.16–3.53)	3.22 (3.03–3.42)	2.67 (2.51–2.83)	2.55 (2.37–2.73)
	Female	2.44 (2.18–2.70)	2.52 (2.29–2.76)	2.31 (2.15–2.46)	2.46 (2.28–2.64)	2.35 (2.16–2.54)	2.49 (2.30–2.68)
Place of living	Rural	2.64 (2.46–2.80)	2.63 (2.42–2.84)	2.88 (2.71–3.05)	2.86 (2.68–3.03)	2.57 (2.41–2.73)	2.56 (2.39–2.73)
	Intermediate	2.87 (2.24-3.50)	2.85 (2.46-3.24)	2.66 (2.25–3.07)	2.68 (2.36–3.00)	2.38 (2.06–2.71)	2.50 (2.20–2.81)
	Urban	2.86 (2.54-3.17)	2.90 (2.60–3.21)	2.73 (2.52–2.94)	2.81 (2.59–3.04)	2.48 (2.24–2.72)	2.44 (2.20–2.69)
Education	No university degree	n.a.	n.a.	2.74 (2.59–2.88)	2.76 (2.61–2.92)	2.43 (2.31–2.55)	2.45 (2.31–2.60)
	University degree			2.94 (2.69–3.19)	2.92 (2.69–3.16)	2.88 (2.47–3.29)	2.85 (2.52–3.17)
Family net income	Below average (≤EUR 1300)	n.a.	n.a.	2.63 (2.40–2.85)	2.71 (2.50–2.93)	2.44 (2.30–2.60)	2.51 (2.33–2.64)
	Above average (>EUR 1300)			2.92 (2.74–3.10)	2.89 (2.71–3.05)	2.68 (2.42–2.94)	2.56 (2.31–2.81)
BMI	Normal	2.95 (2.72–3.16)	2.94 (2.75–3.14)	2.72 (2.52–2.92)	2.81 (2.60–3.01)	2.51 (2.28–2.73)	2.57 (2.31–2.83)
	Overweight and obese	2.34 (2.13–2.56)	2.35 (2.10–2.62)	2.86 (2.69–3.02)	2.82 (2.66–2.98)	2.52 (2.37–2.66)	2.50 (2.36–2.65)
IPAQ	Low intensity	2.77 (2.55–2.99)	2.72 (2.40–3.05)	2.79 (2.60–2.99)	2.78 (2.56–2.99)	2.56 (2.33–2.78)	2.54 (2.32–2.76)
	Moderate	2.64 (2.27–3.02)	2.71 (2.42–3.00)	2.74 (2.48–2.98)	2.82 (2.60–3.05)	2.30 (2.14–2.46)	2.35 (2.13–2.57)
	High intensity	2.79 (2.56–3.01)	2.77 (2.53–2.99)	2.89 (2.65–3.12	2.85 (2.63–3.06)	2.69 (2.45–2.93)	2.67 (2.45–2.89)
Employment	Employed	n.a.	n.a.	2.91 (2.75–3.08)	2.81 (2.64–2.98)	n.a.	n.a.
	Unemployed			2.51 (2.14–2.87)	2.85 (2.46–3.23)		
	Student			2.91 (2.53–3.29)	2.96 (2.47–3.44)		
	Retired			2.54 (2.23–2.86)	2.76 (2.44–3.07)		

Notes: BMI was considered to be normal when it was below 25 kg/m^2, except for adolescents, where sex/age adjusted cut-off points [22,23] were used. Linear regression analysis conducted on samples with excluded missing values (Family net income: $n = 57$ (adults) and 40 (elderly); IPAQ: $n = 5$ (adolescents), 4 (adults), 6 (elderly)); Difference in predicted marginal means of vitamin D intake per different socio-demographic groups within age categories: $p < 0.05$ sex (adolescents), $p < 0.001$ BMI (adolescents); $p < 0.0001$ sex (adults); $p < 0.05$ education (elderly). The energy intake was a significant predictor of vitamin D intake, with β = regression coefficients per different age group as follows: β = +0.0004 ($p \leq 0.001$) for adolescents, β = +0.0006 ($p \leq 0.001$) for adults, β = +0.0006 ($p \leq 0.001$) for the elderly.

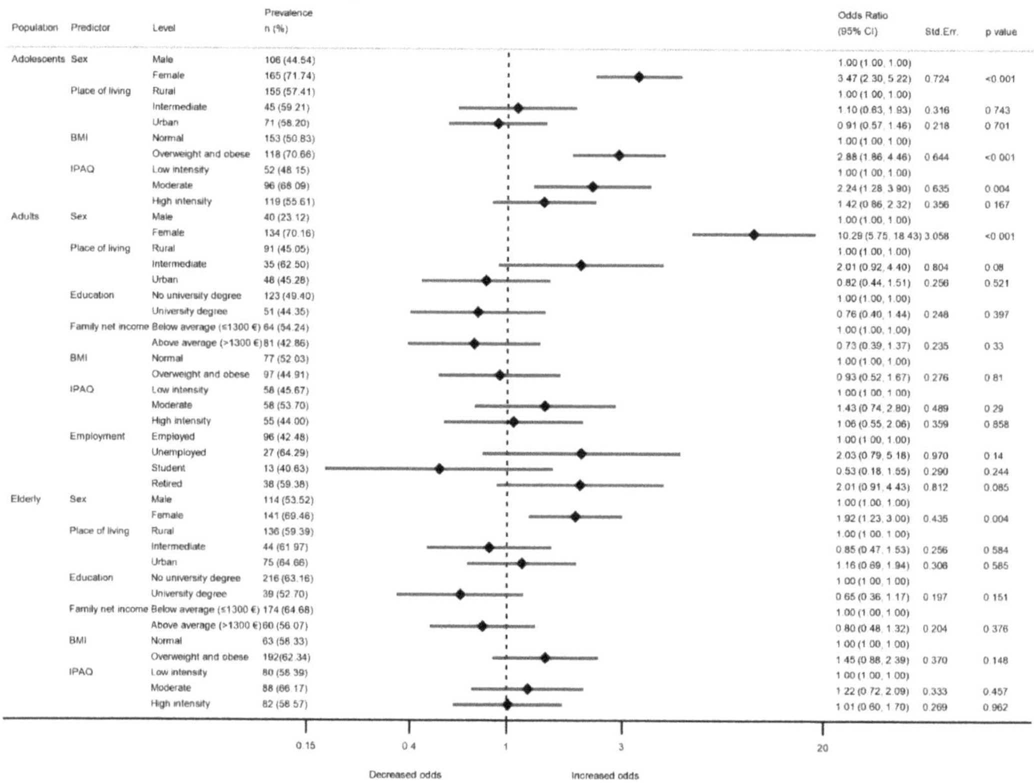

Figure 1. Percentage of the population with a very low vitamin D intake (lower reference nutrient intake; 2.5 μg/day) by sex, place of living, education, family net income, body mass index (BMI), International Physical Activity Questionnaire (IPAQ) score, and employment. **Notes**: BMI was considered to be normal when it was below 25 kg/m2, except for adolescents, where sex/age adjusted cut-off points [22,23] were used. Logistic regression analysis was conducted on samples with excluded missing values (Family net income: n = 57 (adults) and 40 (elderly); IPAQ: n = 5 (adolescents), 4 (adults), 6 (elderly)); Prevalence odds ratio for lower reference nutrient intake (<2.5 μg/day of vitamin D); lower reference nutrient intake prevalence probability test per different socio-demographic groups within age categories: $p < 0.001$ sex (adolescents), $p < 0.001$ BMI (adolescents), $p < 0.05$ IPAQ (adolescents); $p < 0.001$ sex (adults), $p < 0.005$ sex (elderly).

We further investigated the relative contributions of specific food categories to the dietary VitD intakes for the three investigated age groups (Figure 2). Beef, veal and pork meat, sea fish, and eggs were found to be the most important contributors of VitD intake among all age groups, with minor differences among groups. These food groups contributed from 13 to 20% of the dietary VitD intake. Processed meat and fish (fish cans and pate, and sausages, hot dogs, and meat pate) together contributed up to 11% of the VitD intake. In adolescents, processed breakfast cereals (sweetened flakes) were also found as a notable VitD source (~7%), while this was not observed in adults and the elderly population.

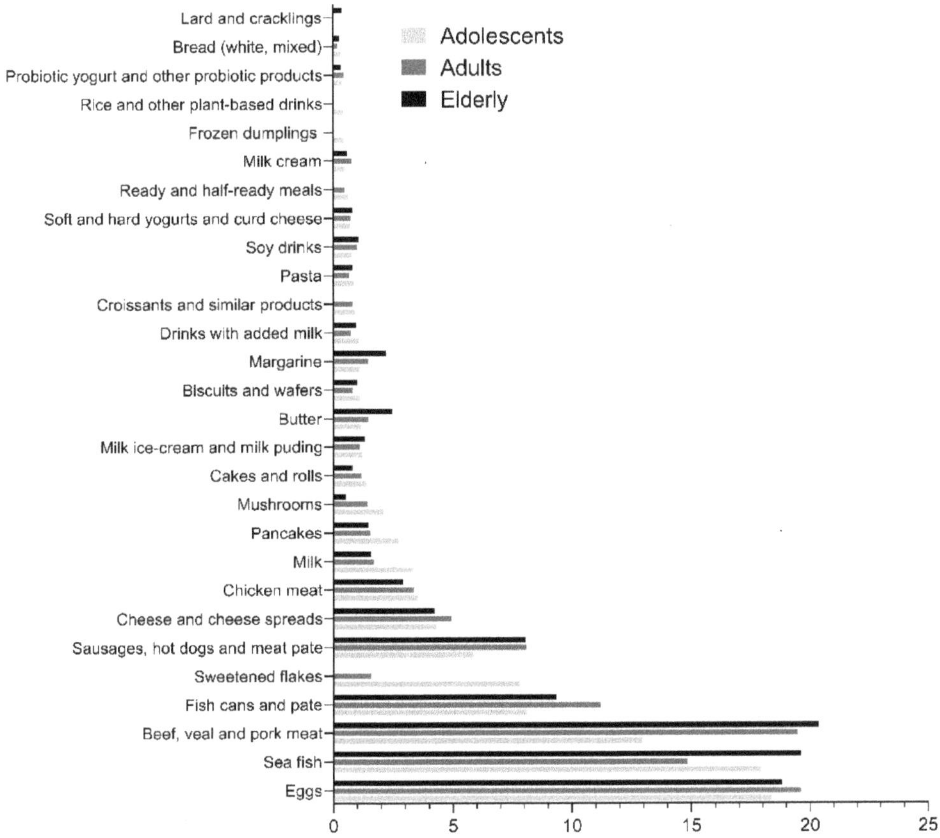

Figure 2. The relative contribution of food categories to Vitamin D intake among different population groups (% of total Vitamin D intake).

For a more comprehensive insight into the importance of specific foods for VitD intake, we also investigated population-weighted consumption patterns for age groups (Supplementary Table S2), with the use of a threshold for LRNI (2.5 µg/day). In all age groups, those with a very low VitD intake reported a lower consumption of sea fish. In adolescents, this was also observed for beef, veal, and pork meat, sweetened flakes, and mushrooms. Among the adult population, those with intakes below LRNI had also lower intakes of fish cans and pate, sausages, hot dogs, and meat pate, milk, cakes and rolls, milk ice-cream and milk pudding, butter, biscuits and wafers, croissants and similar products, and ready and half-ready meals. In the elderly age group, the same applied to fish cans and pate, cheese and cheese products, milk, and mushrooms.

Our study also aimed to estimate changes in dietary VitD intakes in a hypothetical scenario of mandatory milk fortification. Using food intake data from all age groups, we modelled VitD intake in the scenario in which all milk would be fortified with 2 µg of VitD per 100 mL. As presented in Figure 3 and Table 4, the effect of such mandatory fortification would be most notable in adolescents. The projected increase in daily VitD intake due to milk fortification was 2 µg for adolescents and about 1 µg for adults and the elderly population. The estimated total mean VitD intake after accounting for the fortification would be 4.82 µg (SD: 2.51), 4.01 µg (SD: 1.88), and 3.58 µg (SD: 1.79) for adolescents, adults, and the elderly, respectively.

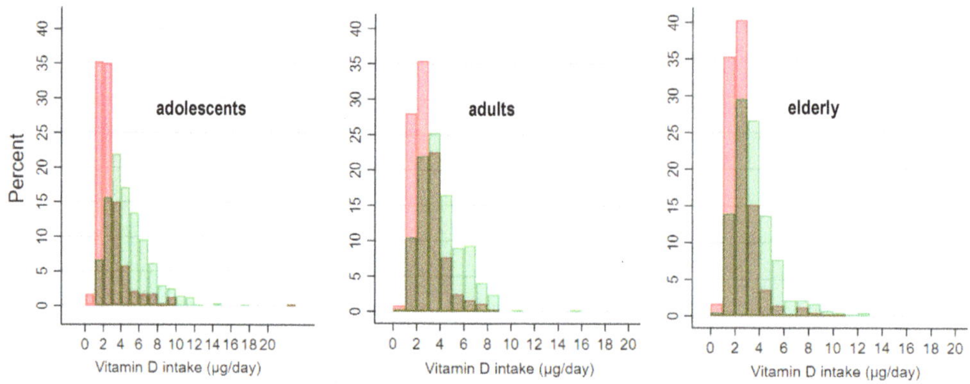

Figure 3. Histograms of vitamin D intakes from the regular diet (red bars) and in a projected fortified diet (scenario of mandatory fortification of milk with 2 μg Vitamin D per 100 mL) (green bars).

Table 4. Comparison of estimated vitamin D intake (μg/day) from a regular diet and projected fortified diet.

	Adolescents (10–17)			Adults (18–64)			Elderly (65–74)		
	All	Male	Female	All	Male	Female	All	Male	Female
	Unadjusted mean vitamin D intake—μg/day (SD)								
Regular diet	2.73 (1.77)	3.01 (2.44)	2.44 (2.00)	2.80 (1.24)	3.35 (1.23)	2.31 (1.03)	2.51 (1.26)	2.67 (1.16)	2.35 (1.35)
Projected fortified diet *	4.82 (2.51)	5.46 (2.29)	4.16 (2.56)	4.01 (1.88)	4.53 (1.90)	3.56 (1.73)	3.58 (1.79)	3.62 (1.83)	3.54 (1.75)
	Median vitamin D intake—μg/day								
Regular diet	2.25	2.66	1.95	2.59	3.02	2.02	2.26	2.38	2.01
Projected increase with fortification *	2.01	4.10	3.15	1.10	1.04	1.23	1.01	0.87	1.25
Projected fortified diet *	4.26	6.76	5.1	3.69	4.06	3.25	3.27	3.25	3.26

Notes: Vitamin D intakes in the study sample, without population weighting. SD: Standard deviation; * Projected fortified diet with the scenario of mandatory fortification of milk with 2 μg Vitamin D per 100 mL.

4. Discussion

While a very high prevalence of VitD deficiency has been reported recently in Slovenia [11], there was no nationally representative data available on the dietary intakes of VitD. Using data collected by the Slovenian national food consumption survey (SI.Menu) we estimated daily VitD intakes using two non-consecutive 24 h dietary recalls and FPQ data. Three different age groups (adolescents, adults, and the elderly) were included in the analyses of VitD intake. That estimated VitD intakes were 2.73 μg/day, 2.85 μg/day, and 2.45 μg/day, respectively. While Lichthammer et al. [52] reported similar daily VitD intakes, they observed the highest intakes in their youngest age group (14–18 years; 3.72 μg/day). It should be noted that in their study, VitD intake was estimated with a different method (food frequency questionnaire; FFQ), and that their sample size was much smaller (n = 434 with 81 subjects from Slovenia). VitD intakes have been investigated in some other Slovenian populations, such as children [42,43], teenagers [44–46], and others [47–51]. VitD intake in adolescents varied from 2 to 4 μg/day [44,46], and was 2.1 μg/day in pregnant women [49]. In the elderly living in residential homes, intakes were particularly concerning (1 μg/day) [51]. When comparing our results to other European countries, we found many similarities. In two studies from neighboring Austria, the VitD intake was 2.1 μg/d (n = 4972; various dietary intake tools; all age groups above 4 years) [67], with the highest intakes in the adult population. The intake was 2.53 μg/d in a study by Kudlacek et al. (n = 1048; 21–76 years) [68]. In girls (11–17 years), the mean VitD intake ranged from 1.5 μg/day in Spain to 3.2 μg/day in Poland, whereas for boys (11–17 years), intakes ranged from 1.9 μg/day in France to 4.8 μg/day in Poland [35]. VitD intake in the elderly population (>65 years) ranged from 0.7 μg/day in Spain to 15 μg/day in Norway [38]. Across Europe, mean VitD intake varied from 1.1 to 6 μg per day [14,31,36–40], with higher intakes in Northern Europe (up to 14 μg/day) and lower intakes in Southern Europe [14,26,37,38,41]. Higher intakes in Northern Europe can be explained by the higher intake of fish and VitD enriched foods.

In adolescents and adults, our results indicate higher intakes of VitD in males; however, this difference was not significant in the elderly population. The same pattern was observed when looking at the odds ratio (OR) for VitD intakes below the lower reference nutrient intake (LRNI; 2.5 μg/d). Adolescent males had an adjusted VitD intake of 2.95 μg/d, in comparison to 2.52 μg/d in females. Meanwhile, the adjusted intake in adult males was 3.21 μg/d in comparison to 2.31 μg/d in adult females. Males usually have higher VitD intakes than women [14,37], which is also related to the higher amount of consumed food. This was also observed in our study, where daily energy intake was a significant predictor of VitD intake. Furthermore, in adolescents, those with normal BMI had a significantly higher intake of VitD (and lower OR for very low VitD intake) in comparison with overweight or obese adolescents. Similar observations were made in a study from Northern Norway [69], but in the adult population. In the elderly, VitD intake was statistically higher among those with a higher education level. Similar observations were reported in two Swiss studies in some population groups [70,71].

To investigate the prevalence of very low VitD intake, we took the cut-off value of 2.5 μg/day. The prevalence of intake below the LRNI was 55.0, 45.8, and 61.0% in adolescents, adults, and the elderly, respectively. The highest prevalence of below LRNI intake was among teenage females (72.6%), and the lowest was among adult males (21.0%). In the population of adolescent males, 38.8% had intakes below 2.5 μg/day, in comparison to 72.6% of females. In European adolescents (11–17 years), the prevalence of intakes of VitD below the LRNI ranged from 17.1% (Netherlands) to 81.7% (France) in males, and 36.6% (Netherlands) to 97.9% in females (Spain) [35]. In the Slovenian adult population, males had a much lower prevalence of below LRNI intakes (21.0%) than females (71.0%). The prevalence data from other European countries were also very varied, ranging from 26.8% (Netherlands; 31–60 years) to 94.7% (Spain; 18–60 years) in adult females, and from 7.3% (Netherlands; 31–60 years) to 87.4% (Spain; 18–60 years) in adult males [35]. In our elderly population, 58.2% of males and 63.5% of females had below LRNI intakes. In other elderly European populations (<60 years) [35], the prevalence ranged from 17.4% (Netherlands) to 100% (Spain) in females, and from 5.7% (Netherlands) to 100% (Spain) in males. We should also note that a notable trend of lower risk for very low VitD intake was observed in those with higher financial status in both adults and elderlies (OR 0.73 and 0.80, respectively), but was not statistically significant. This also corresponded to the observed trend of increased OR for unemployed (OR 2.03, p=0.14).

In the present study, the main contributors to VitD intake across age groups were eggs, sea fish, beef, veal, and pork, fish cans and pate, sweetened flakes (only in adolescents), and sausages, hot dogs, and meat pate. Likewise, in other European countries, the leading food group contributors to VitD intake were fish/shellfish, added fats, meats/meat products, cakes, cereals, and dairy products [31,37,72].

Very low dietary intake of VitD in the Slovenian population, shown in our study, partially explains the previously reported high prevalence of VitD deficiency during winter, when sun-induced biosynthesis of VitD is not sufficient [11]. While the public health outcomes of this epidemiological situation have not been well investigated, it should be mentioned that VitD is a key component in various bodily functions [3–7], also related to the functioning of the immune system and bone health. Although this study did not address this topic, it should be mentioned that Slovenia has been among the countries with the most severe death toll from COVID-19 (currently 2,174 deaths/1 mio. population [73]; the majority of those deaths occurred during the autumn/winter pandemic wave in 2020/21), and that we also have quite a high prevalence of osteoporosis in comparison with some other European countries. For example, in Slovenia, 27.5% of women over 50 years had osteoporosis [74], while reported rates for France and Span are much below 20% [75].

In Slovenia, there is no mandatory fortification of foods with VitD, or even national recommendations. However, rules for the enrichment of foods are defined in the EU Regulation No 1925/2006 on the addition of vitamins and minerals and of certain other substances to foods [21], which enables the enrichment of foods with VitD using either cholecalciferol or ergocalciferol [76]. The fortification of foods with VitD (either mandatory or voluntarily based on national recommendations) is currently in place in North American and some European countries [77–79]. The amount of added VitD varies across countries, as do the food matrixes. Typical matrixes for fortification include milk and dairy products. In countries with implemented VitD fortification policies, the amount of added VitD is increasing [79]. Canada recently proposed a mandatory policy to increase fluid milk fortification from 1 μg to 2 μg/100 mL, due to inadequate VitD intake in the population [63]. Our goal was to investigate a hypothetical scenario in which milk was fortified with 2 μg of VitD per 100 mL of milk, as proposed in Canada [63]. The projected increase in daily dietary VitD intake in this model was 2.0 μg, 1.1 μg, and 1.0 μg in adolescents, adults, and the elderly, respectively. It should be noted that a review of Black et al. [80] showed that each ingested microgram of VitD with fortified food leads to an increase in serum 25(OH)D of 1.2 nmol/L. However, the problem of limiting fortification to milk is in the fact that not everyone in the population consumes milk (i.e., due to lactose intolerance, etc.), and the idea of mandatory fortification is to protect the general population. A preferable fortification approach may therefore be adding VitD in smaller amounts to various food matrixes [81]. Jääskeläinen et al. [77] reported the effects of the implementation of VitD fortification on VitD status in Finland. Their approach was to fortify both fluid milk (1.0 μg/100 g) and spreadable fats (20 μg/100 g). The intervention resulted in higher VitD intakes and a lower prevalence of VitD deficiency, based on serum 25(OH)D levels [77]. We should also mention that in Finland and the United States, fortified foods have the highest contribution to dietary VitD intake [82].

The strength of our study is that VitD intake was estimated with the exploitation of data collected with the Slovenian national SI.Menu food consumption survey, using both 24 h recalls and food propensity questionnaire data. Another strength is that sampling was carried out with three quotas, enabling insights into the more vulnerable populations of adolescents and the elderly. We should also mention that since the SI.Menu study collected a series of sociodemographic and lifestyle indicators, we were able to include those in the regression analyses. Study limitations should be also noted. While the SI.Menu study was designed using a very robust EU Menu methodology and harmonized with the EFSA, the study was not primarily designed to investigate dietary VitD intakes. The objective of the SI.Menu study was to collect data on food consumption in order to inform regulatory risk assessments related to the use of food additives, food contaminants, etc. Nutritional assessment was a secondary objective, and the FPQ was therefore not tailored for this purpose. Because of this, our FPQ dataset did not include the frequency of consumption of some foods (particularly eggs), which are an important source of VitD in the diet. Considering the study design, food composition (VitD content) needed to be estimated using food composition databases, and not with laboratory analyses. A methodological limitation is that in Slovenia, national food composition data do not include the amount of VitD in some foods, and we were therefore forced to use additional food composition databases. An important study limitation is also that we could only estimate daily intake of VitD with foods, without accounting for the use of medicines and food supplements. While the consumption of such products was included in the FPQ of the SI.Menu study, the collected data did not enable assessment of the corresponding VitD intake. The only available data regarding VitD supplements was if a person was taking them. In the present study, 3.6%, 6.0%, and 4.8% of adolescents, adults, and the elderly, respectively, reported the use of VitD supplements. However, supplementation can be an important additional source of VitD [14,83]. For example, a typical dosage of VitD in food supplements is much higher than the expected intake with foods, at 25 μg VitD daily dosage [23]. Furthermore, the COVID-19 pandemic changed people's behaviors considerably [84], and a much higher frequency of VitD supplementation was reported during the 2020 COVID-19 lockdown in Slovenia [85], affecting VitD intakes considerably. However, the long-term effects of the pandemic are not known. Ideally, further research should focus on the assessment of VitD intakes from all sources in the post-COVID-19 period.

5. Conclusions

The estimated dietary intake of VitD in the Slovenian population was very low; 2.73 μg, 2.85 μg, and 2.45 μg in adolescents, adults, and the elderlies, respectively. The VitD intake was well below the recommended intake in all age groups. Most of the population did not even meet the threshold for a lower reference nutrient intake (LRNI; 2.5 μg/day). In all population groups, daily energy intake was a significant predictor of VitD intake. An additional predictor for adolescents and adults was also sex, with higher VitD intakes among males. Higher BMI was also found to be a predictor for lower VitD intake in adolescents, while in the elderly population this was observed in those with lower education levels. The main contributors to VitD intake in the Slovenian diet were eggs, fish and fish products, and meat and meat products. Assessment of the hypothetical scenario of mandatory milk fortification with 2 μg VitD per 100 mL showed a notable increase in the predicted VitD intake, with the highest effect in adolescents, but the expected dietary VitD intake would still be notably below the recommended intake.

Supplementary Materials: The following are available online at https://www.mdpi.com/article/10.3390/nu13103528/s1, Supplementary Table S1. Patterns of observed food consumption related to vitamin D in the diet: comparison of true consumers with the use of FPQ data and 24 h dietary recall. Supplementary Table S2. Patterns of observed FPQ food consumption related to vitamin D in the diet in subjects with intakes lower and higher than 2.5 μg vitamin D per day.

Author Contributions: M.H. wrote the first version of the manuscript; M.H. and H.H. performed the data analyses; B.K.S. was responsible for information technology, and H.H. for the preparation of the database and data analyses. M.H. conducted food-matching to estimate VitD levels. Ž.L. participated in data collection and manuscript review; M.G. and U.B. were responsible for the SI. Menu study design and food consumption data. K.Ž. was responsible for the funding of the study, participated in the study design, collaborated in the data analyses, and reviewed the manuscript. I.P. revised the final draft and prepared the submission. All authors reviewed the manuscript and agreed to the published version of the manuscript.

Funding: This study was conducted within the national research program P3-0395 "Nutrition and Public health" and projects "Challenges in achieving adequate vitamin D status in the adult population" (L7-1849), "Children and adults nutrition as a protective or health-risk factor" (L3-8213) and "Dietary challenges in the population of nursing homes' residents", supported by the Slovenian Research Agency and Ministry of Health of the Republic of Slovenia. The original SI.Menu 2017/18 national food consumption survey was funded by the European Food Safety Authority (EFSA contract No. OC/EFSA/DATA/2014/02-LOT2-CT03), Ministry of Health of the Republic of Slovenia, and Slovenian Research Agency.

Institutional Review Board Statement: The study was conducted according to the guidelines of the Declaration of Helsinki, and approved by the National Medical Ethics Committee, Ljubljana, Slovenia (KME 53/07/16; approval No. 0120-337/2016 issued on 19 July 2016).

Informed Consent Statement: Informed consent was obtained from all subjects involved in the study.

Data Availability Statement: The data presented in this study are available on request from the corresponding author.

Acknowledgments: We acknowledge the support of all SI.Menu project partners and their researchers, particularly Nataša Delfar, Metka Zaletel, Darja Lavtar, Ada Hočevar Grom, and Ivan Eržen (National Institute of Public Health, Ljubljana, Slovenia); Petra Golja (Department of Food Science at Biotechnical Faculty and Group of Anthropology at Biotechnical Faculty of University of Ljubljana, Slovenia); Stojan Kostanjevec (University of Ljubljana, Faculty of Education, Ljubljana,

Slovenia; Majda Pajnkihar (University of Maribor, Faculty of Health Sciences, Maribor, Slovenia); Tamara Poklar Vatovec (University of Primorska, Faculty of Health Sciences, Koper, Slovenia); Eva Valenčič (Jožef Stefan Institute, Ljubljana, Slovenia); Evgen Benedik (University Medical Centre Ljubljana, University Children' s Hospital, Ljubljana, Slovenia); and students: Katarina Žlavs and Naska Kvarantan. We also acknowledge the support of Veronika Belec and her team and interviewers at the GfK Research Institute (Slovenia) for conducting the survey fieldwork.

Conflicts of Interest: The authors declare no conflict of interest. The funders had no role in the design of the study; in the collection, analyses, or interpretation of data; in the writing of the manuscript; or in the decision to publish the results. I.P. has led and participated in various other research projects in the area of nutrition, public health, and food technology that were (co)funded by the Slovenian Research Agency, Ministry of Health of the Republic of Slovenia, the Ministry of Agriculture, Forestry, and Food of the Republic of Slovenia, and, for specific applied research projects, and also by food businesses. I.P., K.Ž, M.G., and U.B are members of a national workgroup responsible for the development of recommendations for adequate vitamin D status among the Slovenian population.

References

1. Cashman, K.D.; Dowling, K.G.; Škrabáková, Z.; Gonzalez-Gross, M.; Valtueña, J.; De Henauw, S.; Moreno, L.; Damsgaard, C.T.; Michaelsen, K.F.; Mølgaard, C.; et al. Vitamin D deficiency in Europe: Pandemic? *Am. J. Clin. Nutr.* **2016**, *103*, 1033–1044. [CrossRef] [PubMed]
2. Holick, M.F. The vitamin D deficiency pandemic: Approaches for diagnosis, treatment and prevention. *Rev. Endocr. Metab. Disord.* **2017**, *18*, 153–165. [CrossRef] [PubMed]
3. Autier, P.; Boniol, M.; Pizot, C.; Mullie, P. Vitamin D status and ill health: A systematic review. *Lancet Diabetes Endocrinol.* **2014**, *2*, 76–89. [CrossRef]
4. Autier, P.; Mullie, P.; Macacu, A.; Dragomir, M.; Boniol, M.; Coppens, K.; Pizot, C.; Boniol, M. Effect of vitamin D supplementation on non-skeletal disorders: A systematic review of meta-analyses and randomised trials. *Lancet Diabetes Endocrinol.* **2017**, *5*, 986–1004. [CrossRef]
5. Holick, M.F. Vitamin D status: Measurement, interpretation, and clinical application. *Ann. Epidemiol.* **2009**, *19*, 73–78. [CrossRef]
6. Zittermann, A. Vitamin D in preventive medicine: Are we ignoring the evidence? *Br. J. Nutr.* **2003**, *89*, 552–572. [CrossRef]
7. Zittermann, A.; Pilz, S.; Hoffmann, H.; Marz, W. Vitamin D and airway infections: A European perspective. *Eur. J. Med. Res.* **2016**, *21*, 1–10. [CrossRef]
8. Kimlin, M.G. Geographic location and vitamin D synthesis. *Mol. Aspects Med.* **2008**, *29*, 453–461. [CrossRef]
9. O'Neill, C.M.; Kazantzidis, A.; Ryan, M.J.; Barber, N.; Sempos, C.T.; Durazo-Arvizu, R.A.; Jorde, R.; Grimnes, G.; Eiriksdottir, G.; Gudnason, V.; et al. Seasonal Changes in Vitamin D-Effective UVB Availability in Europe and Associations with Population Serum 25-Hydroxyvitamin, D. *Nutrients* **2016**, *8*, 533. [CrossRef]
10. Ovsenik Jeglič, T. *Climate of Slovenia 1971–2000—Sunshine Duration*; The Environmental Agency of the Republic of Slovenia: Ljubljana, Slovenia, 2006.
11. Hribar, M.; Hristov, H.; Gregorič, M.; Blaznik, U.; Zaletel, K.; Oblak, A.; Osredkar, J.; Kušar, A.; Žmitek, K.; Rogelj, I.; et al. Nutrihealth study: Seasonal variation in vitamin D status among the Slovenian adult and elderly population. *Nutrients* **2020**, *12*, 1838. [CrossRef]
12. Holick, M.F. Vitamin D deficiency. *N. Engl. J. Med.* **2007**, *357*, 266–281. [CrossRef] [PubMed]
13. World Health Organization/International Research Agency for Cancer (WHO/IARC). *Vitamin D and Cancer*; WHO: Geneva, Switzerland, 2008; p. 210. Available online: https://publications.iarc.fr/Book-And-Report-Series/Iarc-Working-Group-Reports/Vitamin-D-And-Cancer-2008 (accessed on 10 May 2021).
14. Kiely, M.; Black, L.J. Dietary strategies to maintain adequacy of circulating 25-hydroxyvitamin D concentrations. *Scand. J. Clin. Lab. Investig. Suppl.* **2012**, *243*, 14–23.
15. Roseland, J.M.; Phillips, K.M.; Patterson, K.Y.; Pehrsson, P.R.; Taylor, C.L. Vitamin D in Foods: An Evolution of Knowledge. In *Vitamin D*, 4th ed.; Feldman, D., Ed.; Academic Press: Cambridge, MA, USA, 2018; pp. 41–77.
16. Whiting, S.J.; Calvo, M.S. Vitamin D Fortification and Supplementation Policies to Correct Vitamin D Insufficiency/Deficiency Globally. In *Vitamin D*, 4th ed.; Feldman, D., Ed.; Academic Press: Cambridge, MA, USA, 2018; pp. 91–108.
17. Cheney, M.C. Canadian experience with food fortification. *Public Health Rev.* **2000**, *28*, 171–177. [PubMed]
18. Pilz, S.; Marz, W.; Cashman, K.D.; Kiely, M.E.; Whiting, S.J.; Holick, M.F.; Grant, W.B.; Pludowski, P.; Hiligsmann, M.; Trummer, C.; et al. Rationale and Plan for Vitamin D Food Fortification: A Review and Guidance Paper. *Front. Endocrinol.* **2018**, *9*, 373. [CrossRef] [PubMed]
19. Hennessy, Á.; Walton, J.; Flynn, A. The impact of voluntary food fortification on micronutrient intakes and status in European countries: A review. *Proc. Nutr. Soc.* **2013**, *72*, 433–440. [CrossRef] [PubMed]
20. European Commission (EC). Commission Regulation (EC) No 1170/2009 of 30 November 2009 amending Directive 2002/46/EC of the European Parliament and of Council and Regulation (EC) No 1925/2006 of the European Parliament and of the Council as

regards the lists of vitamin and minerals and their forms that can be added to foods, including food supplements. *Off. J. Eur. Union* **2009**, *L314/36*, 336–342.
21. European Commission (EC). Regulation (EC) No 1925/2006 of the European Parliament and of the Council of 20 December 2006 on the addition of vitamins and minerals and of certain other substances to foods. *Off. J. Eur. Union* **2006**, *L404*, 426–438.
22. EFSA Panel on Nutrition; Novel Foods; Food Allergens; Turck, D.; Castenmiller, J.; De Henauw, S.; Hirsch-Ernst, K.I.; Kearney, J.; Maciuk, A.; Mangelsdorf, I.; et al. Safety of extended uses of UV-treated baker's yeast as a Novel Food pursuant to Regulation (EU) 2015/2283. *EFSA J.* **2021**, *19*, e06602.
23. Žmitek, K.; Krušič, S.; Pravst, I. An Approach to Investigate Content-Related Quality of Nutraceuticals Used by Slovenian Consumers: A Case Study with Folate and Vitamin D Supplements. *Foods* **2021**, *10*, 845. [CrossRef]
24. Battelino, T. Zapisnik 56. redne seje RSK za pediatrijo. *Slov. Pediatrija* **2010**, *17*, 241–243.
25. van Schoor, N.; Lips, P. Global Overview of Vitamin D Status. *Endocrinol. Metab. Clin.* **2017**, *46*, 845–870. [CrossRef] [PubMed]
26. Lips, P.; Cashman, K.D.; Lamberg-Allardt, C.; Bischoff-Ferrari, H.A.; Obermayer-Pietsch, B.; Bianchi, M.L.; Stepan, J.; Fuleihan, G.E.-H.; Bouillon, R. Current vitamin D status in European and Middle East countries and strategies to prevent vitamin D deficiency: A position statement of the European Calcified Tissue Society. *Eur. J. Endocrinol.* **2019**, *180*, 23–54. [CrossRef] [PubMed]
27. World Health Organization (WHO). *Vitamin and Mineral Requirements in Human Nutrition*, 2nd ed.; World Health Organization: Geneva, Switzerland, 2005.
28. Institute of Medicine (US) Committee to Review Dietary Reference Intakes for Calcium and Vitamin D. The National Academies Collection: Reports funded by National Institutes of Health. In *Dietary Reference Intakes for Calcium and Vitamin D*; Ross, A.C., Taylor, C.L., Yaktine, A.L., Del Valle, H.B., Eds.; National Academies Press: Washington, DC, USA, 2011.
29. Holick, M.F.; Binkley, N.C.; Bischoff-Ferrari, H.A.; Gordon, C.M.; Hanley, D.A.; Heaney, R.P.; Murad, M.H.; Weaver, C.M. Evaluation, treatment, and prevention of vitamin D deficiency: An Endocrine Society clinical practice guideline. *J. Clin. Endocrinol. Metab.* **2011**, *96*, 1911–1930. [CrossRef] [PubMed]
30. Bresson, J.L.; Burlingame, B.; Dean, T.; Fairweather-Tait, S.; Heinonen, M.; Hirsch-Ernst, K.I.; Mangelsdorf, I.; McArdle, H.; Naska, A.; Neuhauser-Berthold, M.; et al. Dietary reference values for vitamin D. *EFSA J.* **2016**, *14*, e04547.
31. Spiro, A.; Buttriss, J.L. Vitamin D: An overview of vitamin D status and intake in Europe. *Nutr. Bull.* **2014**, *39*, 322–350. [CrossRef]
32. Bouillon, R. Comparative analysis of nutritional guidelines for vitamin D. *Nat. Rev. Endocrinol.* **2017**, *13*, 466–479. [CrossRef]
33. German Nutrition, S. New reference values for vitamin D. *Ann. Nutr. Metab.* **2012**, *60*, 241–246. [CrossRef]
34. European Commission (EC). Regulation (EU) No 1169/2011 of the European Parliament and of the Council of 25 October 2011 on the provision of food information to consumers, amending Regulations (EC) No 1924/2006 and (EC) No 1925/2006 of the European Parliament and of the Council, and repealing Commission Directive 87/250/EEC, Council Directive 90/496/EEC, Commission Directive 1999/10/EC, Directive 2000/13/EC of the European Parliament and of the Council, Commission Directives 2002/67/EC and 2008/5/EC and Commission Regulation (EC) No 608/2004 Text with EEA relevance. *Off. J. Eur. Union* **2011**, *L 304/18*, 18–63.
35. Mensink, G.B.; Fletcher, R.; Gurinovic, M.; Huybrechts, I.; Lafay, L.; Serra-Majem, L.; Szponar, L.; Tetens, I.; Verkaik-Kloosterman, J.; Baka, A.; et al. Mapping low intake of micronutrients across Europe. *Br. J. Nutr.* **2013**, *110*, 755–773. [CrossRef]
36. Novaković, R.; Cavelaars, A.E.; Bekkering, G.E.; Roman-Viñas, B.; Ngo, J.; Gurinović, M.; Glibetić, M.; Nikolić, M.; Golesorkhi, M.; Medina, M.W.; et al. Micronutrient intake and status in Central and Eastern Europe compared with other European countries, results from the EURRECA network. *Public Health Nutr.* **2013**, *16*, 824–840. [CrossRef]
37. Jenab, M.; Salvini, S.; van Gils, C.H.; Brustad, M.; Shakya-Shrestha, S.; Buijsse, B.; Verhagen, H.; Touvier, M.; Biessy, C.; Wallström, P.; et al. Dietary intakes of retinol, beta-carotene, vitamin D and vitamin E in the European Prospective Investigation into Cancer and Nutrition cohort. *Eur. J. Clin. Nutr.* **2009**, *63*, S150–S178. [CrossRef]
38. Roman Viñas, B.; Ribas Barba, L.; Ngo, J.; Gurinovic, M.; Novakovic, R.; Cavelaars, A.; de Groot, L.C.; van't Veer, P.; Matthys, C.; Serra Majem, L. Projected prevalence of inadequate nutrient intakes in Europe. *Ann. Nutr. Metab.* **2011**, *59*, 84–95. [CrossRef]
39. Freisling, H.; Fahey, M.T.; Moskal, A.; Ocke, M.C.; Ferrari, P.; Jenab, M.; Norat, T.; Naska, A.; Welch, A.A.; Navarro, C.; et al. Region-specific nutrient intake patterns exhibit a geographical gradient within and between European countries. *J. Nutr.* **2010**, *140*, 1280–1286. [CrossRef]
40. Rippin, H.L.; Hutchinson, J.; Jewell, J.; Breda, J.J.; Cade, J.E. Adult Nutrient Intakes from Current National Dietary Surveys of European Populations. *Nutrients* **2017**, *9*, 1288. [CrossRef]
41. Elmadfa, I.; Meyer, A.; Nowak, V.; Hasenegger, V.; Putz, P.; Verstraeten, R.; Remaut-DeWinter, A.M.; Kolsteren, P.; Dostálová, J.; Dlouhý, P.; et al. European Nutrition and Health Report 2009. *Forum Nutr.* **2009**, *62*, 12–13.
42. Policnik, R.; Pokorn, D.; Kulnik, D.; Micetic-Turk, D.; Hlastan-Ribic, C. Energy and nutrient intake among pre-school children in central slovenia. *Acta Aliment.* **2013**, *42*, 291–300. [CrossRef]
43. Zdešar Kotnik, K. Smiselnost Uporabe Vitaminskih in Mineralnih Prehranskih Dopolnil Pri Mladostnikih = Advisability of Vitamin and Mineral Dietary Supplement Use Among Adolescents. Ph.D. Thesis, Univerza v Ljubljani, Biotehniška Fakulteta, Ljubljana, Slovenia, 2019.
44. Fidler Mis, N.; Kobe, H.; Štimec, M. Dietary intake of macro-and micronutrients in Slovenian adolescents: Comparison with reference values. *Ann. Nutr. Metab.* **2012**, *61*, 305–313. [CrossRef] [PubMed]

45. Gregorič, M. Ocena Prehranskega Vnosa Pri Mladostnikih Z Vidika Varovanja Zdravja = Assessment of Dietary Intake Among Adolescents from Health Protection Aspect. Ph.D. Thesis, Univerza v Ljubljani, Biotehniška Fakulteta, Ljubljana, Slovenia, 2015.
46. Koch, V.; Gregorič, M. Prehranska kakovost zajtrka slovenskih srednješolcev = Nutritional quality of breakfast eaten by secondary school students in Slovenia. *Zdr. Varst.* **2009**, *48*, 131–142.
47. Jeretina, B. Dejavniki Vpliva Na Pojav Prezgodnje Menopavze = Risk Factors for The Onset of Premature Menopause. Ph.D. Thesis, Univerza v Ljubljani, Zdravstvena Fakulteta, Ljubljana, Slovenia, 2019.
48. Juvan, S. Kakovost vegetarijanske prehrane med študenti = Quality of Vegetarian Nutrition between Students. Ph.D. Thesis, Univerza v Ljubljani, Biotehniška Fakulteta, Ljubljana, Slovenia, 1997.
49. Soltirovska Salamon, A.; Benedik, E.; Bratanič, B.; Velkavrh, M.; Rogelj, I.; Fidler Mis, N.; Bogovič Matijašić, B.; Paro-Panjan, D. Vitamin D status and its determinants in healthy Slovenian pregnant women. *Ann. Nutr. Metab.* **2015**, *67*, 96–103. [CrossRef] [PubMed]
50. Urh, N.; Babnik, K.; Rebec, D.; Poklar Vatovec, T. Ocena prehranskega stanja starejših v socialnovarstvenem zavodu = Assessment of the nutritional status of the elderly in a residential home. *Obz. Zdrav. Nege.* **2017**, *51*, 207–218. [CrossRef]
51. Gregorič, M.; Kotnik, K.Z.; Pigac, I.; Blenkuš, M.G. A Web-Based 24-H Dietary Recall Could Be a Valid Tool for the Indicative Assessment of Dietary Intake in Older Adults Living in Slovenia. *Nutrients* **2019**, *11*, 2234. [CrossRef] [PubMed]
52. Lichthammer, A.; Nagy, B.; Orbán, C.; Tóth, T.; Csajbók, R.; Molnár, S.; Tátrai-Németh, K.; Bálint, M.V. A comparative study of eating habits, calcium and Vitamin D intakes in the population of Central-Eastern European countries. *New Med.* **2015**, *19*, 66–70.
53. European Food Safety Authority. Guidance on the EU Menu methodology. *EFSA J.* **2014**, *12*, 3944.
54. Gregorič, M.; Blaznik, U.; Delfar, N.; Zaletel, M.; Lavtar, D.; Koroušić Seljak, B.; Golja, P.; Zdešar Kotnik, K.; Pravst, I.; Fidler Mis, N.; et al. Slovenian national food consumption survey in adolescents, adults and elderly. *EFSA Supporting Publ.* **1729**, *16*, en-1729. [CrossRef]
55. Craig, C.L.; Marshall, A.L.; Sjöström, M.; Bauman, A.E.; Booth, M.L.; Ainsworth, B.E.; Pratt, M.; Ekelund, U.; Yngve, A.; Sallis, J.F. International physical activity questionnaire: 12-country reliability and validity. *Med. Sci. Sports Exerc.* **2003**, *35*, 1381–1395. [CrossRef] [PubMed]
56. Vede, T. Izdelava in validacija slikovnega gradiva za določanje vnosa živil [Design and Validation of Food Picture Book]. Ph.D. Thesis, University of Ljubljana, Biotechnical Faculty, Ljubljana, Slovenia, 2016.
57. Dodd, K.W.; Guenther, P.M.; Freedman, L.S.; Subar, A.F.; Kipnis, V.; Midthune, D.; Tooze, J.A.; Krebs-Smith, S.M. Statistical methods for estimating usual intake of nutrients and foods: A review of the theory. *J. Am. Diet. Assoc.* **2006**, *106*, 1640–1650. [CrossRef]
58. Korošec, M.; Golob, T.; Bertoncelj, J.; Stibilj, V.; Koroušić Seljak, B. The Slovenian food composition database. *Food Chem.* **2013**, *140*, 495–499. [CrossRef] [PubMed]
59. Harttig, U.; Haubrock, J.; Knüppel, S.; Boeing, H. The MSM program: Web-based statistics package for estimating usual dietary intake using the Multiple Source Method. *Eur. J. Clin. Nutr.* **2011**, *65*, S87–S91. [CrossRef]
60. Nutrition Unit of the National Institute for Health and Welfare. Fineli—The National Food Composition Database. Available online: https://fineli.fi/fineli/en/index (accessed on 10 May 2020).
61. Roe, M.; Pinchen, H.; Church, S.; Finglas, P. McCance and Widdowson's The Composition of Foods Seventh Summary Edition and updated Composition of Foods Integrated Dataset. *Nutr. Bull.* **2015**, *40*, 36–39. [CrossRef]
62. Bodner-Montville, J.; Ahuja, J.K.; Ingwersen, L.A.; Haggerty, E.S.; Enns, C.W.; Perloff, B.P. USDA food and nutrient database for dietary studies: Released on the web. *J. Food Compos.* **2006**, *19*, S100–S107. [CrossRef]
63. Department of Health. Regulations Amending Certain Regulations Made Under the Food and Drugs Act (Nutrition Symbols, Other Labelling Provisions Partially Hydrogenated Oils and Vitamin D.). In *Part I*; No. 6; Canada Gazette: Ottawa, ON, Canada, 2018; Volume 152, p. 256.
64. Zupanič, N.; Hristov, H.; Gregorič, M.; Blaznik, U.; Delfar, N.; Koroušić Seljak, B.; Ding, E.L.; Fidler Mis, N.; Pravst, I. Total and Free Sugars Consumption in a Slovenian Population Representative Sample. *Nutrients* **2020**, *12*, 1729. [CrossRef] [PubMed]
65. Haubrock, J.; Nöthlings, U.; Volatier, J.-L.; Dekkers, A.; Ocké, M.; Harttig, U.; Illner, A.-K.; Knüppel, S.; Andersen, L.F.; Boeing, H.; et al. Estimating Usual Food Intake Distributions by Using the Multiple Source Method in the EPIC-Potsdam Calibration Study. *J. Nutr.* **2011**, *141*, 914–920. [CrossRef] [PubMed]
66. Kolenikov, S. Calibrating Survey Data using Iterative Proportional Fitting (Raking). *Stata J.* **2014**, *14*, 22–59. [CrossRef]
67. Koenig, J.; Elmadfa, I. Status of calcium and vitamin D of different population groups in Austria. *Int. J. Vitam. Nutr. Res.* **2000**, *70*, 214–220. [CrossRef]
68. Kudlacek, S.; Schneider, B.; Peterlik, M.; Leb, G.; Klaushofer, K.; Weber, K.; Woloszczuk, W.; Willvonseder, R. Assessment of vitamin D and calcium status in healthy adult Austrians. *Eur. J. Clin. Invest.* **2003**, *33*, 323–331. [CrossRef] [PubMed]
69. Kamycheva, E.; Joakimsen, R.M.; Jorde, R. Intakes of Calcium and Vitamin D Predict Body Mass Index in the Population of Northern Norway. *J. Nutr.* **2003**, *133*, 102–106. [CrossRef]
70. de Mestral, C.; Marques-Vidal, P.; Gaspoz, J.-M.; Theler, J.-M.; Guessous, I. Independent association between socioeconomic indicators and macro- and micro-nutrient intake in Switzerland. *PLoS ONE* **2017**, *12*, e0174578. [CrossRef]
71. Galobardes, B.; Morabia, A.; Bernstein, M.S. Diet and socioeconomic position: Does the use of different indicators matter? *Int. J. Epidemiol.* **2001**, *30*, 334–340. [CrossRef]

72. Rossum, C.T.M.v.; Fransen, H.P.; Verkaik-Kloosterman, J.; Buurma-Rethans, E.J.M.; Ocké, M.C. *Dutch National Food Consumption Survey 2007–2010: Diet of Children and Adults Aged 7 to 69 Years*; National Institute for Public Health and the Environment: Utrecht, The Netherlands, 2011. Available online: https://www.rivm.nl/bibliotheek/rapporten/350050006.pdf (accessed on 14 May 2021).
73. WORLDOMETERS. Coronavirus in Slovenia. Available online: https://www.worldometers.info/coronavirus/country/slovenia/ (accessed on 23 September 2021).
74. IOS. The Eastern European & Central Asian Regional Audit Epidemiology, Costs & Burden of Osteoporosis in 2010. Available online: https://www.osteoporosis.foundation/sites/iofbonehealth/files/2019-06/2010_Eastern_European_Central_Asian_Audit_English.pdf (accessed on 23 September 2021).
75. Borgström, F.; Karlsson, L.; Ortsäter, G.; Norton, N.; Halbout, P.; Cooper, C.; Lorentzon, M.; McCloskey, E.V.; Harvey, N.C.; Javaid, M.K.; et al. Fragility fractures in Europe: Burden, management and opportunities. *Arch. Osteoporos.* **2020**, *15*, 1–21. [CrossRef]
76. European Commission (EC). Community Register on the Addition of Vitamins and Minerals and of Certain other Substances to Foods. Available online: https://ec.europa.eu/food/system/files/2021-01/labelling_nutrition-vitamins_minerals-comm_reg_en.pdf (accessed on 23 September 2021).
77. Jääskeläinen, T.; Itkonen, S.T.; Lundqvist, A.; Erkkola, M.; Koskela, T.; Lakkala, K.; Dowling, K.G.; Hull, G.L.; Kröger, H.; Karppinen, J.; et al. The positive impact of general vitamin D food fortification policy on vitamin D status in a representative adult Finnish population: Evidence from an 11-y follow-up based on standardized 25-hydroxyvitamin D data. *Am. J. Clin. Nutr* **2017**, *105*, 1512–1520. [CrossRef]
78. Calvo, M.S.; Whiting, S.J. Vitamin D Fortification in North America: Current Status and Future Considerations. In *Handbook of Food Fortification and Health*; Humana Press: Totowa, NJ, USA, 2013; pp. 259–275.
79. Itkonen, S.T.; Erkkola, M.; Lamberg-Allardt, C.J.E. Vitamin D Fortification of Fluid Milk Products and Their Contribution to Vitamin D Intake and Vitamin D Status in Observational Studies-A Review. *Nutrients* **2018**, *10*, 1054. [CrossRef]
80. Black, L.J.; Seamans, K.M.; Cashman, K.D.; Kiely, M. An updated systematic review and meta-analysis of the efficacy of vitamin D food fortification. *J. Nutr.* **2012**, *142*, 1102–1108. [CrossRef] [PubMed]
81. Brown, J.; Sandmann, A.; Ignatius, A.; Amling, M.; Barvencik, F. New perspectives on vitamin D food fortification based on a modeling of 25(OH)D concentrations. *Nutr. J.* **2013**, *12*, 1–12. [CrossRef] [PubMed]
82. O'Mahony, L.; Stepien, M.; Gibney, M.J.; Nugent, A.P.; Brennan, L. The potential role of vitamin D enhanced foods in improving vitamin D status. *Nutrients* **2011**, *3*, 1023–1041. [CrossRef]
83. Whiting, S.J.; Langlois, K.A.; Vatanparast, H.; Greene-Finestone, L.S. The vitamin D status of Canadians relative to the 2011 Dietary Reference Intakes: An examination in children and adults with and without supplement use. *Am. J. Clin. Nutr.* **2011**, *94*, 128–135. [CrossRef]
84. Janssen, M.; Chang, B.P.I.; Hristov, H.; Pravst, I.; Profeta, A.; Millard, J. Changes in Food Consumption During the COVID-19 Pandemic: Analysis of Consumer Survey Data From the First Lockdown Period in Denmark, Germany, and Slovenia. *Front. Nutr.* **2021**, *8*, 60. [CrossRef] [PubMed]
85. Žmitek, K.; Hribar, M.; Lavriša, Ž.; Hristov, H.; Kušar, A.; Pravst, I. Socio-Demographic and Knowledge-Related Determinants of Vitamin D Supplementation in the Context of the COVID-19 Pandemic: Assessment of an Educational Intervention. *Front. Nutr.* **2021**, *8*, 290. [CrossRef]

nutrients

MDPI

Article

Evaluation of Nutrient Intake and Food Consumption among Dutch Toddlers

Elly Steenbergen [1], Anne Krijger [2], Janneke Verkaik-Kloosterman [1], Liset E. M. Elstgeest [2], Sovianne ter Borg [1], Koen F. M. Joosten [2] and Caroline T. M. van Rossum [1,*]

[1] National Institute for Public Health and the Environment, 3720 BA Bilthoven, The Netherlands; elly.steenbergen@rivm.nl (E.S.); janneke.verkaik@rivm.nl (J.V.-K.); sovianne.ter.borg@rivm.nl (S.t.B.)
[2] Erasmus MC, University Medical Center Rotterdam, 3015 GD Rotterdam, The Netherlands; j.j.a.krijger@erasmusmc.nl (A.K.); l.elstgeest@erasmusmc.nl (L.E.M.E.); k.joosten@erasmusmc.nl (K.F.M.J.)
* Correspondence: caroline.van.rossum@rivm.nl; Tel.: +31-(0)6-4686-0803

Abstract: Improving dietary habits at a young age could prevent adverse health outcomes. The aim was to gain insight into the adequacy of the dietary intake of Dutch toddlers, which may provide valuable information for preventive measures. Data obtained from the Dutch National Food Consumption Survey 2012–2016 were used, which included 672 children aged one to three years. Habitual intakes of nutrients were evaluated according to recommendations set by the Dutch Health Council. Specific food groups were evaluated according to the Dutch food-based dietary guidelines. For most nutrients, intakes were estimated to be adequate. High intakes were found for saturated fatty acids, retinol, iodine, copper, zinc, and sodium. No statement could be provided on the adequacy of intakes of alpha-linoleic acids, N-3 fish fatty acids, fiber, and iron. 74% of the toddlers used dietary supplements, and 59% used vitamin D supplements specifically. Total median intakes of vegetables, bread, and milk products were sufficient. Consumption of bread, potatoes and cereals, milk products, fats, and drinks consisted largely of unhealthy products. Consumption of unfavorable products may have been the cause of the observed high and low intakes of several nutrients. Shifting towards a healthier diet that is more in line with the guidelines may positively affect the dietary intake of Dutch toddlers and prevent negative health impacts, also later in life.

Keywords: dietary intake; macronutrients; micronutrients; food groups; dietary guidelines; young children

Citation: Steenbergen, E.; Krijger, A.; Verkaik-Kloosterman, J.; Elstgeest, L.E.M.; ter Borg, S.; Joosten, K.F.M.; van Rossum, C.T.M. Evaluation of Nutrient Intake and Food Consumption among Dutch Toddlers. *Nutrients* **2021**, *13*, 1531. https://doi.org/10.3390/nu13051531

Academic Editor: Murielle Bochud

Received: 16 March 2021
Accepted: 29 April 2021
Published: 1 May 2021

1. Introduction

A healthy diet, characterized by an adequate, safe, and balanced nutritional intake, is pivotal in preserving and promoting overall health throughout the life course [1]. Early childhood is a period of rapid growth and development, and therefore, a time of great opportunity, yet also vulnerability. Hence, nutrition during early life is of special importance and increasingly recognized for its long-term implications [2]. Undernutrition during childhood, defined as insufficient intakes of energy or nutrients, has been linked to short-term consequences, such as impaired growth and development as well as higher infection and mortality risk [3]. In addition, undernutrition is also related to later life health consequences, such as the increased risk of diabetes and hypertension. In addition, an inadequate diet might also have sociodemographic consequences in the long-term, such as lower education level and lower income, due to poorly developed cognitive function [4]. Overnutrition comprises the excess and insufficiency of dietary intake along with overweight and obesity. Childhood obesity is associated with various comorbidities, including childhood manifestations of cardiovascular disease, obstructive sleep apnea, non-alcoholic fatty liver disease, and psychosocial problems [5]. Moreover, childhood obesity has been shown to track into adulthood and increases the risk of type 2 diabetes, hypertension, dyslipidemia, and carotid-artery atherosclerosis in those children with persisting obesity [6,7]. Diet-related health consequences are a major threat in public health in Europe as well as worldwide [8].

Although dietary habits established during childhood likely persist into adulthood [9], diet is considered an important modifiable factor [10]. Hence, improving dietary habits at a young age could sustainably prevent adverse health outcomes. In many countries, food-based dietary guidelines are developed to help consumers eat healthily. A healthy diet provides a sufficient intake of nutrients to maintain or improve people's health. A review on the dietary intake of young children from several European countries has shown potential deficiencies or excess in the intake of nutrients and food groups [11]. However, some of the included studies were conducted more than a decade ago. National food consumption surveys, carried out in several countries, are periodically conducted to provide insight into dietary habits at the population level so that, for example, policymakers and health professionals can implement this in practice by facilitating the shift to more sustainable and safe food for the consumers.

In 2020, a Dutch governmental project was launched on developing a screening tool to assess the nutrition and lifestyle of young children living in the Netherlands, after which measures can be implemented to prevent negative health outcomes. The present study is part of this project and aimed to identify potential nutritional challenges of Dutch children aged one up until three years, which could be considered to be considered in the screening tool. To identify nutritional challenges, the habitual dietary intake, in terms of macronutrients and micronutrients and specific food groups, are described and examined on adequacy by using the most recent food consumption data of the Dutch National Food Consumption Survey (DNFCS 2012–2016) conducted in the general population of the Netherlands [12].

2. Materials and Methods

To assess the dietary intake of Dutch toddlers, data of the DNFCS 2012–2016 were used. A detailed methodological description of the DNFCS has been described elsewhere [12].

2.1. Data Collection and Study Population

In short, the DNFCS 2012–2016 was a cross-sectional survey carried out among the general Dutch population (1–79 years; n = 4313). Data were collected from November 2012 to January 2017. Participants were recruited from representative consumer panels of Kantar Public, for which the sampling was adjusted for characteristics, such as region of residence, degree of urbanization, educational level, and stratified for age and gender.

General data on background and lifestyle factors of participants were collected from questionnaires. Data on food consumption (intake of foods, drinks, and dietary supplements) were obtained during two nonconsecutive multiple-pass 24 h dietary recalls [13], with an interval of about four weeks, carried out by trained dieticians. The dietary recalls were evenly distributed over the days of the week and seasons.

For the present study, data of 672 children aged one to three years were used. The dietary recalls in this age group were completed by their parent(s) or caregiver(s); the first interview was performed during a home visit (including height and weight measurements by the dietician), and the second one was by telephone. To cover any consumptions at the daycare or elsewhere, the parent(s) or caregiver(s) completed a food diary for their child the day before the interviews took place.

To calculate macronutrient and micronutrient intake, food consumption data were combined with an extended version of the Dutch Food Composition Database (NEVO-online 2016) [14] and the Dutch Supplement Database (NES) dated 1 January 2018 [15]. In addition, the foods were classified into food groups according to the "wheel of five", which is substantiated by the Dutch food-based dietary guidelines [16]. Within this classification, products were distinguished into products that meet the Dutch food-based dietary guidelines (within the wheel of five) and products for which it is advised not to consume or to limit the consumption (outside the wheel of five). In addition, the wheel of five provides general recommendations on food consumption [17].

2.2. Data Analyses

Descriptive statistical analyses of participants' general characteristics were performed for the study population, unweighted and weighted for sociodemographic properties for which a weighting factor was applied to the participants in the analyses for results to be representative for children aged one to three years in the Netherlands. These general characteristics included characteristics of the participants' household, supplement use, and fruit and vegetable consumption. Unless otherwise stated, statistical analyses were performed in SAS, version 9.4 [18].

The habitual intake (also referred to as usual intake) distribution of macronutrients, micronutrients, and food groups was estimated from the observed daily intake by correction for the intra-individual (day-to-day) variance, using the Statistical Program to Assess Dietary Exposure (SPADE version 3.2.52 in R, [19]). SPADE analyses were performed age-dependently by gender, using data from all subjects in DNFCS 2012–2016 to predict the model parameters. Results were combined for specific age groups, e.g., children aged one to three years. For most nutrients, the SPADE one-part model was used. Different models were used for folic acid (two-part model) and micronutrients, fiber, and N-3 fish fatty acids (three-part model). If relevant, usual nutrient intakes from food, dietary supplements, and discretionary salt used at the table or during preparation were modeled separately and subsequently combined to total the usual intake (first shrink then add) [20,21]. For iodine and sodium, salt added during preparation or at the table was considered. To estimate the intake from different sources, a multipart model was used. To estimate habitual food consumption, different SPADE models were used for food groups consumed episodically (two-part model) and daily (one-part model). For more details, see the report on the DNFCS 2012–2016 [12].

Results for children aged one to three years are shown in terms of the mean and the distribution of the habitual nutrient intake and food consumption per day (percentiles 5, 25, 50, 75, and 95). 95% confidence intervals were estimated for the mean and the median (50th percentile) using bootstrap analyses.

2.3. Evaluation of Intake and Consumption

The habitual intake distributions of macronutrients and micronutrients from food only and from food and dietary supplements, if relevant, were evaluated by comparison with the ad-interim Dutch dietary reference intakes set by the Health Council in 2014 [22]. The evaluation method differed depending on the type of dietary reference value that was available. The estimated average requirement (EAR) of nutrients was used to estimate the proportion of Dutch toddlers with inadequate intake, using the EAR cut-point method [23]. If the proportion was less than 10%, the nutrient intake was considered adequate by a rule of thumb. When the EAR was not available, the adequate intake (AI) was used, which qualitatively evaluates whether a low prevalence of inadequate nutrient intake could be assumed [24]. If the median intake was at or above the AI, the intake seemed adequate. If the median intake were below the AI, no statement could be provided on the risk of inadequacy and further research on the intake is required. The evaluation with an EAR or AI does not indicate whether the intake is adequate or tolerable but only indicates the probability of adequacy.

For vitamin D, the intake evaluation was performed by comparing the intake with the AI, which was set assuming sufficient exposure to sunlight (i.e., 3 μg). It was assumed that two-thirds of the requirement was covered by vitamin D production in the skin by sunlight exposure with light skin types [25]. The AI for vitamin D intake when sunlight exposure is insufficient is 10 μg. For energy, the intake could not be evaluated by the EAR cut-point method, as one of the underlying assumptions (i.e., intake and requirement are not correlated) was not met. For vitamin K_1, no estimations were made for the intake from food and supplements as no data were available on vitamin K_1 in the NES database.

The tolerable upper intake levels (UL) for nutrients set by European Food Safety Authority (EFSA) [26] were used to estimate the proportion of Dutch toddlers that may be potentially at risk of adverse effects due to excessive intake of a nutrient. If this proportion (whose intake exceeded the UL) was larger than 2.5%, the nutrient intake was considered high at a population level. Otherwise, the intake was considered tolerable by a rule of thumb.

The habitual consumption distribution of food groups was evaluated by the wheel of five and the Dutch food-based dietary guidelines [17]. Recommendations of intakes of vegetables, fruit, and bread were set in terms of a range. For the intake evaluation, it was assessed per food group whether the median intake was equal to or larger than the recommended intake (or higher than the lower bound of the range) for products within the wheel of five ("in") and for all products within and outside the wheel of five ("total"). For the food groups, cheese and meat, the guideline was a maximum consumption, and it was assessed whether the median intake was below that recommendation.

3. Results

3.1. Population Characteristics

The population characteristics of the study population are shown in Table 1. Within the study population, there was an even distribution of boys and girls, of which the majority had a normal BMI. Eight percent of the study population was overweight or obese, and eight percent was (seriously) underweight. The migration background of the children's parents was mostly Dutch, and most of the parents had finished at least a middle education. Household sizes varied (between two to five persons), of which mostly consisted of four persons. Relatively more households were located in the west, corresponding with the most densely populated area of the Netherlands. From the questionnaires, it was observed that 77% of the toddlers had a daily consumption of fruits and 50% a daily consumption of vegetables. Furthermore, 74% of the toddlers used dietary supplements in general. In total, 59% used vitamin D supplements, and 72% used vitamin D-containing supplements (i.e., vitamin D, a combination of calcium and vitamin D, multivitamins, including minerals, and multivitamins without minerals) in winter and/or during the rest of the year.

3.2. Habitual Nutrient Intake

The habitual mean intake and percentiles of the intake distribution of macronutrients and micronutrients are shown in Table 2a,b, respectively.

The intakes of total protein, total fat, polyunsaturated fatty acids, cis-unsaturated fatty acids, trans-fatty acids, linoleic acid, and total carbohydrates met the recommendations of adequate and safe intakes. The total protein intake was adequate as the median protein intake (13.0 En%) was larger than (more than twice) the AI (5.0 En%). Four percent of the toddlers had an intake of saturated fatty acids above the UL. No statement on inadequacy was possible for alpha-linoleic acids and N-3 fish fatty acids (EPA + DHA) as the median intakes were below the AI. The median intake of N-3 fish fatty acids (EPA + DHA) was almost four times slower than the AI. The median fiber intake was below the recommended level.

Table 1. Population characteristics of children aged one to three years in the Netherlands unweighted and weighted for demographic properties (DNFCS 2012–2016; n = 672).

Variable	Categories	Frequency		
		n	% Unweighted	% Weighted
Gender	Male	332	49.4	50.0
	Female	340	50.6	49.9
BMI [1]	Seriously underweight	18	2.7	3.0
	Underweight	37	5.5	5.4
	Normal weight	563	83.8	83.0
	Overweight	38	5.7	6.4
	Obesity	14	2.1	2.0
	Unknown	2	0.3	0.2
Native country of the parents [2]	Dutch	622	92.6	92.0
	Western immigrant	17	2.5	2.3
	Non-Western immigrant	33	4.9	5.7
Size of household	Two or three persons	195	29.0	30.0
	Four persons	294	43.8	43.2
	Five or more persons	183	27.2	26.8
Highest education of the parents [3]	Low	29	4.3	8.0
	Middle	199	29.6	38.0
	High	444	66.1	54.0
Region of household location [4]	West	303	45.1	47.1
	North	78	11.6	9.8
	East	152	22.6	21.8
	South	139	20.7	21.2
Fruit consumption	Zero to four days per week	59	8.8	9.1
	Five to six days per week	97	14.4	14.4
	Every day	516	76.8	76.5
Vegetable consumption	Zero to four days per week	75	11.2	12.6
	Five to six days per week	257	38.2	37.6
	Every day	340	50.6	49.8
Use of dietary supplements	Yes	504	75.0	74.1
	No	168	25.0	25.9
Use of vitamin D supplements in winter and/or rest of the year	Yes	406	60.4	59.1
	No	266	39.6	40.9
Use of vitamin D containing supplements in winter and/or rest of the year [5]	Yes	491	73.1	71.9
	No	181	26.9	28.1

[1] Body mass index (BMI) was calculated per person as the bodyweight divided by the height squared (kg/m^2). For BMI, age and gender-specific values based on the extended international (IOTF) body mass cut-offs were used [27]. [2] Native countries of the parents. Dutch: both parents were born in the Netherlands; Western immigrant: from Europe, United States, Australia; and non-Western immigrant. For Western and non-Western immigrants, at least one parent was born abroad. [3] Highest education of the parents. Low: primary education, lower vocational education, advanced elementary education; middle: intermediate vocational education, higher secondary education; and high: higher vocational education and university. [4] Region of household location was based on Nielsen CBS division and included the three largest cities Amsterdam, Rotterdam, and The Hague. [5] Supplements containing vitamin D: vitamin D only, a combination of calcium and vitamin D, multivitamins, including minerals, and multivitamins without minerals.

Table 2. (**a**). The distribution of habitual macronutrient intake (per day) from food only ("f") and, if relevant, from food and dietary supplements ("f + s") by Dutch children aged one to three years (DNFCS 2012–2016, n = 672, weighted for demographic characteristics, season, and day of the week). (**b**) The distribution of habitual micronutrient intake (per day) from food only ("f") and, if relevant, from food and dietary supplements ("f + s") by Dutch children aged one to three years (DNFCS 2012–2016, n = 672, weighted for demographic characteristics, season, and day of the week).

(a)

Macronutrient	Source	Mean (95% CI)	P5	P25	P50 (95% CI)	P75	P95	EAR	% < EAR	AI	P50 ≥ AI?	UL	% > UL	Evaluation *
Protein (g/kg)	f	3.1 (3.1–3.2)	1.9	2.5	3.0 (3.0–3.0)	3.6	4.8	0.7	0					EAR: adequate intake
Total protein (g)	f	41 (41–42)	25	34	40 (40–41)	48	61	11	0					EAR: adequate intake
Total protein (En%)	f	13.2 (13.1–13.2)	9.8	11.6	13.0 (13.0–13.1)	14.6	17.1			5	Yes	20	0.5	AI: seems adequate; UL: tolerable intake
Total fat (En%)	f	29.4 (29.3–29.5)	22.1	26.4	29.4 (29.3–29.5)	32.5	36.9			25	Yes	40	1.1	AI: seems adequate; UL: tolerable intake
Saturated fatty acids (En%)	f	11.0 (11.0–11.1)	7.6	9.5	10.9 (10.9–11.0)	12.5	14.8					15	4.2	UL: high intake
Polyunsaturated fatty acids (En%)	f	5.6 (5.6–5.6)	3.6	4.7	5.5 (5.5–5.5)	6.4	8.0					12	0	UL: tolerable intake
Cis-unsaturated fatty acids (En%)	f	15.7 (15.6–15.7)	11.1	13.6	15.5 (15.5–15.6)	17.6	20.7					38	0	UL: tolerable intake
Trans fatty acids (En%)	f	0.3 (0.3–0.3)	0.1	0.2	0.3 (0.3–0.3)	0.3	0.5					1	0	UL: tolerable intake
Linoleic acid (En%)	f	4.6 (4.6–4.7)	2.9	3.8	4.5 (4.5–4.6)	5.4	6.8			2	Yes			AI: seems adequate
Alpha linoleic acid (En%)	f	0.6 (0.6–0.6)	0.4	0.5	0.6 (0.6–0.6)	0.7	0.9			1	No			AI: no statement
N-3 fish fatty acids (EPA + DHA, mg)	f	54 (51–58)	8	20	38 (35–40)	68	158			150	No			AI: no statement
	f + s	57 (52–62)	8	20	38 (36–41)	69	167			150	No			AI: no statement
Total carbohydrates (g)	f	174 (172–175)	100	138	169 (167–171)	204	260	92	3					EAR: adequate intake
Total carbohydrates (En%)	f	54.9 (54.8–55.0)	46.6	51.6	55.0 (54.8–55.1)	58.3	63.1			45	Yes			AI: seems adequate
Fiber (g/MJ)	f	2.4 (2.4–2.5)	1.6	2.1	2.4 (2.4–2.4)	2.8	3.4			2.8 [1]	No			Guideline: no statement

(b)

Micronutrient	Source	Mean (95% CI)	P5	P25	P50 (95% CI)	P75	P95	AI	P50 ≥ AI?	UL	% > UL	Evaluation *
Retinol activity equivalents (RAE, μg) [1]	f	558 (534–80)	237	371	508 (481–524)	685	1062	300	Yes			AI: seems adequate
	f + s	594 (579–609)	246	387	533 (520–546)	735	1147	300	Yes			AI: seems adequate
Retinol (μg)	f	422 (405–438)	149	256	373 (353–383)	528	874			800	7.0	UL: high intake
	f + s	469 (456–482)	163	282	411 (400–422)	593	969			800	10.5	UL: high intake
Vitamin B_1 (mg)	f	0.6 (0.6–0.6)	0.4	0.5	0.6 (0.6–0.6)	0.7	1.0	0.3	Yes			AI: seems adequate
	f + s	1.0 (0.6–1.3)	0.4	0.5	0.6 (0.6–0.7)	0.8	1.3	0.3	Yes			AI: seems adequate
Vitamin B_2 (mg)	f	1.1 (1.0–1.1)	0.6	0.8	1.0 (1.0–1.0)	1.3	1.7	0.5	Yes			AI: seems adequate
	f + s	1.4 (1.0–1.9)	0.6	0.9	1.1 (1.1–1.1)	1.4	2.0	0.5	Yes			AI: seems adequate
Vitamin B_3 (mg)	f	8.5 (8.5–8.6)	4.8	6.6	8.2 (8.1–8.2)	10.1	13.6	4	Yes			AI: seems adequate
	f + s	9.9 (9.5–10.4)	4.9	6.8	8.6 (8.5–8.8)	11.1	17.0	4	Yes			AI: seems adequate
Vitamin B_6 (mg)	f	1.0 (1.0–1.0)	0.6	0.8	0.9 (0.9–0.9)	1.1	1.5	0.4	Yes	5	0	AI: seems adequate; UL: tolerable intake
	f + s	1.1 (1.0–1.3)	0.6	0.8	1.0 (1.0–1.0)	1.2	1.8	0.4	Yes	5	0.4	AI: seems adequate; UL: tolerable intake
Folate equivalents (μg) [2]	f	141 (140–142)	82	111	136 (134–137)	165	219	85	Yes			AI: seems adequate
	f + s	172 (164–179)	88	119	149 (146–152)	193	334	85	Yes			AI: seems adequate
Folic acid (μg)	f	9 (8–10)	0	0	4 (3–5)	13	37			200	0	UL: tolerable intake
	f + s	25 (21–28)	0	1	9 (7–10)	29	106			200	0.7	UL: tolerable intake
Vitamin B_{12} (μg)	f	2.7 (2.6–2.7)	1.2	1.9	2.5 (2.5–2.5)	3.2	4.7	0.7	Yes			AI: seems adequate
	f + s	4.9 (2.3–7.6)	1.4	2.0	2.7 (2.7–2.7)	3.6	5.4	0.7	Yes			AI: seems adequate

Table 2. *Cont.*

Micronutrient	Source	Mean (95% CI)	P5	P25	P50 (95% CI)	P75	P95	AI	P50 ≥ AI?	UL	% > UL	Evaluation *
Vitamin C (mg)	f	77 (76–78)	29	50	71 (70–72)	96	145	25	Yes			AI: seems adequate
	f + s	96 (75–118)	32	55	79 (77–80)	108	172	25	Yes			AI: seems adequate
Vitamin D [3] (μg)	f	2.6 (2.6–2.7)	0.9	1.6	2.4 (2.3–2.4)	3.3	5.3	3	No			AI: no statement
	f + s	8.4 (8.0–8.9)	1.3	3.7	7.6 (6.9–8.3)	11.9	17.5	3	Yes			AI: seems adequate
Vitamin E (mg)	f	7.2 (7.1–7.3)	3.6	5.3	6.8 (6.7–6.9)	8.7	12.0	4	Yes	100	0	AI: seems adequate; UL: tolerable intake
	f + s	8.7 (8.1–9.3)	3.9	5.7	7.4 (7.2–7.5)	9.8	15.8	4	Yes	100	0.3	AI: seems adequate; UL: tolerable intake
Vitamin K_1 (μg)	f	39.2 (37.3–41.1)	9.5	19.0	30.4 (29.1–31.8)	49.0	98.9	12	Yes			AI: seems adequate
Calcium (mg)	f	700 (696–705)	361	527	671 (665–676)	841	1144	500	Yes			AI: seems adequate
	f + s	720 (711–729)	371	539	686 (679–693)	860	1185	500	Yes			AI: seems adequate
Copper (mg)	f	0.7 (0.7–0.7)	0.5	0.6	0.7 (0.7–0.7)	0.9	1.1	0.3	Yes	1	10.2	AI: seems adequate; UL: high intake
	f + s	0.8 (0.7–0.8)	0.5	0.6	0.7 (0.7–0.7)	0.9	1.1	0.3	Yes	1	11.5	AI: seems adequate; UL: high intake
Iodine (μg)	f	121 (119–125)	70	96	117 (116–121)	143	186	70	Yes	200	2.7	AI: seems adequate UL: high intake
	f + s	127 (125–131)	73	99	122 (119–125)	149	200	70	Yes	200	5.1	AI: seems adequate UL: high intake
Iron (mg)	f	5.8 (5.7–5.8)	3.4	4.6	5.6 (5.5–5.6)	6.7	8.6	8	No			AI: no statement
	f + s	6.2 (6.0–6.4)	3.6	4.8	5.9 (5.8–5.9)	7.1	9.7	8	No			AI: no statement
Magnesium (mg)	f	182 (180–183)	112	148	177 (175–178)	210	267	85	Yes			AI: seems adequate
	f + s	186 (184–188)	115	151	180 (179–182)	215	275	85	Yes			AI: seems adequate
Phosphorus (mg)	f	851 (845–857)	520	691	829 (823–836)	988	1253	470	Yes			AI: seems adequate
	f + s	848 (841–854)	521	690	826 (819–833)	983	1249	470	Yes			AI: seems adequate
Potassium (mg)	f	1840 (1830–1851)	1141	1509	1799 (1787–1811)	2131	2677	1400	Yes			AI: seems adequate
	f + s	1831 (1818–1843)	1131	1495	1790 (1776–1804)	2125	2660	1400	Yes			AI: seems adequate
Selenium (μg)	f	23 (23–24)	13	18	22 (22–22)	27	37	20	Yes	60	0.1	AI: seems adequate; UL: tolerable intake
	f + s	25 (24–25)	14	19	23 (23–24)	29	41	20	Yes	60	0.5	AI: seems adequate; UL: tolerable intake
Sodium (g)	f	3.1 (3.0–3.1)	1.7	2.3	2.9 (2.9–3.0)	3.6	4.8			3 [4]	47.5	Guideline: high intake
Zinc (mg)	f	5.7 (5.7–5.8)	3.6	4.7	5.6 (5.5–5.6)	6.6	8.4	5	Yes	7	18.6	AI: seems adequate; UL: high intake
	f + s	6.0 (5.9–6.1)	3.7	4.9	5.8 (5.8–5.9)	7.0	8.9	5	Yes	7	24.3	AI: seems adequate; UL: high intake

(**a**) CI = confidence intervals; EAR = estimated average requirement; AI = adequate intake; UL = upper tolerable level. * The habitual intake seemed or was considered to be adequate or tolerable if % < EAR is below 10%, P50 ≥ AI or % > UL is equal to or smaller than 2.5%. [1] This is a guideline rather than an AI [28]. (**b**) CI = confidence intervals; EAR = estimated average requirement; AI = adequate intake; UL = upper tolerable level. * The habitual intake seemed or was considered to be adequate or tolerable if % < EAR is below 10%, P50 ≥ AI or % > UL is equal to or smaller than 2.5%. The EAR was not incorporated in the table as there were no values for EAR for the observed micronutrients. [1] Calculated as μg retinol + μg β-carotene/12 + μg other carotenoids/24 [29]. [2] Calculated using the amount of folate naturally present in foods (in μg) plus 1.7 times the amount of folic acid in enriched foods (in μg) plus 2.0 times the amount of folic acid in dietary supplements (in μg) [14]. [3] Assuming that two-thirds of the requirement is covered by vitamin D production in the skin by sunlight exposure with light skin types [25]. [4] This is a guideline rather than a UL [30].

The intakes of vitamins B_1, B_2, B_3, B_6, B_{12}, C, E, and K_1, as well as folate equivalents, folic acid, calcium, magnesium, potassium, and selenium, met the recommendations. Under the assumption of sufficient sunlight exposure (i.e., two-thirds of the requirement was covered by vitamin D production in the skin by sunlight exposure with light skin types) for the toddlers, the median vitamin D intake from food and supplements was higher than the AI (as shown in Table 2b); thus, the intake met the recommendation. However, when using the AI for vitamin D intake when sunlight exposure is insufficient (i.e., 10 µg), the median vitamin D intake from food and supplements was below that AI. The intake of retinol from food only and from both food and dietary supplements was considered high as the proportion exceeding the UL was 7.9% and 10.5%, respectively. The median intake of retinol activity equivalents (RAE) from both food only (508 µg) and food and supplements combined (533 µg) was above the AI (300 µg). Therefore, there was a low risk of inadequate intakes. For copper and zinc, the intakes seemed to be adequate according to the AI. However, high intakes of copper and zinc from both food only as from food combined with supplements were observed (for copper 10.2% and 11.5%, and zinc 18.6% and 24.3%, respectively had an intake above the UL). For iodine via food combined with dietary supplements, the intake was considered high for a subgroup of the children (5.1% exceeded the UL). For iron, the median intake from food only (4.6 mg) as well as from food and dietary supplements (4.8 mg) was quite below the AI (8 mg); therefore, no statement on inadequacy could be provided. On the contrary, the median intake of vitamin C and magnesium was twice the AI. Sodium intake was considered high as the proportion exceeding the guideline of 6 g per day was 47.5%. Except for vitamin D, no major differences were observed between the intake via food or via food combined with dietary supplements.

Not reported in tables is the habitual intake of energy. The EAR for the energy intake was 5 MJ per day, and the observed median intake was 5.2 MJ per day. However, the energy intake could not be evaluated with the EAR.

3.3. Food Group Consumption

The mean habitual consumption and percentiles of the consumption distribution of food groups mentioned in the wheel of five are shown in Table 3. For each food group, the consumption was compared with recommended consumption levels and evaluated for products that fit the wheel of five (categorized as "in" the wheel of five) and the "total" consumption (in and outside the wheel of five). Evaluation of food groups that do not consist of products that fit the wheel of five ("out") are not shown in Table 3 as there are no recommended consumption levels for these products. However, it is recommended to limit the consumption of products that do not fit the guidelines.

The total median intakes (thus, of products both in and outside the wheel of five) of vegetables, bread, and milk products were larger than the (lower bound of the) recommended consumption levels. The 95th percentile of the consumption of these food groups equaled to or exceeded the (lower bound of the) recommendations. However, the median intake of products that fit the wheel of five of these food groups remained below the recommended consumption levels. For several food groups, less than 25% of the toddlers consumed following the recommendations (legumes and pulses, nuts, fish, eggs, and fats). For the food groups bread, potatoes and cereals, milk products, fats, and drinks, a large part of the total consumption came from products outside the wheel of five, despite the guidelines to minimize the consumption of sugar-sweetened beverages, to replace refined grains with whole wheat and whole-grain products, and to replace solid fats and butter by liquid fats, margarine and plant-based oils.

Of the food groups, of which all products are categorized outside the wheel of five, the daily consumption was the highest for snacks. It is recommended that toddlers do not consume cheese (0 g per day); however, in practice, they do (median intake is 10 g per day). The median intake of meat was 33 g per day, close to the recommended maximum level of 35 g per day.

Table 3. The distribution of habitual consumption of several food groups (in g/day) by Dutch children aged one to three years, compared to the guidelines of the wheel of five (DNFCS 2012–2016, n = 672, weighted for demographic characteristics, season and day of the week).

Food Group	Wheel of Five *	Mean (95% CI)	P5	P25	P50 (95% CI)	P75	P95	Wheel of Five Recommendation (Min-Max)	P50 ≥ Recommendation?
Vegetables	In	47 (45–49)	14	27	41 (39–43)	60	100	75 (50–100)	No
	Total	56 (54–59)	19	35	51 (48–53)	72	113	75 (50–100)	Yes [1]
Fruit	In	123 (119–128)	30	74	114 (110–119)	162	249	150	No
	Total	136 (132–140)	36	83	125 (121–130)	179	269	150	No
Bread	In	59 (57–61)	16	37	55 (54–57)	77	115	88 (70–105)	No
	Total	89 (87–91)	41	64	84 (83–87)	109	154	88 (70–105)	Yes [1]
Potatoes	In	27 (25–29)	7	16	24 (22–26)	35	57	53	No
	Total	38 (36–39)	13	24	34 (32–36)	48	73	53	No
Cereal products	In	4 (3–5)	0	0	0 (0–0)	3	22	38	No
	Total	28 (26–30)	5	13	23 (21–24)	37	69	38	No
Potatoes and cereals [2]	In	31 (29–33)	7	16	26 (25–28)	41	70	120 (60–120)	No
	Total	63 (61–65)	24	42	59 (57–61)	80	116	120 (60–120)	No
Legumes, pulses	In	2 (1–2)	0	0	0 (0–1)	2	8	4	No
	Total	2 (1–2)	0	0	0 (0–1)	2	8	4	No
Nuts	In	0 (0–1)	0	0	0 (0–0)	0	2	15	No
	Total	4 (4–5)	0	0	2 (2–2)	6	15	15	No
Fish	In	5 (4–6)	0	1	3 (2–4)	6	18	7	No
Eggs	In	6 (5–6)	1	3	5 (4–6)	8	14	11	No
Milk products	In	195 (188–203)	17	91	174 (166–183)	274	446	300	No
	Total	343 (333–354)	96	213	319 (309–329)	448	671	300	Yes
Fats	In	7 (7–8)	2	4	6 (6–7)	10	17	30	No
	Total	14 (14–15)	5	9	13 (13–13)	18	29	30	No
Drinks	In	178 (167–189)	10	54	124 (114–134)	245	527	636	No
	Total	560 (552–568)	198	365	521 (510–527)	713	1057	636	No
Meat	In	8 (8–9)	2	4	7 (6–8)	11	21	35 [3]	Yes [4]
	Total	37 (35–38)	13	24	33 (32–35)	46	70	35 [3]	Yes [4]
Cheese	In	2 (2–2)	0	0	0 (0–1)	2	10	0 [3]	No [4]
	Total	12 (11–13)	2	6	10 (10–11)	16	28	0 [3]	Yes [4]
Soups	Out	8 (7–9)	0	0	2 (1–3)	9	38		
Sauces	Out	9 (8–9)	2	4	7 (6–8)	11	22		
Snacks	Out	50 (48–52)	15	29	44 (43–46)	65	105		
Bread toppings	Out	14 (13–14)	2	6	11 (11–12)	18	33		
Other	Out	20 (18–23)	0	1	3 (3–4)	14	91		

* Categorized in the wheel of five ('in"), outside the wheel of five ("out"), and both in as outside the wheel of five ("total"). [1] Within the range of recommendation. [2] This food group, including both potatoes and cereals, was included as their products are interchangeable. [3] Maximum consumption recommendation. [4] In this case, when P50 is equal to or larger than the recommendation, the consumption does not meet the recommended level as it involves a maximum level.

4. Discussion

In the present study, it was observed that for most nutrients, the estimated habitual intake of Dutch children aged one to three years met the recommendations for adequate and safe intakes. However, there are still opportunities for improvement of the nutrient intake and food consumption of these children.

For toddlers in several other European countries, results similar to those of the present study were found. The intakes of N-3 fatty acids, iron, and vitamin D and the consumption of vegetables were consistently below recommended levels, while intakes of saturated fatty acids, sodium, free sugar, and protein were often higher than recommended levels [11].

Compared to a previous study of the DNFCS among young children, conducted in 2005–2006, similar results were found regarding the consumption of vegetables and fruit and the intakes of fiber, retinol, iron, copper, and zinc [31]. The results refer to children aged two to three years rather than to children aged one to three years as in the present study; however, similar conclusions were drawn. Compared to the previous DNFCS, the folate equivalents intake seemed to be improved [32]. A high intake of copper among young children was also observed [33], for which the main source of copper was cereals and cereal products. In the present study, copper intake is still considered high, and cereal products are still the main source [34]. However, also products, which are not needed for a healthy diet contribute to copper intake. For instance, non-alcoholic beverages (waters excluded) contribute 9.2–11.7% of the copper intake among boys and girls in this age group [34]. In addition, as far as we know, there are no indications of health problems in the Netherlands due to high copper intake reported in the literature; therefore, the copper intake is not considered a dietary nutritional challenge, yet this may be further studied. Vitamin D intake from food and dietary supplements did not meet the AI in the previous study, though it did in the present study. However, in the present study, a lower AI was used, as sufficient sunlight exposure was assumed.

In the present study, 74% of the toddlers used dietary supplements in general, and 59% used vitamin D supplements specifically. The median vitamin D intake from food only was 2.4 μg per day, whereas the median vitamin D intake from food and dietary supplements was 7.6 μg per day. For children in the Netherlands aged up to four years, it is advised to take an additional 10 μg of vitamin D supplements daily [29]. This advice was based on the dietary reference values for adults whose levels below 25 nmol/L were estimated to result in vitamin D deficiency [35]. In 2019, a study on the vitamin D status of Dutch children concluded that one-third of the children were vitamin D deficient in winter, which was likely due to low adherence to the supplementation advice [36]. However, vitamin D deficiency was defined as <50 nmol/L, which is twice the threshold level used by the Dutch Health Council. Nevertheless, more emphasis could be put on compliance with the supplementation advice. Therefore, the intake of vitamin D is a potential nutritional challenge in the dietary habits of Dutch toddlers, depending on the sufficiency of sunlight exposure. In addition, studies on the status of other nutrients, for example, of those of which no statement could be done or of which low intakes were observed in the present study, could be useful in identifying potential nutritional challenges.

For toddlers in the present study, the total protein intake was adequate. However, even the 5th percentile (10 En%) of the protein intake was above the AI (5 En%). Currently, an upper intake level of protein is not yet set. However, a high intake of protein during early childhood is reported to be associated with higher BMI in childhood and a higher risk of obesity in later life [37]. Eight percent of the toddlers in the present study were overweight or obese.

For 50% of the toddlers, it was reported that they ate vegetables every day. The median habitual consumption of vegetable products categorized in the wheel of five was below the recommended level. However, the total consumption of vegetable products (both favorable and unfavorable products categorized in and outside the wheel of five) did meet the recommended consumption level. Toddlers also consumed unfavorable products from several other food groups, especially from bread, potatoes and cereals, milk products, fats, and drinks, which contrasts with the guidelines. The guidelines specifically mention limiting sugar-sweetened beverages, increasing the consumption of whole wheat and whole grain products instead of refined grains, and replacing solid fats and butter with liquid fats, margarine, and plant-based oils. Those products that do not fit the wheel of five are low in fiber or high in unfavorable fats, sugar or salt. The relatively high consumption of unfavorable products may have been the cause for the observed high intake of saturated fatty acids and the median intake of fiber far below the guideline.

As far as we know, no indications of health problems were observed (as it was not examined in the present study) and of insufficient intakes of nutrients. A potential nutritional challenge in the dietary intake of Dutch toddlers is the vitamin D intake, which has been found to be similar for other countries. Therefore, supplementation advice exists for this age group in the Netherlands. However, it remains difficult to assess the adequacy of vitamin D with dietary assessment due to the substantial effect of sunlight exposure. For alpha-linoleic acids, N-3 fish fatty acids, and iron, no statement on adequacy could be provided, though the median intakes were not close to the AI; therefore, these nutrients may be potential nutritional challenges. To gain more in-depth knowledge on potential nutritional challenges and the causal associations between the dietary habits of Dutch toddlers and the impact on their health, further (additional, long-term follow-up) research should be done concerning growth and neuro-development. Insight into the nutrient intake, of which no statement could be done or of which low estimations were observed in the present study, could be provided by additional research, such as on nutritional status. This could be valuable for listing potential nutritional challenges, as was done by studying vitamin D status in Dutch children [36]. In addition, additional analyses within subgroups of this population could potentially provide insight into more class-specific dietary habits related to, for example, age group or socioeconomic status.

There were a few limitations in this study, as in a study involving (self-reporting of) dietary intake, misreporting (underreporting or overreporting) of dietary intake was likely. With self-reporting of dietary intake, misreporting cannot be fully avoided. This is possibly even more the case when the recall day is known. For energy intake, the average level of misreporting than the expected energy intake was estimated as underreported by about ten percent on average, with 2% of the study participants who reported an unlikely low-energy intake [12]. Based on this, the underreporting seems limited. However, bias in the intakes can still not be fully excluded. To estimate the intake of macronutrients and micronutrients, data were combined with the databases NEVO and NES. It is evaluated that the NEVO database is complete though not all products and their declarations are listed and/or available, for which a comparable food product was selected. In the end, the average percentage of missing values for the nutrients presented in this study was only 3% [12]. For the data on supplements, NES uses the nutrient declaration available on the packaging rather than data available through laboratory analyses, which involves average compositions and may lead to overestimation and underestimation of nutrient intake via supplements [38]. In addition, the reference values used for the comparison with the habitual intake of children are ad interim values of the Dutch Health Council, which may be adjusted, as they are working on new reference values for children [39].

For evaluating the intake of food groups, the Dutch food-based dietary guidelines (presented in the wheel of five) were used [16]. However, no compliance with the guidelines does not necessarily mean that the food pattern is inadequate because consumption of various foods and food groups can still lead to adequate intakes of nutrients, as was shown in the present study. The guidelines are set as guidance for individuals rather than for populations. Because the individual requirement is unknown in individual nutritional advice, the recommended daily intake (RDI) is used for guidelines rather than the EAR. The RDI is a value that meets the requirement of 97.5% of the population; thus, for most individuals, it will be more than their individual requirement [29]. For this reason, the EAR cut-point method is usually applied to evaluate the adequacy of intake in populations [23]. Unfortunately, the food-based dietary guidelines are not available in an EAR-like measure. Therefore, in the present study, we made a qualitative comparison of the median consumption of a food group with the guidelines to gain knowledge at a population level rather than assuming that every individual must meet the guidelines.

One of the strengths of the present study was that due to sampling and weighing the results on small deviances on the sociodemographic characteristics. It was possible to obtain results that are representative of the target population. Data were retrieved by using food diaries and repeated 24 h-recalls conform the European guidance for harmonized food consumption data in EU member states by EFSA [40], of which the habitual intake could be estimated and compared with reference values. In addition, of all nutrients from food only as well as from food combined with dietary supplements, the habitual intake was estimated rather than the reported intake on two individual days; therefore, the day-to-day (intraindividual) variation was accounted for, and a better estimate of the proportion with inadequate intakes could be made.

5. Conclusions

The dietary intake of Dutch children aged one to three years seems adequate for most nutrients. Vitamin D is a potential nutritional challenge, and several nutrients need to be further looked at for potential nutritional challenges: alpha-linoleic acids, N-3 fish fatty acids, and iron. The dietary pattern of the toddlers consists partially of unfavorable products that may have been the cause of the high intakes of several nutrients, such as sodium and saturated fatty acids, and the low intake of fiber.

Therefore, for young children, shifting to and following a healthy diet, which is (more) in line with the guidelines, may improve the nutrient intake, of which in the present study was found to be low or for which no statement on adequacy could be done. This is important as early-life dietary habits affect health, also later in life. Further research or potential intervention studies on indicators and predictors of a healthy diet for children aged one to three years may be useful to prevent negative health impacts and encourage a healthy life in the future. This knowledge could be incorporated into the screening tool that is being developed for toddlers in The Netherlands.

Author Contributions: Conceptualization, C.T.M.v.R., J.V.-K., and K.F.M.J.; formal analysis, E.S.; writing—original draft preparation, E.S. and A.K.; writing—review and editing, E.S., A.K., J.V.-K., L.E.M.E., S.t.B., K.F.M.J., and C.T.M.v.R. All authors have read and agreed to the published version of the manuscript.

Funding: This study was conducted as a part of the research project "Nutrition and lifestyle screening tool for youth healthcare 2019–2022", which was funded by the Dutch Ministry of Health, Welfare and Sport (VWS).

Institutional Review Board Statement: Ethical review and approval were waived for this study. The Utrecht University Medical Ethical Review Committee evaluated that the study from which the data were used was not subject to the Medical Research Involving Human Subjects Act (WMO) of the Netherlands (reference number 12–359/C).

Informed Consent Statement: Informed consent was obtained from all subjects involved in the study before an interview during a home visit. For interviews by telephone, written informed consent was not necessary according to the Dutch regulations at the time of data collection.

Data Availability Statement: The data used in this study are available on request from https://www.rivm.nl/en/dutch-national-food-consumption-survey/data-on-request or from the corresponding author (data accessed on 29 July 2020).

Conflicts of Interest: The authors declare no conflict of interest.

References

1. Taylor-Baer, M.; Herman, D. From Epidemiology to Epigenetics: Evidence for the Importance of Nutrition to Optimal Health Development Across the Life Course. In *Handbook of Life Course Health Development*; Halfon, N., Forrest, C.B., Lerner, R.M., Faustman, E.M., Eds.; Springer: Cham, Switzerland, 2018; pp. 431–462.
2. Georgieff, M.K.; Brunette, K.E.; Tran, P.V. Early life nutrition and neural plasticity. *Dev. Psychopathol.* **2015**, *27*, 411–423. [CrossRef] [PubMed]
3. Black, R.E.; Victora, C.G.; Walker, S.P.; Bhutta, Z.A.; Christian, P.; de Onis, M.; Ezzati, M.; Grantham-McGregor, S.; Katz, J.; Martorell, R.; et al. Maternal and child undernutrition and overweight in low-income and middle-income countries. *Lancet* **2013**, *382*, 427–451. [CrossRef]
4. Martins, V.J.B.; Florêncio, T.M.M.; Grillo, L.P.; Do Carmo, P.; Franco, M.; Martins, P.A.; Clemente, A.P.G.; Santos, C.D.L.; de Fatima, A.; Vieira, M.; et al. Long-lasting effects of undernutrition. *Int. J. Environ. Res. Public Health* **2011**, *8*, 1817–1846. [CrossRef] [PubMed]
5. Kumar, S.; Kelly, A.S. Review of Childhood Obesity: From Epidemiology, Etiology, and Comorbidities to Clinical Assessment and Treatment. *Mayo Clin. Proc.* **2017**, *92*, 251–265. [CrossRef] [PubMed]
6. Juonala, M.; Magnussen, C.G.; Berenson, G.S.; Venn, A.; Burns, T.L.; Sabin, M.A.; Srinivasan, S.R.; Daniels, S.R.; Davis, P.H.; Chen, W.; et al. Childhood adiposity, adult adiposity, and cardiovascular risk factors. *N. Engl. J. Med.* **2011**, *365*, 1876–1885. [CrossRef] [PubMed]
7. Ryder, J.R.; Jacobs, D.R.J.; Sinaiko, A.R.; Kornblum, A.P.; Steinberger, J. Longitudinal Changes in Weight Status from Childhood and Adolescence to Adulthood. *J. Pediatr.* **2019**, *214*, 187–192.e2. [CrossRef] [PubMed]
8. World Health Organization (WHO). Malnutrition. Available online: https://www.who.int/news-room/fact-sheets/detail/malnutrition (accessed on 16 February 2021).
9. Movassagh, E.Z.; Baxter-Jones, A.D.G.; Kontulainen, S.; Whiting, S.J.; Vatanparast, H. Tracking Dietary Patterns over 20 Years from Childhood through Adolescence into Young Adulthood: The Saskatchewan Pediatric Bone Mineral Accrual Study. *Nutrients* **2017**, *9*, 990. [CrossRef] [PubMed]
10. Kim, J.; Lim, H. Nutritional Management in Childhood Obesity. *J. Obes. Metab. Syndr.* **2019**, *28*, 225–235. [CrossRef] [PubMed]
11. Alles, M.S.; Eussen, S.R.; van der Beek, E.M. Nutritional challenges and opportunities during the weaning period and in young childhood. *Ann. Nutr. Metab.* **2014**, *64*, 284–293. [CrossRef] [PubMed]
12. Van Rossum, C.; Buurma-Rethans, E.; Dinnissen, C.; Beukers, M.; Brants, H.; Dekkers, A.; Ocké, M. *The Diet of the Dutch: Results of the Dutch National Food Consumption Survey 2012–2016*; RIVM Report 2020-0083; RIVM: Bilthoven, The Netherlands, 2020.
13. Slimani, N.; Ferrari, P.; Ocké, M.; Welch, A.; Boeing, H.; Liere, M.; Pala, V.; Amiano, P.; Lagiou, A.; Mattisson, I.; et al. Standardization of the 24-hour diet recall calibration method used in the european prospective investigation into cancer and nutrition (EPIC): General concepts and preliminary results. *Eur. J. Clin. Nutr.* **2000**, *54*, 900–917. [CrossRef] [PubMed]
14. NEVO-Online Version 2019/6.0. Available online: https://nevo-online.rivm.nl/ (accessed on 16 February 2021).
15. Buurma-Rethans, E.; Fransen, H.; Ghameshlou, Z.; de Jong, N. Een databestand voor supplementen: Behoeftes en acties. *Voeding Nu* **2008**, *10*, 21–24.
16. Brink, E.; van Rossum, C.; Postma-Smeets, A.; Stafleu, A.; Wolvers, D.; van Dooren, C.; Toxopeus, I.; Buurma-Rethans, E.; Geurts, M.; Ocké, M. Development of healthy and sustainable food-based dietary guidelines for the Netherlands. *Public Health Nutr.* **2019**, *22*, 2419–2435. [CrossRef] [PubMed]
17. Netherlands Nutrition Centre. Food Based Dietary Guidelines. Available online: https://www.voedingscentrum.nl/Assets/Uploads/voedingscentrum/Documents/Professionals/Schijf%20van%20Vijf/Richtlijnen%20Schijf%20van%20Vijf.pdf (accessed on 9 February 2021).
18. *SAS*, Version 9.4; SAS Institute Inc.: Cary, NC, USA, 2014.
19. Dekkers, A.L.; Verkaik-Kloosterman, J.; van Rossum, C.T.; Ocké, M.C. SPADE, a new statistical program to estimate habitual dietary intake from multiple food sources and dietary supplements. *J. Nutr.* **2014**, *144*, 2083–2091. [CrossRef] [PubMed]
20. Verkaik-Kloosterman, J.; Buurma-Rethans, E.J.M.; Dekkers, A.L.M. *The Iodine Intake of Children and Adults in the Netherland: Results of the Dutch National Food Consumption Survey 2007–2010*; RIVM Report 350090012/2012; RIVM: Bilthoven, The Netherlands, 2012.
21. Van Rossum, C.T.M.; Buurma-Rethans, E.J.M.; Fransen, H.P.; Verkaik-Kloosterman, J.; Hendriksen, M.A.H. *Salt Consumption of Children and Adults in the Netherlands: Results from the Dutch National Food Consumption Survey 2007–2010*; RIVM report 350050007/2012; RIVM: Bilthoven, The Netherlands, 2012.
22. Health Council of the Netherlands. Ad-Interim Dietary Health Intakes. 2014. Available online: https://www.gezondheidsraad.nl/documenten/adviezen/2018/09/18/tijdelijke-voedingsnormen-2014 (accessed on 16 February 2021).
23. Carriquiry, A.L.; Camaño-Garcia, G. Evaluation of dietary intake data using the tolerable upper intake levels. *J. Nutr.* **2006**, *136*, 507S–513S. [CrossRef] [PubMed]
24. Institute of Medicine (US) Subcommittee on Interpretation and Uses of Dietary Reference Intakes; Institute of Medicine (US) Standing Committee on the Scientific Evaluation of Dietary Reference Intakes. *DRI Dietary Reference Intakes: Applications in Dietary Assessment*; National Academies Press (US): Washington, DC, USA, 2000. [CrossRef]
25. Health Council of the Netherlands. *File Evaluation of the Dietary Reference Intakes for Vitamin D*; Publication No. 2012/15; Health Council of the Netherlands: The Hague, The Netherlands, 2012.

26. European Food Safety Authority (EFSA). Tolerable Upper Intake Levels for Vitamins and Minerals. Available online: https://www.efsa.europa.eu/sites/default/files/efsa_rep/blobserver_assets/ndatolerableuil.pdf (accessed on 16 February 2021).
27. Cole, T.J.; Lobstein, T. Extended international (IOTF) body mass index cut-offs for thinness, overweight and obesity. *Pediatr. Obes.* **2012**, *7*, 284–294. [CrossRef] [PubMed]
28. Health Council of the Netherlands. *Guideline for the Consumption of Fiber*; Publication No. 2006/03; Health Council of the Netherlands: The Hague, The Netherlands, 2006.
29. Health Council of the Netherlands. *Dietary Reference Values for Vitamins and Minerals for Adults*; Publication No. 2018/19; Health Council of the Netherlands: The Hague, The Netherlands, 2018.
30. Netherlands Nutrition Centre. How Much Salt Can I Eat? Available online: https://www.voedingscentrum.nl/nl/service/vraag-en-antwoord/gezonde-voeding-en-voedingsstoffen/hoeveel-zout-mag-ik-eten.aspx (accessed on 16 February 2021).
31. Ocké, M.C.; van Rossum, C.T.M.; Fransen, H.P.; Buurma, E.J.M.; de Boer, E.J.; Brants, H.A.M.; Niekerk, E.M.; van der Laan, J.D.; Drijvers, J.J.M.M.; Ghameshlou, Z. *Dutch National Food Consumption Survey Young Children 2005/2006*; RIVM Report 350070001/2008; RIVM: Bilthoven, The Netherlands, 2018.
32. Westenbrink, S.; Jansen-van der Vliet, M.; Van Rossum, C. Updated folate data in the Dutch Food Composition Database and implications for intake estimates. *Food Nutr. Res.* **2012**, *56*, 5449. [CrossRef] [PubMed]
33. Büchner, F.L.; van der Vliet, M.J.; van Rossum, C.T.M. *Copper Intake in Young Children*; RIVM Letter Report 350023001/2010; RIVM: Bilthoven, The Netherlands, 2010.
34. National Institute for Public Health and the Environment (RIVM). DNFCS2012–2016; Mean Contribution of Food Sources to the Intake of Nutrients. Available online: https://statline.rivm.nl/#/RIVM/nl/ (accessed on 16 February 2021).
35. Health Council of the Netherlands. *An. Evaluation of the EFSA's Dietary Reference Values (DRVs), Part 1*; Publication No. 2018/19A; Health Council of the Netherlands: The Hague, The Netherlands, 2018.
36. Hoevenaar-Blom, M.P.; Wielders, J.P.; Groeneveld, H.; de Leeuw, E.; Schmits, R.J.; Pepermans, C.; de Jong, P.C.P.; Hogeman, P.H. Prevalence and determinants of vitamin D deficiency in infants and toddlers in the Netherlands: A pilot study. *Ann. Clin. Biochem.* **2019**, *56*, 613–618. [CrossRef] [PubMed]
37. Hörnell, A.; Lagström, H.; Lande, B.; Thorsdottir, I. Protein intake from 0 to 18 years of age and its relation to health: A systematic literature review for the 5th Nordic Nutrition Recommendations. *Food Nutr. Res.* **2013**, *57*, 21803. [CrossRef] [PubMed]
38. Verkaik-Kloosterman, J.; Seves, S.M.; Ocké, M.C. Vitamin D concentrations in fortified foods and dietary supplements intended for infants: Implications for vitamin D intake. *Food Chem.* **2017**, *221*, 629–635. [CrossRef] [PubMed]
39. Health Council of the Netherlands. Committee 2021 Health Council. Available online: https://www.gezondheidsraad.nl/documenten/overige/2020/09/15/werkprogramma-2021 (accessed on 16 February 2021).
40. European Food Safety Authority (EFSA). Guidance on the EU Menu methodology. *EFSA J.* **2014**, *12*, 3944. [CrossRef]

nutrients MDPI

Article

Dietary Changes among Adults in The Netherlands in the Period 2007–2010 and 2012–2016. Results from Two Cross-Sectional National Food Consumption Surveys

Ceciel S. Dinnissen [1,*], Marga C. Ocké [1,2], Elly J. M. Buurma-Rethans [1] and Caroline T. M. van Rossum [1]

1 National Institute for Public Health and the Environment (RIVM), 3721 BA Bilthoven, The Netherlands; marga.ocke@rivm.nl (M.C.O.); ellybuurma@planet.nl (E.J.M.B.-R.); caroline.van.rossum@rivm.nl (C.T.M.v.R.)
2 Division of Human Nutrition and Health, Wageningen University and Research, 6708 WE Wageningen, The Netherlands
* Correspondence: ceciel.dinnissen@rivm.nl; Tel.: +31-30-2742834

Abstract: Insight into dietary trends is important for the development and evaluation of dietary policies. The aim of this study is to describe changes in dietary intakes of Dutch adults and to evaluate these changes by age, gender, and education. In 2007–2010 and 2012–2016, two national food consumption surveys were conducted including 2106 and 1540 adults, respectively. Data collection included two non-consecutive 24 h dietary recalls. Mean habitual intakes of foods and nutrients relevant for a healthy diet of both surveys were estimated. Between the two periods the mean consumption of red or processed meat, dairy, sodium and alcohol and the ratio of whole-grain to cereal products decreased by 4–30% and the consumption of fibre and unsaturated fatty acids increased by about 3% and 6%, respectively. For most food groups, changes in consumption were comparable for both sexes and in all age groups. A healthier consumption pattern and several favorable changes were observed among higher-educated people. Most, but not all, changes in food consumption are favorable from a public health point of view. However, there is still a large potential for further improvements. A healthier consumption pattern was observed in adults with a higher educational level which calls for attention to social disparities when developing dietary policies.

Keywords: dietary change; food consumption; food monitoring; educational level

Citation: Dinnissen, C.S.; Ocké, M.C.; Buurma-Rethans, E.J.M.; van Rossum, C.T.M. Dietary Changes among Adults in The Netherlands in the Period 2007–2010 and 2012–2016. Results from Two Cross-Sectional National Food Consumption Surveys. *Nutrients* **2021**, *13*, 1520. https://doi.org/10.3390/nu13051520

Academic Editors: Murielle Bochud and Igor Pravst

Received: 26 March 2021
Accepted: 27 April 2021
Published: 30 April 2021

1. Introduction

Diabetes, cardiovascular diseases and some types of cancers are chronic diseases that are, for the most part, related to poor diet and obesity [1,2]. To help achieve an optimal diet and reduce preventable diet-related illnesses, there is a need to develop effective nutrition and food policies. Based on this knowledge, health councils stipulate dietary guidelines for a healthy diet for an entire population [3–6]. These guidelines are evidence-based and aimed to reduce the risk of chronic diseases for the general population.

National food consumption surveys provide the opportunity to get insight into the dietary behavior of a population [7]. These surveys can therefore be used to help formulate and evaluate dietary policies. Repeated dietary surveys can help to gain insight into the development of dietary intake over time. To be suitable for this purpose it is essential that the repeated food consumption surveys are standardized.

Alongside population diet quality, surveys can identify disparities between subgroups. Diet-related risk factors are not distributed equally across population subgroups leading to dietary- and, possibly subsequently, health inequalities. There are gender differences in eating behavior, women tend to eat healthier foods than men do [8]. Additionally, with increasing age, people tend to eat less and make different food choices [9,10]. Poorer health and unhealthier lifestyles are more prevalent in individuals with a lower socio-economic position (SEP) than that for individuals with a higher SEP [11–13]. SEP is usually assessed by determining education, occupation and income. Education is considered to be most related to dietary behavior [14]. In line with this, groups who are less educated and have a lower income appear to consume a less healthy diet [15]. Few studies have described time trends in food consumption for a large variety of food groups and nutrients and studied if the observed trends differed by socio-economic groups.

The aim of the current paper is to describe dietary changes in the period 2007–2010 versus 2012–2016 of Dutch adults overall and by age group, gender, and educational level with data of the Dutch national food consumption survey (DNFCS) from 2007–2010 and 2012–2016.

2. Materials and Methods

Food consumption data from participants aged 19–69 years in two DNFCSs were used for the present analyses. Data from the first survey was collected between March 2007 and April 2010 among 3819 6–69-year-olds including 2106 Dutch adults aged 19–69 years. Data from the second survey was collected between November 2012 and January 2017 amongst 4313 1–79-year-olds including 1540 adults aged 19–69 years [16]. Both surveys largely followed the EU Menu guidelines [17].

In both surveys, participants were recruited via a consumer panel. Exclusion criteria were involvement in a food consumption survey during the previous four years, pregnancy or lactation, being institutionalized or inadequate command of the Dutch language. Sampling was stratified by age, gender, region, degree of urbanization and educational level. A questionnaire was used to obtain information on relevant background and lifestyle factors including educational level. The dietary assessment was based on two telephone-administered 24 h dietary recalls on non-consecutive days with at least 2–6 weeks in between. Interviews were conducted by trained dietitians using the computer-based interview program EPIC–Soft, now called GloboDiet [18]. The recalls were distributed over all days of the week and seasons.

Mean habitual intake of foods and nutrients, associated with foods groups mentioned in the Dutch dietary guidelines set by the Health Council of the Netherlands were selected to be presented [3]. These guidelines recommend following a more plant-based and less animal-based dietary pattern. Daily or weekly dietary guidelines were created for fruits, vegetables, brown bread, wholemeal bread or other wholegrain, unsalted nuts, legumes, dairy, (oily) fish and tea. Consumption of salt and that of red meat, particularly processed meat should be limited, and the consumption of sugar-containing beverages should be minimized. It is advised to drink no alcohol or no more than one glass daily. We looked at intake of sodium and the nutrient alcohol rather than salt intake and the consumption of alcoholic beverages, respectively. For three food groups, the guidelines advise a replacement of specific foods rather than a consumed amount. It is advised to replace refined cereal products with whole-grain products and to replace butter, hard margarines and cooking fats with soft margarines, liquid cooking fats and vegetable oils. For those foods the habitual consumption of the advised subgroup was calculated as a percentage of the total food group, by dividing the mean intake of the subgroup by the mean intake of the total food group. It is also advised to replace unfiltered coffee with filtered coffee, however, our data cannot differentiate between these two states and is therefore not calculated.

Energy and nutrient intakes per person per day were calculated using the Dutch Food Composition Databases (NEVO). The NEVO table of 2011 was used for the 2007–2010 data and NEVO-online 2016 was used for the 2012–2016 data [19,20]. Vegetable protein and animal protein intake to study if a shift towards a more plant-based eating pattern was seen. The other nutrients were mono- and disaccharides (because of the limitation of sugar-containing beverages), fiber (because of the increase in whole-grain, vegetable and fruit consumption), unsaturated fatty acids (because of the replacement of solid fats to liquid fats and oils), n-3 fish fatty acids intake (because of the guideline on (oily) fish consumption), sodium intake (because of the limitation of salt) calcium and vitamin D intake (because of the dairy consumption) and alcohol (because of the limitation of alcoholic beverages).

Habitual daily energy-, nutrient- and food group intake distribution for each survey was estimated using RStudio software (version 4.0.2, RStudio, PBC, Boston, MA, USA) and the package 'SPADE' (Statistical Program to Assess Dietary Exposure, RIVM, version 3.2.55) [21]. SPADE is an R package which includes modeling options to estimate habitual consumption based on two consumption days for foods or nutrients that are consumed daily and by almost all participants (all nutrients and for bread and cereals, dairy and total fat) or episodically (all other food groups). If consumed episodically, SPADE took information on never-consumers available from the general questionnaire into account for alcohol. The habitual intake distributions were modeled as a function of age using data of all survey participants, after which intake distribution estimates were requested for adults aged 19–69 years. 95% Confidence intervals of the estimated mean food group consumptions and nutrient intakes were obtained by bootstrapping [21]. When the means of the two surveys had non-overlapping confidence intervals, the intakes were considered statistically significantly different. Due to low consumption frequency of unsalted nuts and legumes, we could not calculate their habitual consumption and consequently could not study the change in the consumption of those foods.

Data were analyzed for the total population and stratified by educational level, age and gender. Educational level was categorized into low (primary education, lower vocational education, advanced elementary education), middle (intermediate vocational education, higher secondary education) and high (higher vocational education and university). Age was classified into three groups (19–30, 31–50, and 51–69 years).

SAS software, (Version 9.4, SAS Institute Inc., Cary, NC, USA) was used to calculate frequency distributions to describe the study population. Weighing factors were applied in all analyses to correct for deviances from the Dutch population in sociodemographic characteristics (gender, age groups, region, level of education, urbanization) and for day of the week and season at the time of each survey.

3. Results

In the 2012–2016 survey, a lower proportion of participants were categorized in the low educational level group and a higher proportion of participants were categorized in a high educational level group compared to the previous survey. The proportion of overweight or obese participants and participants who indicated drinking alcoholic beverages was slightly higher in 2012–2016 than in the 2007–2010 survey. A lower proportion of 31–50-year-olds was seen in the more recent survey. The proportion of the two other age classes was higher. Smoking status, compliance with the physical activity guideline and dietary supplement use was comparable in both surveys (Table 1).

Between 2007–2010 and 2012–2016, the mean consumption of red- or processed meat and dairy decreased (Table 2). The consumption of whole-grain products in gram and as a percentage of cereals and cereal products also decreased. For most food groups, the direction of the changes in food consumption was similar for both men and women. However, the change in mean consumption of sugar-containing beverages (excluding coffee and tea) and the consumption of whole-grain products as a percentage of cereals and cereal products did differ by gender. The consumption of the sugar-containing beverages only significantly decreased in women; and the decrease in the proportion of whole-grain products was only significant in men, while in women the consumed amount of whole-grain products decreased but not the proportion of total cereals. In both surveys, women consumed more fruit and tea, and less red or processed meat, dairy and sugar-containing beverages compared to men. The gender differences in fruit, red or processed meat and dairy consumption decreased in 2012–2016 compared to those in 2007–2010, while for sugar-containing beverages and tea the gender differences increased.

Table 1. Characteristics of adults (19–69 years) participating in the Dutch National Food Consumption Surveys in 2007–2010 (n = 2106) and 2012–2016 (n = 1540).

Characteristics	2007–2010		2012–2016	
	n	Weighted %	n	Weighted %
Total	2106		1540	
Men	1055	50.2	770	50.0
Women	1051	49.8	770	50.0
Educational level				
Low [1]	708	32.1	323	25.6
Middle [2]	935	43.6	655	43.4
High [3]	463	24.3	562	31.0
Age				
19–30 years	703	21.5	516	22.4
31–50 years	699	44.7	523	40.7
51–69 years	704	33.9	501	36.9
BMI classes				
Normal and underweight [4]	1010	45.6	714	42.8
Overweight/obese [5]	1095	54.4	826	57.2
Current smoker				
Yes	544	24.2	362	23.9
No	1521	75.8	1167	75.7
Alcohol use				
Yes	1456	69.5	1155	73.6
No	649	30.5	385	26.4
Meeting guideline for physical activity				
No	1577	74.6	1155	76.6
Yes [6]	529	24.4	353	23.4
Dietary supplement use				
Yes	916	44.6	663	43.1
No	1189	55.4	877	56.9

[1] primary education, lower vocational education, advanced elementary education; [2] intermediate vocational education, higher secondary education; [3] higher vocational education and university; [4] BMI $\leq$ 25 kg/m^2; [5] BMI > 25 kg/m^2, [6] Physical activity >5 day and >30 min.

The mean consumption of red or processed meat and dairy decreased in all age groups (Table 3). Only in the youngest age group, the mean consumption of fruit increased and sugar-containing beverages decreased. The consumption of whole-grain products in total and as a percentage of cereals and cereal products decreased in the two oldest age groups. The percentage of soft fats to total fats increased in the youngest age group but decreased in the oldest age group.

Table 2. Habitual mean [¥] (95%CI) food consumption (g/day) * for Dutch adults aged 19–69 years in 2007–2010 (n = 2106; DNFCS 2007–2010) and 2012–2016 (n = 1540; DNFCS 2007–2010) for the total population and by gender.

Food Group	Total		Men		Women	
	2007–2010	2012–2016	2007–2010	2012–2016	2007–2010	2012–2016
	n = 2106	n = 1540	n = 1055	n = 1051	n = 770	n = 770
Vegetables	137 (134–139)	140 (136–144)	138 (134–142)	142 (136–148)	135 (131–139)	138 (133–142)
Fruit	102 (98–106)	106 (102–110)	90 (85–96)	96 (90–102)	114 (109–119)	116 (110–122)
Red or processed meat	**93 (91–95)**	**80 (77–82)** [1]	**112 (109–116)**	**97 (93–101)** [1]	**74 (71–77)**	**63 (59–66)** [1]
Dairy	**372 (369–375)**	**333 (330–335)** [1]	**412 (408–416)**	**364 (361–367)** [1]	**332 (329–336)**	**301 (298–304)** [1]
Fish	17 (16–18)	17 (16–19)	19 (17–20)	18 (16–21)	15 (14–17)	16 (14–18)
Sugar-containing beverages	328 (317–340)	307 (294–321)	376 (357–395)	372 (351–392)	**281 (268–294)**	**243 (228–259)** [1]
Tea	230 (219–241)	245 (231–260)	164 (151–177)	165 (148–183)	296 (278–314)	325 (302–348)
Cereals and cereal products	**208 (207–209)**	**203 (202–205)** [1]	234 (232–236)	237 (235–239)	**182 (181–183)**	**170 (169–171)** [1]
Whole-grain products	**101 (99–103)**	**93 (90–95)** [1]	114 (110–118)	106 (102–110)	**88 (86–91)**	**79 (76–82)** [1]
Whole-grain products. Cereals and cereal products (%)	**49 (48–49)**	**46 (45–46)** [1]	**49 (47–50)**	**45 (43–46)** [1]	49 (48–50)	46 (45–48)
Fats	**28 (28–28)**	**23 (23–23)** [1]	**33 (33–33)**	**27 (27–27)** [1]	**23 (22–23)**	**18 (18–18)** [1]
Soft margarines, liquid cooking fats, and vegetable oils	**21 (21–22)**	**18 (17–18)** [1]	**26 (25–26)**	**22 (21–23)** [1]	**17 (17–18)**	**14 (13–15)** [1]
Soft fats **/total fats (%)	77 (76–79)	79 (78–81)	78 (77–80)	80 (78–82)	76 (74–78)	77 (75–79)

[1] Sig. difference in mean consumption (95%CI) between DNFCS 2007–2010 and DNFCS 2012–2016, in bold; * except when indicated with (%) for substitution guidelines; ** Soft margarines, liquid cooking fats and vegetable oils; [¥] weighted for socio-demographic characteristics, season and day of the week.

In both surveys, adults aged 51–69 years consumed more vegetables, fruit, fish, tea and a higher proportion of whole grain products and less sugar-containing beverages than the 19–30-year-olds. The difference in fruit and sugar-containing drinks consumption and proportion of whole-grain products between the youngest age group and the other groups decreased in 2012–2016 compared to 2007–2010. In 2007–2010 the 19–30-year-olds consumed the same amount of dairy compared to the 51–69-year-olds, in 2012–2016 the oldest group consumed more dairy than the youngest age group.

The decrease in dairy consumption was observed in all educational level groups (Table 4). Other changes were only observed in one or two educational level subgroups. Only in the group with the highest educational level, the mean consumption of vegetables increased. Only in the middle educational level group, the proportion of whole-grain products of total cereals decreased and the percentage of soft margarines, liquid cooking fats and vegetable oils to total fat increased between the two time periods. Additionally, the consumption of red or processed meat decreased in the low- and high-educated groups, but not in the middle group.

Table 3. Habitual mean[¥] (95%CI) food consumption (g/day) * for Dutch adults aged 19–69 years in 2007 to 2010 (n = 2106; DNFCS 2007–2010) and 2012 to 2016 (n = 1540; DNFCS 2007–2010) by age groups (19–30, 31–50, 51–69).

Food Group	19–30 Years		31–50 Years		51–69 Years	
	2007–2010	2012–2016	2007–2010	2012–2016	2007–2010	2012–2016
	n = 703	n = 516	n = 699	n = 523	n = 704	n = 501
Vegetables	117 (113–120)	121 (118–125)	135 (132–139)	139 (134–144)	151 (146–156)	152 (147–157)
Fruit	**80 (76–84)**	**92 (87–98)** [1]	97 (92–102)	100 (95–106)	123 (117–129)	121 (115–127)
Red or processed meat	**90 (87–94)**	**76 (73–79)** [1]	**92 (89–95)**	**81 (78–85)** [1]	**96 (92–99)**	**80 (77–83)** [1]
Dairy	**376 (372–381)**	**327 (323–331)** [1]	**372 (368–376)**	**330 (327–333)** [1]	**370 (366–374)**	**339 (336–341)** [1]
Fish	13 (11–14)	13 (11–15)	17 (15–18)	17 (14–19)	20 (17–22)	21 (18–23)
Sugar-containing beverages	**545 (527-563)**	**491 (471–511)** [1]	330 (313–346)	313 (295–330)	189 (178–200)	190 (178–203)
Tea	184 (170–198)	194 (178–209)	231 (214–247)	268 (246–289)	259 (243–276)	252 (232–273)
Cereals and cereal products	**230 (227–232)**	**218 (214–221)** [1]	215 (213–217)	212 (209–215)	185 (184–187)	185 (183–188)
Whole-grain products	92 (89–95)	85 (81–89)	**103 (100–106)**	**95 (91–99)** [1]	**105 (101–108)**	**95 (91–98)** [1]
Whole-grain products/Cereals and cereal products (%)	40 (39–41)	39 (38–40)	**48 (47–49)**	**45 (44–46)** [1]	**56 (55–58)**	**51 (50–52)** [1]
Fats	**26 (25–26)**	**20 (20–21)** [1]	**28 (27–28)**	**23 (22–23)** [1]	**29 (29–29)**	**24 (24–25)** [1]
Soft margarines, liquid cooking fats, and vegetable oils	**20 (19–20)**	**17 (17–18)** [1]	**21 (21–22)**	**18 (17–19)** [1]	**23 (22–23)**	**18 (18–19)** [1]
Soft fats **/total fats (%)	**76 (75–77)**	**84 (83–85)** [1]	77 (76–79)	81 (79–82)	**78 (77–80)**	**75 (74–76)** [1]

[1] Sig. difference in mean consumption (95%CI) between DNFCS 2007–2010 and DNFCS 2012–2016, in bold; * except when indicated with (%) for substitution guidelines; ** Soft margarines, liquid cooking fats and vegetable oils; [¥] weighted for socio-demographic characteristics, season and day of the week.

In both surveys, people with a high educational level consumed more vegetables, fruit and tea and less red or processed meat and sugar-containing beverages than people with a lower educational level. The difference between the highest educational level group and the lower educational level group, seen in 2007–2010, increased in 2012–2016 for vegetables, fruit, tea and the proportion of whole grains but decreased for fish and the percentage of soft margarines, liquid cooking fats and vegetable oils to total fat. In 2007–2010, dairy consumption was highest in people with a high educational level. The larger decrease in consumption in the subgroup caused this group to become the lowest consumers of dairy in 2012–2016.

For most nutrients, the changes in mean habitual intake between 2007–2010 and 2012–2016 were in the same direction for both sexes as for the total Dutch population (Table 5). The mean habitual intake of fiber and unsaturated fatty acids increased and intake of animal protein, mono- and disaccharides, sodium, calcium, vitamin D and alcohol decreased significantly in the total population and in both men and women. Some differences between men and women were observed. Energy intake decreased in women and stayed stable in men. Vegetable protein intake decreased slightly in women, whereas an increase in intake was seen in men. Furthermore, n-3 fish fatty acids intake increased in men and was not significantly increased in women.

Table 4. Habitual mean ¥ (95%CI) food consumption (g/day) * for Dutch adults aged 19–69 years in 2007–2010 (n = 2106; DNFCS 2007–2010) and 2012–2016 (n = 1540; DNFCS 2007–2010) by educational level (low, middle, high).

Food Group	Low		Middle		High	
	2007–2010	2012–2016	2007–2010	2012–2016	2007–2010	2012–2016
	n = 708	n = 323	n = 935	n =655	n =463	n =526
Vegetables	126 (121–131)	124 (117–130)	131 (126–135)	128 (124–133)	**144 (137–150)**	**167 (160–175)** [1]
Fruit	95 (89–101)	96 (87–104)	102 (97–108)	98 (92–104)	118 (110–126)	133 (126–141)
Red or processed meat	**98 (94–102)**	**86 (81–92)** [1]	93 (90–96)	89 (85–93)	**82 (78–86)**	**72 (67–76)** [1]
Dairy	**361 (356–366)**	**339 (333–345)** [1]	**370 (367–374)**	**344 (340–348)** [1]	**381 (377–386)**	**316 (313–320)** [1]
Fish	14 (12–17)	19 (15–23)	16 (14–18)	15 (13–17)	21 (18–24)	21 (18–24)
Sugar-containing beverages	332 (310–353)	320 (287–354)	357 (339–375)	354 (333–374)	280 (259–301)	258 (239–276)
Tea	206 (188–224)	212 (182–242)	235 (218–252)	217 (196–238)	264 (236–292)	312 (280–344)
Cereals and cereal products	**196 (193–199)**	**190 (187–193)** [1]	**216 (215–218)**	**202 (200–204)** [1]	210 (207–213)	213 (211–216)
Whole-grain products	92 (88–97)	82 (75–88)	**105 (101–109)**	**88 (84–93)** [1]	105 (100–110)	102 (98–107)
Whole-grain products/Cereals and cereal products (%)	47 (46–49)	43 (40–46)	**49 (47–50)**	**44 (42–45)** [1]	50 (49–52)	48 (46–49)
Fats	**29 (28–29)**	**23 (22–23)** [1]	**28 (28–29)**	**22 (22–23)** [1]	**25 (25–25)**	**22 (22–23)** [1]
Soft margarines, liquid cooking fats and vegetable oils	**21 (21–22)**	**17 (16–18)** [1]	**21 (20–22)**	**18 (17–18)** [1]	**19 (19–20)**	**17 (16–18)** [1]
Soft fats **/total fats (%)	75 (73–78)	77 (73–80)	**75 (73–77)**	**79 (78–81)** [1]	77 (74–80)	77 (74–80)

[1] Sig. difference in mean consumption (95% CI) between DNFCS 2007–2010 and DNFCS 2012–2016, in bold; * except when indicated with (%) for substitution guidelines; ** Soft margarines, liquid cooking fats and vegetable oils; ¥ weighted for socio-demographic characteristics, season and day of the week.

In both surveys, only the intake of mono- and disaccharides and fibers was higher in women than in men. The gender differences for the intake of animal protein intake, unsaturated fatty acids, calcium and alcohol decreased in 2012–2016 compared to that of 2007–2010.

Divided by age group, the mean habitual intake changed in a similar direction for the majority of the nutrients (Table 6). The intake of fiber, energy percentage unsaturated fatty acid and n-3 fish fatty acids increased and the intake of animal protein, mono and disaccharides, sodium, calcium and vitamin D decreased in all age groups.

Table 5. Habitual mean ¥ (95%CI) * nutrient intake per day from food sources only for Dutch adults aged 19–69 years in 2007–2010 (*n* = 2106; DNFCS 2007–2010) and 2012–2016 (*n* = 1540; DNFCS 2007–2010) and by gender.

Nutrient	Total		Men		Women	
	2007–2010	2012–2016	2007–2010	2012–2016	2007–2010	2012–2016
	n = 2106	*n* =1540	*n* = 1055	*n* = 1051	*n* = 770	*n* = 770
Energy (MJ)	9.5 (9.5–9.6)	9.4 (9.3–9.4) [1]	10.9 (10.9–11)	10.9 (10.9–11.0)	8.2 (8.1–8.2)	7.8 (7.8–7.9) [1]
Vegetable protein (g)	31.7 (31.6–31.8)	32.1 (31.9–32.3) [1]	35.8 (35.6–36.0)	36.9 (36.7–37.0) [1]	27.7 (27.6–27.8)	27.3 (27.2–27.4) [1]
Animal protein (g)	53.9(53.8–54.2)	51.0 (50.8–51.3) [1]	61.5 (61.2–61.8)	58.1 (57.8–58.5) [1]	46.5 (46.3–46.7)	43.9 (43.7–44.2) [1]
Mono- and disaccharide (en%)	20.1 (20.0–20.1)	19.4 (19.3–19.5) [1]	18.9 (18.8–19.0)	18.3 (18.2–18.4) [1]	21.2 (21.1–21.3)	20.5 (20.4–20.5) [1]
Fiber (g/MJ)	2.25 (2.25–2.26)	2.31 (2.30–2.31) [1]	2.13 (2.12–2.13)	2.20 (2.19–2.20) [1]	2.38 (2.37–2.39)	2.42 (2.41–2.43) [1]
Unsaturated fatty acids (en%)	18.4 (18.3–18.4)	19.4 (19.4–19.5) [1]	18.8 (18.7–18.8)	19.8 (19.7–19.8) [1]	18.0 (17.9–18.0)	19.1 (19.1–19.2) [1]
n-3 fish fatty acids (EPA and DHA) (mg)	132 (125–140)	162 (152–171) [1]	137 (127–146)	176 (162–189) [1]	128 (117–139)	148 (137–160)
Sodium ** (mg)	2745 (2734–2755)	2582 (2570–2595) [1]	3124 (3107–3142)	2972 (2957–2986) [1]	2367 (2358–2375)	2193 (2182–2203) [1]
Calcium (mg)	1062 (1059–1066)	1007 (1003–1011) [1]	1146 (1141–1150)	1081 (1076–1086) [1]	979 (974–983)	933 (928–938) [1]
Vitamin D (μg)	3.5 (3.5–3.6)	3.1 (3.1–3.2) [1]	4.0 (4.0–4.1)	3.6 (3.5–3.6) [1]	3.1 (3.0–3.1)	2.7 (2.6–2.7) [1]
Alcohol (g)	14.0 (13.3–14.8)	10.7 (10.0–11.4) [1]	19.5 (18.2–20.9)	15.7 (14.5–16.9) [1]	8.6 (7.7–9.5)	5.6 (4.9–6.3) [1]

[1] Sig. difference between mean intake (95%CI), DNFCS 2007–2010 vs. DNFCS 2012–2016, in bold; * unless indicated with (en%) for percentage of energy; ** Added salt is not included; ¥ weighted for socio-demographic characteristics, season, and day of the week.

Energy intake only decreased in the youngest age group whereas, vegetable protein intake increased only in the oldest age group. Furthermore, alcohol intake only decreased in the 31–50-year-olds and in the 51–69-year-olds.

In both surveys, adults aged 51–69 years intake of energy, vegetable protein, mono- and disaccharides, unsaturated fatty acids and sodium was lower and animal protein, fiber and n-3 fish fatty acids, calcium, vitamin D and alcohol was higher than that of 19–30-year-olds. The age difference declined slightly between the two periods for most of the nutrients. The difference remained the same only for unsaturated fatty acids and vitamin D and increased for calcium.

For all educational level groups, the mean habitual intake showed a similar direction of change for almost all nutrients (Table 7). The energy percentage of unsaturated fatty acid in the diet increased and the intakes of animal protein, mono- and disaccharides, sodium, calcium, vitamin D and alcohol decreased.

Changes in the intakes of energy, vegetable protein and fiber intake differed across educational levels. Energy intake increased in the group with a high educational level, while it decreased in the other two educational level groups. Vegetable protein intake decreased in the group with a middle educational level and increased in the group with a high educational level. Fiber intake increased only in the groups with a middle and high educational level.

Table 6. Habitual mean [¥] (95%CI) * nutrient intake per day from food sources only for Dutch adults aged 19–69 years in 2007–2010 (*n* = 2106; DNFCS 2007–2010) and 2012–2016 (*n* = 1540; DNFCS 2007–2010) by age groups.

Nutrient	19–30 Years		31–50 Years		51–69 Years	
	2007–2010	2012–2016	2007–2010	2012–2016	2007–2010	2012–2016
	n = 703	*n* = 516	*n* = 699	*n* = 523	*n* = 704	*n* = 501
Energy (MJ)	**10.1 (10.0–10.2)**	**9.6 (9.4–9.7)** [1]	9.7 (9.6–9.8)	9.7 (9.5–9.8)	8.9 (8.9–9.0)	9.0 (8.9–9.1)
Vegetable protein (g)	32.9 (32.5–33.2)	32.6 (32.2–33.1)	32.6 (32.4–32.9)	33.2 (32.8–33.6)	**29.8 (29.5–30.0)**	**30.6 (30.2–30.9)** [1]
Animal protein (g)	**50.8 (50.2–51.4)**	**47.6 (46.9–48.3)** [1]	**53.9 (53.5–54.4)**	**51.9 (51.4–52.5)** [1]	**56.0 (55.5–56.5)**	**52.1 (51.6–52.6)** [1]
Mono- and disaccharides (en%)	**22.3 (22.2–22.4)**	**21.5 (21.4–21.7)** [1]	**19.9 (19.8–20.0)**	**18.9 (18.8–19.0)** [1]	**18.9 (18.8–18.9)**	**18.6 (18.5–18.7)** [1]
Fiber (g/MJ)	**2.12 (2.11–2.13)**	**2.20 (2.19–2.21)** [1]	**2.24 (2.23–2.25)**	**2.29 (2.28–2.30)** [1]	**2.36 (2.35–2.37)**	**2.39 (2.38–2.40)** [1]
Unsaturated fatty acids (en%)	**18.4 (18.3–18.4)**	**19.4 (19.3–19.4)** [1]	**18.5 (18.5–18.5)**	**19.7 (19.6–19.7)** [1]	**18.2 (18.1–18.2)**	**19.2 (19.2–19.3)** [1]
n-3 fish fatty acids (EPA and DHA) (mg)	**110 (103–116)**	**136 (128–144)** [1]	**129 (122–136)**	**165 (156–175)** [1]	**151 (143–159)**	**174 (164–184)** [1]
Sodium ** (mg)	**2872 (2837–2907)**	**2598 (2561–2635)** [1]	**2821 (2798–2844)**	**2676 (2647–2706)** [1]	**2562 (2541–2583)**	**2469 (2442–2495)** [1]
Calcium (mg)	**1028 (1018–1037)**	**942 (932–952)** [1]	**1068 (1062–1075)**	**1023 (1016–1029)** [1]	**1076 (1071–1081)**	**1029 (1023–1034)** [1]
Vitamin D (µg)	**3.2 (3.1–3.2)**	**2.8 (2.8–2.9)** [1]	**3.5 (3.5–3.6)**	**3.1 (3.1–3.2)** [1]	**3.8 (3.8–3.9)**	**3.4 (3.3–3.4)** [1]
Alcohol (g)	9.1 (8.0–10.1)	8.5 (7.3–9.7)	**13.1 (11.9–14.3)**	**9.9 (8.9–11)** [1]	**18.4 (17.1–19.8)**	**12.9 (11.7–14.2)** [1]

[1] Sig. difference between mean intake (95%CI), DNFCS 2007–2010 vs. DNFCS 2012–2016, in bold; * unless indicated with (en%) for percentage of energy; ** Added salt is not included; [¥] weighted for socio-demographic characteristics, season, and day of the week.

In both surveys, people with a high educational level consumed more vegetable protein (and less animal protein), fiber, n-3 fish fatty acids, calcium, and alcohol than people with a low educational level. For fiber and protein intake the changes resulted in a larger gap between the group with a low and a high educational level. For some nutrients, the association with educational level changed. In the 2012–2016 survey, the high educational level group had a higher energy intake and lower contribution of mono-disaccharides to the energy intake than the lower educational level group, while it was comparable between these two groups in the 2007–2010 survey. For unsaturated fatty acids and vitamin D, the intakes were comparable in the most recent survey while in 2007–2010 the intakes were lower in the higher-educated groups than in the lower-educated groups.

Table 7. Habitual mean [¥] (95%CI) * nutrient intake per day from food sources only for Dutch adults aged 19–69 years in 2007–2010 (*n* = 2106; DNFCS 2007–2010) and 2012–2016 (*n* = 1540; DNFCS 2007–2010) by educational level.

Nutrient	Low		Middle		High	
	2007–2010	2012–2016	2007–2010	2012–2016	2007–2010	2012–2016
	n = 708	*n* = 323	*n* = 935	*n* =655	*n* =463	*n* =526
Energy (MJ)	**9.4 (9.4–9.5)**	**9.1 (9.0–9.2)** [1]	**9.7 (9.7–9.8)**	**9.4 (9.3–9.5)** [1]	**9.3 (9.3–9.4)**	**9.5 (9.4–9.6)** [1]
Vegetable protein (g)	30.4 (30.2–30.6)	30.1 (29.8–30.4)	**32.3 (32.1–32.4)**	**31.8 (31.6–32.0)** [1]	**32.2 (31.9–32.5)**	**34.0 (33.7–34.3)** [1]
Animal protein (g)	**54.2 (53.8–54.6)**	**53.0 (52.6–53.5)** [1]	**54.1 (53.7–54.4)**	**51.8 (51.3–52.3)** [1]	**52.1 (51.8–52.5)**	**50.5 (50.1–50.9)** [1]

Table 7. *Cont.*

Nutrient	Low		Middle		High	
	2007–2010	2012–2016	2007–2010	2012–2016	2007–2010	2012–2016
	n = 708	*n* = 323	*n* = 935	*n* =655	*n* =463	*n* =526
Mono- and disaccharides (en%)	**20.0 (19.9–20.2)**	**19.6 (19.4–19.8)** [1]	**20.4 (20.3–20.5)**	**20.0 (19.9–20.1)** [1]	**20.1 (20.0–20.2)**	**18.6 (18.4–18.7)** [1]
Fiber (g/MJ)	2.20 (2.19–2.22)	2.22 (2.20–2.23)	**2.22 (2.22–2.23)**	**2.25 (2.24–2.26)** [1]	**2.37 (2.36–2.38)**	**2.42 (2.41–2.43)** [1]
Unsaturated fatty acids (en%)	**18.5 (18.5–18.5)**	**19.4 (19.3–19.4)** [1]	**18.3 (18.3–18.3)**	**19.2 (19.2–19.3)** [1]	**17.9 (17.8–17.9)**	**19.5 (19.4–19.5)** [1]
n–3 fish fatty acids (EPA and DHA) (mg)	132 (117–146)	156 (129–183)	129 (118–141)	142 (131–153)	170 (151–189)	187 (171–203)
Sodium ** (mg)	**2703 (2683–2723)**	**2545 (2523–2566)** [1]	**2813 (2797–2828)**	**2596 (2576–2616)** [1]	**2661 (2639–2683)**	**2558 (2535–2581)** [1]
Calcium (mg)	**1032 (1024–1039)**	**968 (960–976)** [1]	**1067 (1061–1072)**	**1013 (1006–1021)** [1]	**1101 (1094–1108)**	**1030 (1025–1035)** [1]
Vitamin D (µg)	**3.7 (3.6–3.7)**	**3.2 (3.1–3.3)** [1]	**3.6 (3.6–3.7)**	**3.1 (3.1–3.2)** [1]	**3.3 (3.2–3.4)**	**3.0 (3.0–3.1)** [1]
Alcohol (g)	**12.8 (11.4–14.2)**	**8.8 (7.1–10.4)** [1]	**14.5 (13.3–15.7)**	**10.9 (9.8–12)** [1]	**16.0 (14.4–17.5)**	**12.3 (11.0–13.6)** [1]

[1] Sig. difference between mean intake (95%CI), DNFCS 2007–2010 vs. DNFCS 2012–2016, in bold; * unless indicated with (en%) for percentage of energy; ** Added salt is not included; ¥ weighted for socio-demographic characteristics, season, and day of the week.

4. Discussion

This study found several positive changes in health-related dietary intake between the periods 2007–2010 and 2012–2016 among Dutch adults. The consumption of fiber and unsaturated fatty acids increased by about 3% and 6%, respectively. The mean consumption of red or processed meat, animal protein, mono-and disaccharides, sodium and alcohol decreased by about 4–30%. Some less favorable changes were also observed; the intake of calcium and vitamin D decreased, potentially due to the decrease in dairy, as did the ratio of whole-grain products to cereals and cereal products. In both surveys, it was observed that people with a high educational level consumed more vegetables, fruit and tea and consumed less red or processed meat, and sugar-containing beverages than people with a lower educational level. However, for most food groups, changes in consumption over time were similar across the educational level groups, as well as for both sexes, and all age groups. Our findings represent the most comprehensive evaluation of the change in relevant dietary habits among Dutch adults in the previous decades.

The favorable decrease in consumption of red or processed meat is reflected in a decreased intake of animal protein. This is not an issue as total protein intake was still sufficient in the Dutch population [7]. The decrease in sodium intake is probably not related to a change in dietary habits but a result of the efforts by food manufacturers to reduce the salt content of their products to comply with the "National agreement to Improve Product Composition: Salt, Saturated Fat, Sugar (Calories)" [22]. The intake of n-3 fish fatty acid also developed favorably. However, its higher intake was not associated with an increase in fish consumption. This suggests that the types of fish consumed changed to kinds richer in n-3 fish fatty acids. Similar changes were found in England in their surveys of 2001 and 2009 [23,24]. Another positive development was the decrease in intake of mono- and disaccharides. As the total consumption of sugar-containing beverages kept stable in most subgroups, this decrease might be explained by the result of several other changes in the food pattern. For instance, changes in the sugar content of sugar-containing drinks or in consumption of some types of beverages (less fruit juices; data not shown) or in consumption of sweet snacks (data not shown), fruit and dairy. It would be interesting to monitor the contribution of sugar-containing drinks to the mono-disaccharides as the food industry made agreements for reformulation of these products.

Not all observed changes between the two surveys were beneficial from a health perspective. The consumption of dairy decreased and with that, the intake of vitamin D and calcium. Since the mean habitual vitamin D intake recommended by the Health Council is not met, a decrease in the intake of vitamin D is therefore unfavorable [25]. Because one of the principal functions of vitamin D is that it stimulates the absorption of calcium from foods [26], a combined decrease with calcium is unfavorable, even though the habitual intake of calcium in the Dutch population is quite high and should be monitored to prevent further decrease [3]. A second unfavorable change was the decrease in the consumption of whole-grain products. This was not reflected in a decreased intake of fiber per MJ of energy, partly because energy intake decreased and partly because fiber intake from other sources was larger. As in both surveys, the intake is far below the adequate intake, an increase in the consumption of whole-grain products is important for reducing the risk of chronic diseases.

Our observed changes in food group consumption of the total Dutch population differ from reported changes in the period 1987–1998. The decrease in the consumption of vegetables and fruit in that 10-year period has stopped, and the increase in tea consumption before 2000 was not significant anymore when comparing 2007–2010 versus 2012–2016. Because of differences in food classification, time trends in the consumption of other food groups could not be compared [12]. Additionally, for other European countries time trends in consumption were reported. Similar to our results, several countries found no change in vegetable [27–30], fruit [27,28] or fish [27,29] consumption and a decrease in (red- or processed) meat consumption [27,28] in the total population. However, contrary to the results of this study, most countries did not find a change in dairy consumption [28,29] or found a decrease in bread and cereal consumption [28,30]. Observed differences could be different trends in other countries but might also be explained by different time periods studied.

Our results show only minimal dietary disparities in changes by age and sex. However, comparing the food consumption by age, most of the differences are in favor of adults in the oldest age group. This group consumed more of the favorable food groups such as vegetables, fruit and tea and a higher proportion of whole grain products and they consumed less sugar-containing beverages. This difference was already apparent in 2007–2010 and continued in 2012–2016. Furthermore, compared to men, women had a more preferable diet since they consumed more fruit and tea, and less red or processed meat, and sugar-containing beverages in both surveys. However, they consumed less dairy, this gender difference decreased in 2012–2016 compared to that in 2007–2010, while for sugar-containing beverages and tea the gender differences increased. Similar but not exactly the same age and sex differences were observed in other Western European countries such as Germany [29], the UK [31] and France [32].

Dietary disparities by educational strata were also observed with a diet in favor of adults in the high educational level group. In both surveys, people with a high educational level consumed more vegetables, fruit and tea and less red or processed meat and sugar-containing beverages than people with a lower educational level. Between the two surveys, the mean consumption of vegetables significantly increased in people with a higher-educational level and the consumption of red or processed meat significantly decreased in both the lower- and the higher-educational level group. Consequently, the gap between people with a low educational level and a high educational level broadened in 2012–2016 for vegetables and decreased slightly for red or processed meat. Across Europe, a higher education level is associated with higher fruit and vegetable consumption [33–35]. In accordance, the Helsinki Health Study survey found that the difference in the consumption of fresh vegetables and fish, increased in favor of people with a higher education level, between 2000–2002 and 2007 [36]. Outside of Europe, disparities by income and educational level also widened for the purchases of fruits, vegetables and the percentage of calories from sugar between 2008 and 2018 [37].

Even though some of the dietary changes seen are a step in the right direction, the Dutch population still does not reach the recommended dietary guidelines set by the Health Council for any of the describe food groups [7]. The guidelines are developed based on scientific evidence and examine the relationship between diet and the risk of disease. Additional health gains can be achieved when the Dutch adopt a diet more in accordance with the Dutch Dietary Guidelines. This would lower the risk of stroke, coronary heart disease, diabetes, colorectal cancer and mortality from any cause [38,39]. Life expectancy of people with a lower educational level is 6 years lower and they live 15 years longer in poorer health than people with a higher educational level. Contributing factors could be a difference in lifestyle and food consumption. The ecological aspects of the Dutch dietary guidelines have also been considered by the Dutch Health Council. In general, a more plant-based and less animal-based diet is associated with a lower ecological footprint [40]. The observed declines in the consumption of meat and dairy are also expected to result in lower greenhouse gas emissions of the diet [13,41]. More research on this topic should be done to get more insight and identify potentially favorable changes for the ecological impact.

Altering food consumption in people is known to be challenging. This is even more the case in subgroups with a lower educational level since they tend to have a poorer attitude toward healthy eating [42]. People with a higher educational level usually have more knowledge about healthier food items compared to people with a lower educational level [43]. Increasing dietary information could therefore be helpful. However, other policies, such as economic interventions, such as taxes aimed at reducing the cost for healthy foods) and also more upstream policies focused on access to decent wages and social support might be even more effective [44,45]. Insight needs to be gained in what types of dietary policies are most effective in altering food consumption in people with lower educational levels. Changes in dietary policies can be made accordingly and help cater to the specific needs of this group [46].

Potential limitations of our study should be considered. We were not able to evaluate the dietary trends on all recommendations of the Dutch food-based dietary guidelines because the data did not allow this (filtered coffee), or because consumption levels were too low to estimate habitual intake for the Dutch population (e.g., legumes, unsalted nuts). The differences in consumption could potentially be caused by differences in participant characteristics between the two surveys. However, we assume that the observed changes are hardly affected by these changes in the Dutch population. In addition, our study also showed trends stratified by gender, age group, and educational level. As with any population measure, dietary information is subject to random and systematic error. At the group level, underestimation of energy intake is common when self-reported methods are used among adults [47,48]. The mean underreporting increased by 3% between the two surveys [7] and might explain part of observed decreases in intake. Most of our observed differences were however larger than 3%. Another potential limitation is that only two consecutive surveys are used. It would be of added value to repeat these analyses when the results of the 2019–2021 survey become available.

Our study also has several strengths. This is the first study to investigate the change in food consumption and nutrient intake in the 21st century in Dutch adults and different subgroups. Multiple food groups and corresponding nutrients were evaluated using similar validated methods across the two DNFCS, making valid comparisons possible. We used detailed and representative data for the Dutch population during the period of data collection. Both surveys had a large sample size. The characteristics of participants in the survey are consistent with the Dutch population demographics at the moment of the surveys. With the exception of the number of subjects with a very low educational level and those with non-Western migration background, they might not be as well represented in the DNFCS in the low educational level group [7]. In any survey, the most vulnerable groups are usually difficult to reach to participate [49,50].

5. Conclusions

Several changes in food consumption and nutrient intake occurred between 2007–2010 and 2012–2016 in Dutch adults. Various favorable changes were observed, such as an increase in fiber (g/MJ) and unsaturated fatty acids (en%) intake and a decrease in red- or processed meat consumption, in intake of animal protein, mono-and disaccharides (en%), sodium and alcohol intake. Some less favorable changes were also observed; the consumption of dairy, the type of cereals calcium and vitamin D decreased.

A healthier food consumption pattern was still more observed in adults with a higher educational level. Future assessment of trends in food consumption and nutrient intake -in the Dutch population is necessary to evaluate whether the observed changes will continue over the following years or whether they were temporary and to monitor if disparities between educational level groups will change over time.

Author Contributions: Conceptualisation, C.S.D., M.C.O., E.J.M.B.-R. and C.T.M.v.R.; formal analysis, C.S.D. and E.J.M.B.-R.; writing—original draft preparation, C.S.D.; writing—review and editing, M.C.O. and C.T.M.v.R.; All authors have read and agreed to the published version of the manuscript.

Funding: This research received no external funding.

Institutional Review Board Statement: Ethical review and approval were waived for this study. The Utrecht University Medical Ethical Review Committee evaluated that the study from which the data were used was not subject to the Medical Research Involving Human Subjects Act (WMO) of the Netherlands (reference number 12–359/C).

Informed Consent Statement: Informed consent was obtained from all subjects involved in the study before an interview during a home visit. For interviews by telephone, written informed consent was not necessary according to the Dutch regulations at the time of data collection.

Data Availability Statement: The data used in this study are available on request from https://www.rivm.nl/en/dutch-national-food-consumption-survey/data-on-request (accessed on 25 March 2021).

Acknowledgments: We would like to thank Arnold Dekkers for his support with the SPADE analyses and Marja Beukers for her help with the statistical analyses.

Conflicts of Interest: The authors declare no conflict of interest.

References

1. Gakidou, E.; Afshin, A.; Abajobir, A.A.; Abate, K.H.; Abbafati, C.; Abbas, K.M.; Abd-Allah, F.; Abdulle, A.M.; Abera, S.F.; Aboyans, V.; et al. Global, regional, and national comparative risk assessment of 84 behavioural, environmental and occupational, and metabolic risks or clusters of risks, 1990–2016: A systematic analysis for the Global Burden of Disease Study 2016. *Lancet* **2017**, *390*, 1345–1422. [CrossRef]
2. Fardet, A.; Boirie, Y. Associations between food and beverage groups and major diet-related chronic diseases: An exhaustive review of pooled/meta-analyses and systematic reviews. *Nutr. Rev.* **2014**, *72*, 741–762. [CrossRef] [PubMed]
3. Kromhout, D.; Spaaij, C.J.K.; De Goede, J.; Weggemans, R.M. The 2015 Dutch food-based dietary guidelines. *Eur. J. Clin. Nutr.* **2016**, *70*, 869–878. [CrossRef] [PubMed]
4. Nissen, S.E. U.S. Dietary Guidelines. *Ann. Intern. Med.* **2016**, *165*, 605–606. [CrossRef] [PubMed]
5. Hercberg, S.; Chat-Yung, S.; Chaulia, M. The French National Nutrition and Health Program: 2001–2006–2010. *Int. J. Public Health* **2008**, *53*, 68–77. [CrossRef] [PubMed]
6. Gezondheidsraad, H. (Ed.) *Voedingsaanbevelingen voor België-2016*; HGR: Brussel, Belgium, 2016.
7. van Rossum, C.T.M.; Buurma-Rethans, E.J.M.; Dinnissen, C.S.; Beukers, M.H.; Brants, H.A.M.; Ocké, M.C. *The Diet of the Dutch; Results of the Dutch National Food Consumption Survey 2012–2016*; RIVM: Bilthoven, The Netherlands, 2020.
8. Grzymisławska, M.; Puch, E.A.; Zawada, A.; Grzymisławski, M. Do nutritional behaviors depend on biological sex and cultural gender? *Adv. Clin. Exp. Med.* **2020**, *29*, 165–172. [CrossRef] [PubMed]
9. Drewnowski, A.; Shultz, J.M. Impact of aging on eating behaviors, food choices, nutrition, and health status. *J. Nutr. Health Aging* **2001**, *5*, 75–79. [PubMed]
10. Hiza, H.A.; Casavale, K.O.; Guenther, P.M.; Davis, C.A. Diet Quality of Americans Differs by Age, Sex, Race/Ethnicity, Income, and Education Level. *J. Acad. Nutr. Diet.* **2013**, *113*, 297–306. [CrossRef] [PubMed]
11. Volaco, A.; Cavalcanti, A.M.; Filho, R.P.; Precoma, D.B. Socioeconomic Status: The Missing Link between Obesity and Diabetes Mellitus? *Curr. Diabetes Rev.* **2018**, *14*, 321–326. [CrossRef] [PubMed]

12. Hulshof, K.F.A.M.; Brussaard, J.H.; Kruizinga, A.G.; Telman, J.; Löwik, M.R.H. Socio-economic status, dietary intake and 10 y trends: The Dutch National Food Consumption Survey. *Eur. J. Clin. Nutr.* **2003**, *57*, 128–137. [CrossRef] [PubMed]
13. van Bussel, L.M.; van Rossum, C.T.; Temme, E.H.; Boon, P.E.; Ocké, M.C. Educational differences in healthy, environmentally sustainable and safe food consumption among adults in the Netherlands. *Public Health Nutr.* **2020**, *23*, 57–67. [CrossRef] [PubMed]
14. Vlismas, K.; Stavrinos, V.; Panagiotakos, D.B. Socio-economic Status, Dietary Habits and Health-Related Outcomes in Various Parts of the World: A Review. *Cent. Eur. J. Public Health* **2009**, *17*, 55–63. [CrossRef]
15. Darmon, N.; Drewnowski, A. Does social class predict diet quality? *Am. J. Clin. Nutr.* **2008**, *87*, 1107–1117. [CrossRef]
16. van Rossum, C.T.; Fransen, H.P.; Verkaik-Kloosterman, J.A.; Buurma-Rethans, E.J.; Ocke, M.C. *Dutch National Food Consumption Survey 2007–2010; Diet of Children and Adults Aged 7 to 69 Years*; RIVM: Bilthoven, The Netherlands, 2011.
17. European Food Safety Authority. *General Principles for the Collection of National Food Consumption Data in the View of a Pan-European Dietary Survey*; European Food Safety Authority: Parma, Italy, 2009; Volume 7, p. 1435.
18. Crispim, S.P.; Nicolas, G.; Casagrande, C.; Knaze, V.; Illner, A.-K.; Huybrechts, I.; Slimani, N. Quality assurance of the international computerised 24 h dietary recall method (EPIC-Soft). *Br. J. Nutr.* **2013**, *111*, 506–515. [CrossRef] [PubMed]
19. RIVM. NEVO-Online Versie 2016/5.0. 2016. Available online: https://nevo-online.rivm.nl/ (accessed on 25 March 2021).
20. RIVM/Voedingscentrum. *NEVO-Tabel; Nederlands Voedingsstoffenbestand 2011, in NEVO-Tabel*; RIVM/Voedingscentrum: The Hague, The Netherlands, 2011.
21. Dekkers, A.L.M.; Verkaik-Kloosterman, J.; van Rossum, C.T.M.; Ocké, M.C. SPADE, a New Statistical Program to Estimate Habitual Dietary Intake from Multiple Food Sources and Dietary Supplements. *J. Nutr.* **2014**, *144*, 2083–2091. [CrossRef]
22. Akkoord Verbetering Productsamenstelling Zout, Verzadigd Vet, Suiker. 2014. Available online: http://www.rijksoverheid.nl/documenten-en-publicaties/convenanten/2014/01/23/akkoord-verbetering-productsamenstelling-zout-verzadigd-vet-suiker.html (accessed on 23 April 2021). (In Dutch).
23. Whitton, C.; Nicholson, S.K.; Roberts, C.; Prynne, C.J.; Pot, G.K.; Olson, A.; Fitt, E.; Cole, D.; Teucher, B.; Bates, B.; et al. National Diet and Nutrition Survey: UK food consumption and nutrient intakes from the first year of the rolling programme and comparisons with previous surveys. *Br. J. Nutr.* **2011**, *106*, 1899–1914. [CrossRef] [PubMed]
24. Pot, G.K.; Prynne, C.J.; Roberts, C.; Olson, A.; Nicholson, S.K.; Whitton, C.; Teucher, B.; Bates, B.; Henderson, H.; Pigott, S.; et al. National Diet and Nutrition Survey: Fat and fatty acid intake from the first year of the rolling programme and comparison with previous surveys. *Br. J. Nutr.* **2011**, *107*, 405–415. [CrossRef]
25. Health Council of The Netherlands. *Evaluation of the Dietary Reference Values for Vitamin D*; Health Council of The Netherlands: The Hague, The Netherlands, 2012.
26. Christakos, S.; Dhawan, P.; Porta, A.; Mady, L.J.; Seth, T. Vitamin D and intestinal calcium absorption. *Mol. Cell. Endocrinol.* **2011**, *347*, 25–29. [CrossRef] [PubMed]
27. Public Health England. *National Diet and Nutrition Survey: Years 1 to 9 of the Rolling Programme (2008/2009–2016/2017): Time Trend and Income Analyses in National Diet and Nutrition Survey*; Public Health England: London, UK, 2019.
28. Bel, S.; de Ridder, K.A.A.; Lebacq, T.; Ost, C.; Teppers, E.; Cuypers, K.; Tafforeau, J. Habitual food consumption of the Belgian population in 2014–2015 and adherence to food-based dietary guidelines. *Arch. Public Health* **2019**, *77*, 14. [CrossRef]
29. Gose, M.; Krems, C.; Heuer, T.; Hoffmann, I. Trends in food consumption and nutrient intake in Germany between 2006 and 2012: Results of the German National Nutrition Monitoring (NEMONIT). *Br. J. Nutr.* **2016**, *115*, 1498–1507. [CrossRef]
30. Varela-Moreiras, G.; Ruiz, E.; Valero, T.; Ávila, J.M.; del Pozo, S. The Spanish diet: An update. *Nutr. Hosp.* **2013**, *28*, 13–20.
31. Yau, A.; Adams, J.; Monsivais, P. Time trends in adherence to UK dietary recommendations and associated sociodemographic inequalities, 1986–2012: A repeated cross-sectional analysis. *Eur. J. Clin. Nutr.* **2019**, *73*, 997–1005. [CrossRef] [PubMed]
32. Dubuisson, C.; Lioret, S.; Touvier, M.; Dufour, A.; Calamassi-Tran, G.; Volatier, J.-L.; Lafay, L. Trends in food and nutritional intakes of French adults from 1999 to 2007: Results from the INCA surveys. *Br. J. Nutr.* **2009**, *103*, 1035–1048. [CrossRef] [PubMed]
33. Stea, T.H.; Nordheim, O.; Bere, E.; Stornes, P.; Eikemo, T.A. Fruit and vegetable consumption in Europe according to gender, educational attainment and regional affiliation—A cross-sectional study in 21 European countries. *PLoS ONE* **2020**, *15*, e0232521. [CrossRef] [PubMed]
34. de Irala-Estévez, J.; Groth, M.; Johansson, L.; Oltersdorf, U.; Prättälä, R.; A Martínez-González, M. A systematic review of socio-economic differences in food habits in Europe: Consumption of fruit and vegetables. *Eur. J. Clin. Nutr.* **2000**, *54*, 706–714. [CrossRef]
35. Prättälä, R.; Hakala, S.; Roskam, A.-J.R.; Roos, E.; Helmert, U.; Klumbiene, J.; van Oyen, H.; Regidor, E.; Kunst, A.E. Association between educational level and vegetable use in nine European countries. *Public Health Nutr.* **2009**, *12*, 2174–2182. [CrossRef]
36. Seiluri, T.; Lahelma, E.; Rahkonen, O.; Lallukka, T. Changes in socio-economic differences in food habits over time. *Public Health Nutr.* **2011**, *14*, 1919–1926. [CrossRef]
37. Lacko, A.M.; Maselko, J.; Popkin, B.; Ng, S.W. *Socio-Economic and Racial/Ethnic Disparities in the Nutritional Quality of Packaged Food Purchases in the USA, 2008–2018*; Cambridge University Press: Cambridge, UK, 2021; pp. 1–13.
38. Voortman, T.; Jong, J.C.K.-D.; Ikram, M.A.; Stricker, B.H.; Van Rooij, F.J.A.; LaHousse, L.; Tiemeier, H.; Brusselle, G.G.; Franco, O.H.; Schoufour, J.D. Adherence to the 2015 Dutch dietary guidelines and risk of non-communicable diseases and mortality in the Rotterdam Study. *Eur. J. Epidemiol.* **2017**, *32*, 993–1005. [CrossRef]

39. Boer, J.M.A.; Buurma-Rethans, E.J.M.; van Kranen, H.J.; Milder, I.E.J.; Ocke, M.C.; Verkaik-Kloosterman, J.; van Raaij, J. *Health Aspects of the Dutch Diet, in Background Report to 'What Is on Our Plate? Safe, Healthy and Sustainable Diets in the Netherlands*; RIVM: Bilthoven, The Netherlands, 2017.
40. Gezondheidsraad. *Richtlijnen Goede Voeding Ecologisch Belicht*; Gezondheidsraad: The Hague, The Netherlands, 2011.
41. Vellinga, R.E.; van de Kamp, M.; Toxopeus, I.B.; van Rossum, C.T.M.; de Valk, E.; Biesbroek, S.; Hollander, A.; Temme, E.H.M. Greenhouse gas emissions and blue water use of Dutch diets and its association with health. *Sustainability* **2019**, *11*, 6027. [CrossRef]
42. Lê, J.; Dallongeville, J.; Wagner, A.R.; Arveiler, D.; Haas, B.; Cottel, D.; Simon, C.; Dauchet, L. Attitudes toward healthy eating: A mediator of the educational level–diet relationship. *Eur. J. Clin. Nutr.* **2013**, *67*, 808–814. [CrossRef]
43. Spronk, I.; Kullen, C.; Burdon, C.; O'Connor, H. Relationship between nutrition knowledge and dietary intake. *Br. J. Nutr.* **2014**, *111*, 1713–1726. [CrossRef]
44. McGill, R.; Anwar, E.; Orton, L.; Bromley, H.; Lloyd-Williams, F.; O'Flaherty, M.; Taylor-Robinson, D.; Guzman-Castillo, M.; Gillespie, D.; Moreira, P.; et al. Are interventions to promote healthy eating equally effective for all? Systematic review of socioeconomic inequalities in impact. *BMC Public Health* **2015**, *15*, 1–15.
45. Aggarwal, A.; Monsivais, P.; Cook, A.J.; Drewnowski, A. Does diet cost mediate the relation between socioeconomic position and diet quality? *Eur. J. Clin. Nutr.* **2011**, *65*, 1059–1066. [CrossRef]
46. PHAROS. *Sociaaleconomische Gezondheidsverschillen (SEGV)*; PHAROS: Utrecht, The Netherlands, 2019.
47. Novotny, J.A.; Rumpler, W.V.; Riddick, H.; Hebert, J.R.; Rhodes, D.; Judd, J.T.; Baer, D.J.; McDowell, M.; Briefel, R. Personality characteristics as predictors of underreporting of energy intake on 24-hour dietary recall interviews. *J. Am. Diet. Assoc.* **2003**, *103*, 1146–1151. [CrossRef]
48. Park, Y.; Dodd, K.W.; Kipnis, V.; E Thompson, F.; Potischman, N.; A Schoeller, D.; Baer, D.J.; Midthune, D.; Troiano, R.P.; Bowles, H.; et al. Comparison of self-reported dietary intakes from the Automated Self-Administered 24-h recall, 4-d food records, and food-frequency questionnaires against recovery biomarkers. *Am. J. Clin. Nutr.* **2018**, *107*, 80–93. [CrossRef] [PubMed]
49. Rockliffe, L.; Chorley, A.J.; Marlow, L.A.V.; Forster, A.S. It's hard to reach the "hard-to-reach": The challenges of recruiting people who do not access preventative healthcare services into interview studies. *Int. J. Qual. Stud. Health Well Being* **2018**, *13*, 1479582. [CrossRef] [PubMed]
50. Bonevski, B.; Randell, M.; Paul, C.; Chapman, K.; Twyman, L.; Bryant, J.; Brozek, I.; Hughes, C. Reaching the hard-to-reach: A systematic review of strategies for improving health and medical research with socially disadvantaged groups. *BMC Med. Res. Methodol.* **2014**, *14*, 42. [CrossRef]

nutrients MDPI

Article

Can Healthy and Sustainable Dietary Patterns That Fit within Current Dutch Food Habits Be Identified?

Samantha N. Heerschop [1], Sander Biesbroek [1], Elisabeth H. M. Temme [1,2] and Marga C. Ocké [1,2,*]

[1] Division of Human Nutrition and Health, Wageningen University, P.O. Box 17, 6700 AA Wageningen, The Netherlands; samantha.heerschop@wur.nl (S.N.H.); sander.biesbroek@wur.nl (S.B.); liesbeth.temme@rivm.nl (E.H.M.T.)
[2] National Institute for Public Health and the Environment (RIVM), Postbox 1, Bilthoven, 3720 BA Utrecht, The Netherlands
* Correspondence: marga.ocke@rivm.nl; Tel.: +31-(0)317-48-0363

Abstract: This study investigated major healthy and sustainable dietary patterns in the Dutch population. Two 24-hour dietary recalls were collected in 2078 participants aged 19–79 years in the Dutch National Food Consumption Survey 2012–2016. Dietary patterns were identified using reduced rank regression. Predictor variables were food groups and response variables were Dutch Healthy Diet index 2015 (DHD15-index) score, greenhouse gas emissions (GHGE), and blue water use. Three patterns were discovered, including a "high fruit and vegetable dietary pattern", a "low meat dietary pattern", and a "high dairy, low fruit juices dietary pattern". Diets in the highest quartile of these patterns had higher DHD15-index score than the average population. However, diets of the "high fruit and vegetable dietary pattern" were associated with higher dietary GHGE (14%) and blue water use (69.2%) compared to the average population. Diets of the "low meat dietary pattern" were associated with lower GHGE (19.6%) and higher blue water use (7.7%). Concluding, the "low meat dietary pattern" was the most healthy and sustainable dietary pattern in this population. The addition of blue water use as an environmental impact indicator shows the difficulty of finding existing dietary patterns that have low environmental impact in all determinants.

Keywords: sustainable diets; dietary pattern; reduced rank regression; greenhouse gas emissions; blue water use; acceptability

Citation: Heerschop, S.N.; Biesbroek, S.; Temme, E.H.M.; Ocké, M.C. Can Healthy and Sustainable Dietary Patterns That Fit within Current Dutch Food Habits Be Identified? *Nutrients* **2021**, *13*, 1176. https://doi.org/10.3390/nu13041176

Academic Editors: Murielle Bochud and Igor Pravst

Received: 15 March 2021
Accepted: 31 March 2021
Published: 2 April 2021

1. Introduction

Ongoing climate change emphasizes the need for new strategies to improve sustainability, as stated by the Paris Climate Agreement and the United Nations Sustainable Development Goals [1,2]. Greenhouse gas emissions (GHGE) are a major driver of global warming, and hence, reduction is key [3]. Food production and consumption contribute 20–30% to the total GHGE [4]. In addition, food production is a major determinant in biodiversity loss, land use, and fresh water use [5,6]. Therefore, shifting towards more sustainable diets is important and urgent.

According to the Food and Agriculture Organization (FAO) of the United Nations, sustainable diets have low environmental impacts, are culturally acceptable, and are nutritionally adequate, safe, and healthy for present life and future generations [7]. To identify healthy and sustainable diets, several studies modelled a priori dietary patterns based on nutritional guidelines and environmental impact data. For example, the EAT-Lancet diet was introduced as the healthy and sustainable reference diet that enables us to feed the world without exceeding planetary boundaries [8]. The EAT-Lancet dietary recommendations include high consumption of vegetables, fruit, whole grains, legumes, nuts, and unsaturated oils, moderate consumption of seafood and poultry, and minimal consumption of red meat, processed meat, added sugar, refined grains, and starchy vegetables. Large differences are observed when comparing the current "Western diet", such as the Dutch diet, to this reference diet [9], and consequently, large changes need to be implemented to meet recommendations of the EAT-Lancet diet. As a first step towards such a healthy and sustainable diet, data-driven or a posteriori methods can be used to derive example dietary patterns present within a population [10]. These dietary patterns have proven acceptability by at least part of the population.

An often-used posteriori method to derive dietary patterns is the principal component analysis [11,12]. Another method, the reduced rank regression (RRR), is able to extract dietary patterns that are stronger associated with several particular effect measures (response variables), such as disease risk factors, by combining both a priori and posteriori techniques [13,14]. A previous study used the RRR to search for dietary patterns in a Dutch cohort, using as response variables the Dutch Healthy Diet 2015 index (DHD15-index) as a proxy for healthiness of the diet and GHGE as an environmental impact indicator [15]. In this study, the derived "plant-based diet" included high consumption of fruit, vegetables, and legumes and benefited health as well as the environment. However, the second derived "dairy-based diet", including high consumption of dairy and nuts and seeds, was somewhat healthier but at the expense of higher GHGE.

Since the effects of sustainable diets should not exceed any planetary boundary, the focus should not be only on GHGE. Vellinga et al. (2019) showed that GHGE is highly correlated with acidification, eutrophication, and land use, but not with blue water use. Therefore, including blue water use in the analysis could potentially provide a more complete picture of environmental impact of a diet. Furthermore, to find healthy and environmentally sustainable dietary patterns that are achievable for the Dutch population, a representative study population with most recent dietary information is needed. Participants that completed the latest Dutch National Food Consumption Survey (DNFCS) (2012–2016) would be such a study population [16].

To gain insight in healthy and environmentally sustainable dietary patterns that are realistic and achievable for the Dutch population, this study investigated which dietary patterns were present in the study population that might be beneficial for health and the environment.

2. Materials and Methods

2.1. Study Population

The study population consisted of participants of the DNFCS [16]. This survey consisted of 4313 participants aged 1–79 years and was conducted between 2012–2016 by the National Institute for Public Health and the Environment (RIVM), the Netherlands. Participants were drawn from a representative consumer panel of the market research agency KANTAR TNS. Panel members participate in all types of studies. An age–gender random sampling strategy was applied. Furthermore, representativeness of region, address density, and education was taken into account. The response rate was 65%. The DNFCS was conducted according to the guidelines of the Helsinki Declaration. Because of non-invasive measurements in this survey, the Medical Ethical Committee of the University Medical Centre Utrecht, the Netherlands, concluded that the study did not need to be evaluated according to the "Medical research on human act" (WMO).

Children younger than 19 years were excluded from this study (n = 2235). The total population for analyses was aged 19–79 and consisted of 2078 participants.

2.2. Dietary Assessment

Trained dieticians collected food consumption data by two non-consecutive 24-hour dietary recalls. Standardized interviews were conducted using the GloboDiet (former EPIC-soft©) computer program, provided by the International Agency for Research on Cancer, Lyon, France [17]. Participants aged >70 years received an additional food recording booklet to be kept at the day before the call. To obtain consumption information representative for the whole year, the 24-hour dietary recalls were spread over seasons and both week and weekend days.

The energy of consumed products was derived from the Dutch Food Composition Database (NEVO—online version 2016/5.0) [18]. Food items from the DNFCS were grouped in 21 food groups adapted from the GloboDiet food group categorization. For the present study, consumptions were presented in g/2000 kcal. This standardization was to account for differences in energy intake by age and gender. In this way of standardization, food items that contain no calories, such as water, coffee, and tea, could still be taken into account in the analysis.

2.3. Assessment of Healthiness of Diets

The healthiness of a diet was scored using the DHD15-index [19]. This index distinguishes fifteen components, each representing one of the fifteen Dutch dietary guidelines of 2015 (Table A1) [20]. A score between 0, indicating no adherence, and 10, indicating complete adherence, was attributed to each component [15]. Since there was no information available on type of coffee consumed (filtered or not), this item was not taken into account. DHD15-index scores in this study could, therefore, potentially range between 0 and 140 points, where a score of 140 indicates maximal adherence to the guidelines.

2.4. Assessment of the Environmental Impact of Diets

GHGE (kg CO_2 equivalents/2000 kcal) and blue water use (m^3/2000 kcal) of food products consumed by the Dutch population were determined using Life Cycle Assessment (LCA) [21]. LCAs take into account the process of production, transportation, preparation, and waste or losses of a product at all stages of the life cycle. Blonk Consultants provided Life Cycle Inventories to estimate environmental impact of a product [22]. Vellinga et al. (2019) provided a more extensive description of definitions of GHGE and blue water use and of the usage of LCA in the DNFCS [23].

Table A2 shows median GHGE and blue water use per kg products within food groups. Calculations were based on consumption of participants, excluding non-consumers in each food group.

2.5. Lifestyle and Anthropometric Variables

A general questionnaire was used to derive information on covariates. For participants aged 19–70, height and weight were self-reported. From this information, body mass index (BMI) was calculated (kg/m^2). For people aged >70, weight was measured and height was not reported; therefore, BMI could not be calculated. Educational level was classified as low (primary education, lower vocational education, advanced elementary education), moderate (intermediate vocational education, higher secondary education), or high (higher vocational education and university).

2.6. Statistical Analysis

Characteristics of the study population are presented as mean and standard deviation for continuous, normally distributed variables. For continuous, skewed data, median and interquartile range are shown. Categorical variables are shown as percentages.

RRR was used to extract dietary patterns that might benefit health and the environment. This method determines linear functions of predictors by maximizing the explained variation in various response variables [14]. To perform this analysis, the PROC PLS procedure in SAS was used. Predictor variables were food groups in gram per 2000 kcal. Response variables were DHD15-index score, daily dietary GHGE per 2000 kcal, and daily dietary blue water use per 2000 kcal. The number of dietary patterns derived is equal to the number of response variables. Participants received a pattern score for each pattern. Pattern scores were split in quartiles, and participants in quartile 4 (Q4) were the highest adherents to the pattern. If the highest adherents to a pattern showed lower DHD15-index than lowest adherents, pattern scores were multiplied by (−1) to obtain the healthier diet in Q4. Dietary patterns were labelled based on the two food groups that had the strongest association with the pattern. A factor loading of $> | 0.20 |$ was considered important.

All statistical analyses were performed using SAS software (version 9.4, SAS Institute Inc., Cary, NC, USA). A two-sided p value of < 0.05 was considered statistically significant.

3. Results

3.1. Population Characteristics and Food Consumption

The adult population from the DNFCS consisted of 2078 participants of which 50.2% were male (Table 1). The median age was 51 years, and the median energy intake was 2064 (interquartile range (IQR): 1699–2552) kcal/day. The mean DHD15-index score was 59.4 (standard deviation (SD): 18.6) out of the potentially maximum score of 140. The median dietary GHGE per 2000 kcal was 4.7 (IQR: 4.02–5.62) kg CO_2 equivalents, and the median dietary blue water use was 0.13 (IQR: 0.10–0.19) m^3 per 2000 kcal.

Table 1. Characteristics and consumption (g/2000 kcal) per food group of the total adult population (n = 2078) and diets in quartile one and four of the "high fruit and vegetable dietary pattern", the "low meat dietary pattern", and the "high dairy, low fruit juices dietary pattern", derived by the reduced rank regression.

	Total Population from DNFCS	"High Fruit and Vegetable Dietary Pattern"		"Low Meat Dietary Pattern" [1]		"High Dairy, Low Fruit Juices Dietary Pattern"	
		Quartile 1	Quartile 4	Quartile 1	Quartile 4	Quartile 1	Quartile 4
Age (years) (median, IQR)	51 (31–70)	41 (27–56)	59 (38–72)	56 (36–71)	47 (30–68)	49 (30–68)	55 (31–72)
Males (n (%))	1043 (50.2)	336 (64.7)	143 (27.6)	263 (50.7)	250 (48.2)	278 (53.6)	240 (46.2)
Body mass index (kg/m^2) [2] (median, IQR)	25.5 (22.7–29.0)	25.2 (22.2–29.0)	25.6 (22.9–29.4)	27.2 (24.2–30.6)	24.0 (21.7–27.1)	25.3 (22.7–28.4)	25.9 (23.3–29.5)
Smokers (n (%))	413 (20.0)	136 (26.6)	79 (15.2)	121 (23.4)	74 (14.3)	132 (25.6)	93 (18.0)
Energy intake (kcal/day) (median, IQR)	2064 (1699–2552)	2459 (1968–2956)	1715 (1421–2020)	1922 (1562–2421)	2151 (1772–2660)	2170 (1809–2679)	1857 (1547–2300)
Education (n (%)) [3]							
Low	602 (29.0)	145 (27.9)	160 (30.8)	171 (33.0)	126 (24.3)	133 (25.6)	172 (33.1)
Moderate	789 (38.0)	234 (45.1)	166 (32.0)	207 (39.9)	184 (35.5)	210 (40.5)	217 (41.8)
High	687 (33.1)	140 (27.0)	193 (37.2)	141 (27.2)	209 (40.3)	176 (33.9)	130 (25.1)
Dietary consumption (gram/2000 kcal)							
Animal-based products							
Meat							
Processed meat	35.5 (11.7–67.5)	49.4 (24.6–84.4)	20.8 (0–55.4)	62.43 (26.0–103.6)	12.9 (0–31.3)	44.9 (16.3–77.4)	21.8 (7.6–46.1)
Red unprocessed meat	20.3 (0–53.0)	18.3 (0–46.1)	18.0 (0–57.2)	71.3 (29.9–104.7)	0 (0–16.4)	9.6 (0–38.1)	36.1 (0–75.8)
White unprocessed meat	0 (0–23.4)	0 (0–17.4)	0 (0–33.6)	0 (0–24.3)	0 (0–16.7)	0 (0–16.1)	0 (0–37.7)
Dairy	255.0 (120.0–421.8)	191.1 (65.8–336.4)	303.4 (145.0–485.5)	275.6 (122.2–458.2)	221.8 (95.7–365.2)	148.7 (46.0–270.6)	453.0 (292.3–613.4)
Cheese	28.2 (12.8–48.2)	22.3 (7.5–38.7)	32.8 (17.1–55.0)	28.4 (11.8–49.9)	23.3 (10.8–42.1)	25.7 (8.5–44.6)	32.8 (16.6–56.9)
Fish	0 (0–14.9)	0 (0–0)	0 (0–52.0)	0 (0–0)	0 (0–23.6)	0 (0–0)	0 (0–54.0)
Eggs	0 (0–22.9)	0 (0–17.3)	0 (0–26.2)	0 (0–24.4)	0 (0–21.1)	0 (0–24.1)	0 (0–16.6)
Plant-based foods							
Potatoes and cereals [4]	256.4 ± 85.0	247.7 ± 78.3	249.7 ± 92.8	242.9 ± 89.1	263.1 ± 84.6	227.8 ± 80.5	293.7 ± 89.2
Vegetables	125.6 (73.8–204.1)	65.6 (34.6–100.0)	237.8 (164.5–329.5)	152.7 (90.0–234.7)	109.3 (57.2–189.4)	112.2 (56.8–190.1)	150.5 (94.3–236.0)

Table 1. *Cont.*

	Total Population from DNFCS	"High Fruit and Vegetable Dietary Pattern"		"Low Meat Dietary Pattern" [1]		"High Dairy, Low Fruit Juices Dietary Pattern"	
		Quartile 1	Quartile 4	Quartile 1	Quartile 4	Quartile 1	Quartile 4
Legumes	0 (0–0)	0 (0–0)	0 (0–0)	0 (0–0)	0 (0–0)	0 (0–0)	0 (0–0)
Fruit	95.2 (13.5–193.7)	13.5 (0–65.2)	223.9 (136.59–346.8)	76.7 (0–171.1)	112.4 (37.1–231.2)	83.0 (0–179.9)	113.1 (23.5–214.4)
Nuts and seeds	0 (0–13.1)	0 (0–6.1)	0 (0–20.9)	0 (0–1.4)	7.9 (0–26.1)	0 (0–24.9)	0 (0–0)
Beverages							
Non-alcoholic beverages							
Fruit and vegetables juice	0 (0–81.7)	0 (0–153.0)	0 (0–0)	0 (0–96.9)	0 (0–68.7)	125.6 (3.0–223.5)	0 (0–0)
Soft drinks	121.3 (0–360.3)	324.3 (87.7–653.3)	0 (0–171.4)	127.7 (0–406.6)	74.4 (0–254.3)	152.8 (0–390.3)	84.5 (0–276.9)
Coffee and tea	735.9 (452.2–1146.8)	430.9 (218.5–681.8)	1216.7 (777.6–1742.6)	715.1 (442.9–1141.9)	778.1 (479.9–1223.8)	706.4 (390.8–1106.8)	779.1 (493.4–1189.3)
Water	464.1 (149.4–956.1)	291.7 (65.1–704.6)	730.2 (318.0–1269.3)	504.1 (156.4–1151.4)	436.3 (144.1–918.7)	367.9 (127.5–816.1)	560.8 (201.1–1265.8)
Alcoholic beverages	0 (0–211.2)	62.8 (0–301.4)	0 (0–111.5)	10.8 (0–261.7)	0 (0–131.6)	94.4 (0–351.4)	0 (0–68.5)
Miscellaneous							
Sweets and snacks	71.8 (40.6–109.7)	85.0 (47.7–129.0)	56.8 (25.4–87.1)	48.5 (24.4–80.3)	99.5 (61.1–142.4)	68.8 (39.8–107.0)	64.0 (33.9–103.7)
Fat and oils	20.3 (12.6–28.8)	12.0 (12.5–29.0)	19.0 (11.1–28.3)	19.3 (11.6–27.3)	22.0 (13.7–31.7)	18.9 (11.6–27.2)	21.0 (12.7–29.2)
Broth, sauces, and condiments	50.0 (19.2–118.8)	52.9 (21.9–115.0)	46.2 (14.6–126.9)	47.0 (17.6–113.1)	46.1 (15.5–115.7)	48.7 (20.7–116.8)	39.2 (14.4–111.4)
Other	0 (0–0.2)	0 (0–0)	0 (0–1.6)	0 (0–0.7)	0 (0–0.1)	0 (0–0.2)	0 (0–0.1)

Median and interquartile ranges are shown for continuous variables. Categorical variables are presented as percentages. DNFCS: Dutch National Food Consumption Survey 2012–2016. IQR: interquartile range. [1] Pattern scores of participants are multiplied by (−1) to obtain the healthier and more sustainable diet in quartile 4. [2] From participants aged >70 years, no information on height was obtained. Therefore, body mass index is missing for these participants. Number of missing data on body mass index: total population of the DNFCS: 516. "High fruit and vegetable dietary pattern" quartile 1: 60, quartile 4: 172, "low meat dietary pattern" quartile 1: 147, quartile 4: 114, "high dairy, low fruit juices dietary pattern" quartile 1: 118, quartile 4: 150. [3] Level of education: low (primary education, lower vocational education, advanced elementary education), moderate (intermediate vocational education, higher secondary education), high (higher vocational education and university). [4] Presented as mean ± standard deviation.

3.2. *Dietary Patterns Derived by RRR*

Using RRR, three dietary patterns were derived. The first pattern, "high fruit and vegetable dietary pattern", explained 37.5% of the variation in DHD15-index, dietary GHGE, and dietary blue water use and 8.9% of the variation in food consumption. The second pattern, "low meat dietary pattern", explained 21.3% of the variation in the dependent variables and 4.7% in the predictor variables. The third pattern, "high dairy, low fruit juices dietary pattern", explained 7.7 and 5.3% in the dependent and predictor variables, respectively.

3.2.1. Healthiness and Sustainability of the Three Dietary Patterns

Diets in Q4 of the pattern scores of the "high fruit and vegetable dietary pattern" had a 27.4% higher DHD15-index score than the diets of the average Dutch population (75.6, SD: 15.3) versus 59.4 (SD: 18.6) (Table 2). These diets had a 14.0% higher dietary GHGE and 69.2% higher dietary blue water use than diets of the average Dutch population (5.36 (IQR: 4.54–6.33) kg CO_2 eq/2000 kcal versus 4.70 (IQR: 4.02–5.62) CO_2 eq/2000 kcal and 0.22 (IQR: 0.17–0.28) m^3/2000 kcal versus 0.13 (IQR: 0.10–0.19) m^3/2000 kcal, respectively).

Diets in Q4 of the "low meat dietary pattern" had a 17.0% higher DHD15-index score, a 19.6% lower dietary GHGE, and a 7.7% higher dietary blue water use compared to diets of the average population (69.5 (SD: 17.8) versus 59.4 (SD: 18.6), 3.78 (IQR: 3.35–4.22) kg CO_2 eq/2000 kcal versus 4.70 (IQR: 4.02–5.62) CO_2 eq/2000 kcal and 0.14 (IQR: 0.09–0.20) m^3/2000 kcal versus 0.13 (IQR: 0.10–0.19) m^3/2000 kcal, respectively) (Table 2). With increasing adherence to the "high dairy, low fruit juices dietary pattern", a 13.0% higher DHD15-index score, a 9.1% higher dietary GHGE, and a 7.7% lower dietary blue water use were observed (Table 2).

Table 2. Mean ± standard deviation Dutch Healthy diet index 2015 (DHD15-index) score and median (interquartile range) dietary greenhouse gas (GHG) emissions and dietary blue water use of participants in quartile 1 and 4 of the three derived dietary patterns compared to the total adult population of the Dutch National Food Consumption Survey (DNFCS) 2012–2016.

	Total Population of the DNFCS	**"High Fruit and Vegetable Dietary Pattern"**		**"Low Meat Dietary Pattern"** [a]		**"High Dairy, Low Fruit Juices Dietary Pattern"**	
		Quartile 1	**Quartile 4**	**Quartile 1**	**Quartile 4**	**Quartile 1**	**Quartile 4**
DHD15-index score [b]	59.4 ± 18.6	42.1 ± 13.1	75.6 ± 15.3	51.5 ± 17.3	69.5 ± 17.8	52.2 ± 19.0	67.1 ± 16.9
GHGE (kg CO_2 equivalents/2000 kcal)	4.70 (4.02–5.62)	4.26 (3.70–4.98)	5.36 (4.54–6.33)	5.98 (5.20–6.96)	3.78 (3.35–4.22)	4.52 (3.78–5.49)	5.13 (4.42–6.05)
Blue water use (m^3/2000 kcal)	0.13 (0.10–0.19)	0.09 (0.07–0.11)	0.22 (0.17–0.28)	0.13 (0.10–0.19)	0.14 (0.09–0.20)	0.18 (0.12–0.24)	0.12 (0.09–0.17)

[a] Pattern scores of participants are multiplied by (−1) to obtain the healthier and more sustainable diet in quartile 4. [b] DHD15-index score: score out of 140 points.

3.2.2. Dietary Characterization of the Three Dietary Patterns

In the "high fruit and vegetable dietary pattern", Q4 was characterized by high consumption of vegetables (factor loading (Fl): 0.51), fruit (Fl: 0.48), coffee and tea (Fl: 0.40), water (Fl: 0.22), and fruit and vegetable juices (Fl: 0.20), and low consumption of soft drinks (Fl: −0.22) (Figure 1). The high fruit and vegetable consumption and the low consumption of soft drinks within dietary pattern caused the high DHD15-index score compared to the average population. The relatively high consumption of dairy, fruit, vegetables, and coffee and tea in Q4 of the "high fruit and vegetable dietary pattern" caused the increased GHGE. Besides, the consumption of fruit, vegetables, and coffee and tea caused the high dietary blue water use in this pattern. See Table 1 for intake per food group in Q1 and Q4 per pattern and Table A2 for GHGE and blue water use per kg food group.

The "low meat dietary pattern" was defined as high consumption of sweets and snacks (Fl: 0.36) and nuts and seeds (Fl: 0.26), and the low consumption of red unprocessed meat (Fl: −0.65) and processed meat (Figure 1). The increased nuts and seeds consumption and the decreased red unprocessed and processed meat consumption caused the increased DHD15-index score compared to the average population. However, due to the increased consumption of sweets and snacks, a smaller increase in DHD15-index score is observed than in the "high fruit and vegetable dietary pattern". The "low meat dietary pattern" had the lowest dietary GHGE, caused by relatively low processed and unprocessed red meat, dairy, vegetable, and fruit consumption (Table A2). Fruit, nuts and seeds, sweets and snacks, and coffee and tea consumption caused the slightly increased dietary blue water use compared to the average population (Table A2).

Q4 of the "high dairy, low fruit juices dietary pattern" was characterized as high consumption of dairy (Fl: 0.43) and potatoes and cereals (Fl: 0.23), and the low consumption of fruit and vegetable juices (Fl: −0.70), alcoholic beverages (Fl: −0.23), and nuts and seeds (Fl: −0.20) (Figure 1). The slightly increased dairy, fruit, and vegetable consumption and slightly decreased soft drinks and sweets and snacks consumption caused the increased DHD15-index score compared to the average population. The unprocessed red meat and dairy consumption in this diet was responsible for the slightly increased GHGE (Table A2), and the relatively low vegetable, fruit, and coffee and tea consumption caused the slightly decreased dietary blue water use compared to the average population and the other two dietary patterns (Table A2).

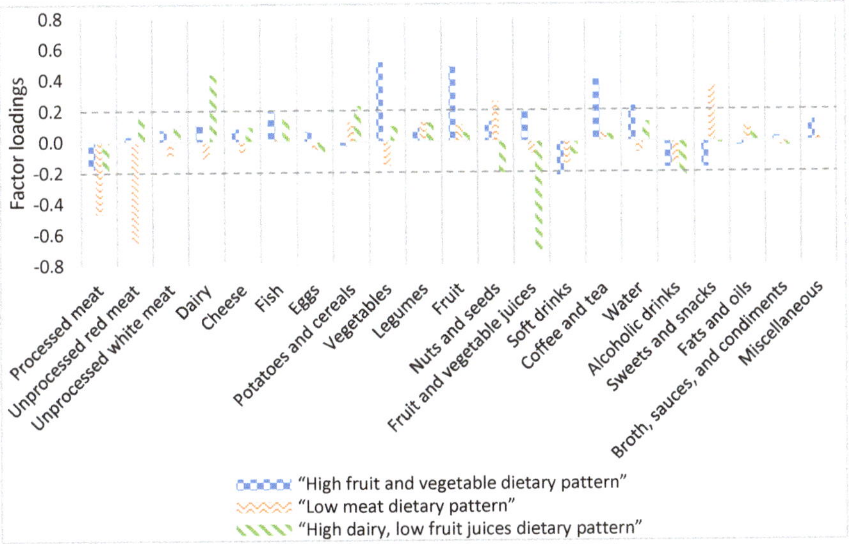

Figure 1. Factor loadings of the food groups on three dietary patterns derived by reduced rank regression explaining the variation in the Dutch Healthy Diet 2015 index scores, dietary greenhouse gas emission, and dietary blue water use. Factor loadings > | 0.20 | were considered important contributors to a dietary pattern. Note: of the "low meat dietary pattern", pattern scores of participants and factor loadings were multiplied by (−1) to obtain the healthier and more sustainable diet in quartile 4.

Fruit intake in Q4 of the "low meat dietary pattern" and the "high dairy, low fruit juices dietary pattern" differed only 0.8% (112 g versus 113 g, respectively), but blue water use of the food group "fruit" of diets in Q4 of the "low meat dietary pattern" was 36.4% higher than that of diets in Q4 of the "high dairy, low fruit juices dietary pattern" (0.015 m^3 versus 0.011 m^3, respectively). This reveals that types of fruit eaten differs between the patterns. The same is true for the food group coffee and tea in these two dietary patterns.

3.2.3. Characteristics of Adherents of the Three Dietary Patterns

Adherents of the "high fruit and vegetable dietary pattern" (participants in Q4 of the pattern scores) were more likely to be older, to be female, and to be higher educated compared to the average population (59 vs. 51 years; 72.4 vs. 49.8% females; 37.0 vs. 33.1% higher educated, respectively) (Table 1). Besides, these adherents had lower energy intake and smoked less compared to the average population (1715 vs. 2064 kcal/day; 15.2 vs. 20.0% smokers). Adherents of the "low meat dietary pattern" tended to be higher educated (40.3 vs. 33.1% higher educated), to be younger (47 vs. 51 years), smoked less (14.3 vs. 20.0% smokers), had a slightly lower BMI (24.0 vs. 25.5 kg/m^2), and had higher energy intake compared to the average population (2151 vs. 2064 kcal/day). Adherents of the "high dairy, low fruit juices dietary pattern" were older and had a lower energy intake compared to the average population (55 vs. 51 years; 1857 vs. 2064 kcal/day).

3.3. Differences in Pattern Scores per Level of Education

Table 3 shows pattern scores per level of education, stratified for age and gender and corrected for BMI. For the "high fruit and vegetable dietary pattern", higher educated males and females of all ages had higher pattern scores than the lower educated groups.

Besides, with increasing age, pattern scores for every educational level increased for both males and females. Females had higher pattern scores on this dietary pattern at every age and every level of education compared to males.

Table 3. Mean pattern scores per level of education, stratified for age and gender. Participants from the Dutch National Food Consumption Survey 2012–2016.

Dietary Pattern	Gender	Age	Level of Education		
			Low	Middle	High
"High fruit and vegetable dietary pattern" (pattern scores range between −2.913 and 7.789)	Male	<40	−1.033 [a] (*N* = 38)	−0.748 [a] (*N* = 141)	−0.436 [b] (*N* = 159)
		40–59	−0.508 (*N* = 53)	−0.479 (*N* = 134)	−0.250 (*N* = 104)
		>59	−0.323 [ab] (*N* = 47)	−0.412 [a] (*N* = 52)	0.031 [b] (*N* = 56)
		>70 *	*N* = 104	*N* = 79	*N* = 76
	Female	<40	−0.350 [a] (*N* = 39)	−0.216 [a] (*N* = 154)	0.434 [b] (*N* = 160)
		40–59	0.120 [a] (*N* = 81)	0.118 [a] (*N* = 136)	0.892 [b] (*N* = 65)
		>59	0.371 [a] (*N* = 240)	0.604 [a] (*N* = 93)	1.631 [b] (*N* = 67)
		>70 *	*N* = 167	*N* = 50	*N* = 40
"Low meat dietary pattern" (pattern scores range between −5.689 and 2.633) **	Male	<40	−0.039 [ab]	−0.063 [a]	0.247 [b]
		40–59	−0.043	−0.146	0.076
		>59	0.003	−0.103	−0.081
	Female	<40	−0.210 [a]	0.073 [ab]	0.229 [b]
		40–59	−0.048	0.011	0.042
		>59	−0.022	−0.045	0.303
"High dairy, low fruit juices dietary pattern" (pattern scores range between −5.638 and 2.834) **	Male	<40	−0.170	−0.014	−0.125
		40–59	−0.193	−0.083	−0.118
		>59	0.069	−0.174	−0.280
	Female	<40	−0.064	−0.130	0.013
		40–59	0.194	0.044	−0.032
		>59	0.262	0.094	0.085

Analysis of covariance: different superscript letters showing significant differences between pattern scores within a row (stratum). Pattern scores are adjusted for body mass index. * From participants aged > 70 years, no information on height was obtained. Therefore, body mass index is missing for these participants, and they are left out of this analysis. ** Number of participants per age, gender, and level of education is similar to the "high fruit and vegetable dietary pattern".

For the "low meat dietary pattern", higher educated males and females younger than 40 years had higher patterns scores than the lower educated groups. Pattern scores for males and females of the ages 40–59 years and >59 years did not differ significantly per level of education.

Pattern scores for males and females of all age groups did not differ significantly per level of education for the "high dairy, low fruit juices dietary pattern".

4. Discussion

In this study, three dietary patterns were derived from the Dutch National Food Consumption Survey 2012–2016 using RRR: a "high fruit and vegetable dietary pattern", a "low meat dietary pattern", and a "high dairy, low fruit juices dietary pattern". The "low meat dietary pattern" was the most sustainable pattern with diets in Q4 having 19.6% lower GHGE and 7.7% higher blue water use compared to diets of the average population. Since these patterns are derived from food consumption information of the Dutch population, it may be assumed that these patterns are socially acceptable for at least part of the population. In any pattern, as observed in this study, a shift is possible towards healthier and environmentally sustainable diets. As yet, none of the patterns showed the optimal combination of increased DHD15-index score and decreased dietary GHGE and dietary blue water use.

Our results are generally in line with dietary patterns that were found in the EPIC-NL cohort using the RRR with DHD15-index and GHGE as dependent variables [15]. The "plant-based diet" derived from the cohort was healthier and had lower dietary GHGE compared to the average diet of that population. This pattern is a combination of the "high fruit and vegetable dietary pattern" and the "low meat dietary pattern", with blue water as the additional separating environmental impact indicator in the current study. The "dairy-based diet" of the cohort was somewhat healthier and had higher dietary GHGE, which corresponds to our "high dairy, low fruit juices dietary pattern". The same was observed in another study that searched for healthy and environmentally sustainable dietary patterns in five European countries, using a multiple factor analysis, focusing on GHGE and mean adequacy ratio, mean excess ratio, and solid energy density as proxies for nutritional value of the diet [24]. They found a diet that was healthier and more sustainable, in which significantly larger quantities of plant-based products and smaller quantities of meat, soft drinks, and alcoholic beverages were consumed. This is similar to a combination of our first two patterns with blue water use a separator. Another study that used a similar data-driven method and eight response variables, indicating health, environment, and affordability derived a dietary pattern, which included relatively low amounts of animal origin products, especially red meat, sweets, and fatty products, and substantial consumption of soy-based and whole products [25]. This pattern matches our "low meat dietary pattern" regarding low meat consumption. However, sweets and snacks consumption does not correspond, which might be caused by the different response variables used. Similar to our derived "low meat dietary pattern", a review on the impact of dietary changes on the environment concluded that a reduction in animal-based foods would result in substantial reductions in diet-related GHGE, land use, and water use [26]. Another review states that reducing the amount of meat and changing the type of meat mainly affects the environmental improvement potential regarding GHGE and land use [27].

Comparing the dietary patterns derived from the Dutch population to the healthy and sustainable reference diet of the EAT-Lancet Commission, several similarities and differences are observed. [8]. Diets in the highest quartile of the "low meat dietary pattern" match the reference diet in the low red and processed meat consumption and the high consumption of nuts and seeds. However, these diets also show high consumption of sweets and snacks (adding in a limited way to environmental impact), which is in contrast with the EAT-Lancet reference diet. The high fruit and vegetable intake in diets in the highest quartile of the "high fruit and vegetable dietary pattern" do correspond with the reference EAT-Lancet diet, but the relatively high meat and dairy intake do not. The low intake of fruit and vegetable juices in the "high dairy, low fruit juices dietary pattern" is in line, but the high dairy consumption is not. Summarizing, the dietary patterns found in this study still show a distance from this EAT-Lancet reference diet. Even for the most beneficial "low meat dietary pattern", there is much to gain regarding health and environmental impact aspects. A study about dietary changes that are needed to reach a healthy and environmentally sustainable diet in different European countries showed that GHGE could be theoretically decreased by 62–78%, while still being nutritionally adequate, but this is at a strong risk of compromising cultural acceptability of the diets [28]. Concluding, other European countries also still show a large distance from a healthy and environmentally sustainable dietary pattern. However, given the presence (21.3% of the variation in response variables explained) of the "low meat dietary pattern" in the Dutch population, aspects of this diet are achievable for at least part of the population. Despite Dutch diets being far from environmental sustainability yet, the "low meat dietary pattern" is a good starting point for developing realistic environmentally sustainable dietary patterns. To improve the health and sustainability of the "low meat dietary pattern", guidelines may focus on decreasing sweets and snacks consumption and moderating nuts and seeds consumption, in order to reduce blue water use. As an alternative, consumption of, for example, legumes may be promoted, as they benefit both health and environment. Furthermore, a study that used data of the DNFCS 2007–2010 showed that choosing low GHGE foods from each food group within a healthy diet results in reductions in dietary GHGE that are comparable to reductions achieved in healthy diets without meat [29].If consumers besides low meat consumption choose low GHGE products from each food group, large reductions in GHGE of a healthy diet can be achieved, whether or not the same concept holds for blue water use can be a subject for future studies.

Diets in the highest quartile of the "high fruit and vegetable dietary pattern" have highest DHD15-index scores, but also show very high blue water use. It can be worthwhile to study which products cause the high blue water use in this pattern. For example, raspberries have a twelve times higher water use than apples. When using more specific food groups that are more homogeneous in dietary GHGE and blue water use, dietary patterns that optimize health and both environmental indicators might be revealed.

Observed differences when comparing pattern scores of dietary patterns per level of education were comparable to other studies. Biesbroek et al. (2018) found that the higher educated group had higher pattern scores on the "prudent dietary pattern", which is comparable to our "high fruit and vegetable dietary pattern" [30]. This is in line with existing literature, which says that higher educated people have healthier diets according to the DHD15-index score and the consumption of energy, fat, fiber, fruit, vegetables, and energy-rich drinks, respectively, but the diets are less environmentally friendly [31,32]. When developing interventions or dietary guidelines, policy makers may take into account that higher educated people adhere more to a "high fruit and vegetable dietary pattern" and that young (<40 years) higher educated people adhere more to the "low meat dietary pattern". E.g., in young (<40 years) higher educated consumers, the focus could be more on lower consumption of sweets and snacks of which consumption is high in the "low meat dietary pattern". More specific interventions or dietary guidelines for subgroups might increase cultural acceptance and thereby compliance [33–35].

This study has several strengths. The first one is the usage of a second environmental impact indicator to provide a more complete picture of environmental impact of diets. The second strength is the hybrid approach of the RRR. This method allows us to find dietary patterns that are associated with the response variables of interest, if healthy and environmentally sustainable dietary patterns are present in the population, compared to the older and often-used principal component analysis [11]. Another strength is that food consumption information was based on two non-consecutive 24-hour dietary recalls. This method provides detailed food consumption information and is less subject to bias than food frequency questionnaires [36]. Lastly, the used food consumption information is from the most recent DNFCS, which is stratified for age, gender, region, address density, and level of education. This reflects the most current and representative diets of Dutch inhabitants.

Additionally, some limitations should be mentioned. Firstly, because of the addition of a second environmental impact indicator, it might be argued that environmental impact has a larger influence on the derived dietary patterns than health. However, intake in the highest quartile of all patterns that were found are healthier than the average diet. Secondly, despite a large percentage of the variation in the response variables was explained by the dietary patterns, namely 37.5, 21.3, and 7.7%, respectively, only 8.9, 4.7, and 5.3% of the variation in the predictor variables was explained. These percentages are comparable to other studies using RRR to identify dietary patterns [11,14,15]. Data envelopment analyses methods are currently under development and might be used in the future to derive dietary patterns, since this method can maybe explain a larger percentage of the variation in predictor variables [37,38]. A third limitation is the fact that not all environmental impact indicators are included in the study. An important missing indicator is biodiversity loss [39]. Because of the lack of data, we were not able to include this in our analysis. The addition of data on biodiversity loss will improve completeness of the impact of the diet on the environment. As GHGE is highly correlated (rho > 0.7) with acidification, fresh water eutrophication, marine eutrophication, and land use, a wide range of environmental impact indicators is indirectly taken into account in this study [23]. Another limitation is the large amount of non-consumers and the possible misreporting of the self-reported recalls [36,40]. However, assuming that misreporting is independent of specific food groups, due to the standardization for energy intake, part of the misreporting is corrected for [41]. Besides, to obtain a perfect representation of the Dutch population, a weighing factor was desirable to add to the RRR model, but SAS did not provide an option to add this weighing factor. However, with only small deviances from the real population, our study population was a good representation for the Dutch population [16]. Furthermore, despite the fact that the most recent food consumption survey data were used in this study, eating habits might already have changed between 2016 and 2021. The most recent trends in dietary pattern are not taken into account as food consumption surveys are time-bounded. A study on the acceptance of alternative protein sources for meat concludes that Dutch consumers have a higher acceptance towards all alternative proteins in 2019 compared to 2015. However, self-reported consumption of alternative proteins shows no differences across years [42]. Therefore, using data that are representative of the Dutch population at 2012–2016 is still insightful. Another limitation is that food groups and the response variables dietary GHGE and blue water use were standardized for energy intake, but not the DHD15-index. The latter measure is based on absolute consumption, and standardization would violate the true score. Since two out of three response variables were standardized, results might be slightly distorted. However, using only unstandardized variables would result in dietary patterns based on variations in diet quantity, and not in diet quality. The last limitation of this study is the unresolved uncertainty in LCA data. Unresolved problems of LCAs are, for example, spatial variation and local environmental uniqueness [43]. Primary LCA data were available for 242 food products and cover 71% of the quantity consumed in the DNFCS. Remaining food products (29%) are based on extrapolated data. However, due to the extrapolations, our LCA data are complete [23]. Besides, LCAs are the best estimates available for environmental impact of foods, though they always include uncertainties.

5. Conclusions

To the best of our knowledge, this is the first study that provides insight in existing dietary patterns in a representative study population for the Dutch society, using the reduced rank regression analysis. Three socially acceptable dietary patterns were extracted: the "high fruit and vegetable dietary pattern", the "low meat dietary pattern", and the "high dairy, low fruit juices dietary pattern". In any of these patterns, a shift is possible towards healthier and environmentally sustainable diets. However, none of the patterns showed the optimal combination of increased DHD15-index score and decreased dietary GHGE and dietary blue water use. The "low meat dietary pattern" was the healthiest and most environmentally sustainable pattern with diets in the highest quartile having 17.0% higher DHD15-index score, 19.6% lower GHGE, and 7.7% higher blue water use. The addition of blue water use as an environmental impact indicator in this study shows the difficulty of finding existing dietary patterns that have low environmental impact in all determinants. Future research might focus on the role of foods or food groups in dietary patterns where health and/or different environmental impact indicators do align to optimize dietary patterns that are socially acceptable, healthy, and sustainable.

Author Contributions: Conceptualization, S.N.H., S.B., and M.C.O.; data curation, S.N.H.; formal analysis, S.N.H.; methodology, S.N.H., S.B., and M.C.O.; resources, M.C.O.; visualization, S.N.H.; writing—original draft, S.N.H.; writing—review and editing, S.N.H., S.B., E.H.M.T., and M.C.O. All authors have read and agreed to the published version of the manuscript.

Funding: This research received no external funding. The APC was funded by Wageningen University Research Centre Department of Agrotechnology and Food Sciences, Division of Human Nutrition and Health.

Institutional Review Board Statement: Ethical review and approval were waived for this study as the Utrecht University Medical Ethical Review Committee evaluated that the study was not subject to the Medical Research Involving Human Subjects Act (WMO) of the Netherlands (reference number 12-359/C). In line with this, written informed consent was not required according to the regulations at the time of data collection.

Informed Consent Statement: Not applicable.

Data Availability Statement: Data of the Dutch National Food Consumption Survey 2012–2016 can be requested for at https://www.rivm.nl/en/dutch-national-food-consumption-survey/data-on-request(accessed on 16 October 2019). Primary environmental data of 250 food products can be found at https://www.rivm.nl/voedsel-en-voeding/duurzaam-voedsel/database-milieubelasting-voedingsmiddelen (accessed on 16 October 2019).

Conflicts of Interest: The authors declare no conflict of interest.

Appendix A

Table A1. Components and scoring criteria of the Dutch Healthy Diet index 2015 (DHD15-index), cited from Looman et al., 2017 [19].

DHD15-Index	Maximum Score [a] (10 Points)	Minimum Score [a] (0 Points)
1. Vegetables (g)	≥200	0
2. Fruit (g)	≥200	0
3a. Wholegrain products (g)	≥90 (5 points)	0
3b. Replace refined with wholegrain products	No consumption of refined products or ratio wholegrain/refined ≥11 (5 points)	No consumption of wholegrain products or ratio wholegrain/refined ≤0.7
4. Legumes (g)	≥10	0
5. Nuts (g)	≥15	0
6. Dairy products [b] (g)	300–450	0 or ≥750
7. Fish [c] (g)	≥15	0
8. Tea (g)	≥450	0
9. Replace butter and hard fats with margarines and oils	No consumption of fats or ratio oils/fats ≥13	No consumption of oils or ratio ≤0.6
10. Replace unfiltered coffee with filtered coffee	Consumption of only filtered coffee or no coffee consumption	Any consumption of unfiltered coffee
11. Red meat (g)	<45	≥100
12. Processed meat (g)	0	≥50
13. Sweetened beverages and fruit juices (g)	0	≥250
14. Alcohol (g)	≤10	Men: ≥30 Women: ≥20
15. Sodium (g)	<1.9	≥3.8

Abbreviations: g—grams. [a] A score above the recommended intake is 10 points, whereas an intake below is given a proportional score between 0 and 10 points. [b] A maximum of 40 g cheese per day could be included. [c] A maximum of 4 g lean fish per day could be included.

Table A2. Categorization of 21 food groups adapted from GloboDiet, and their greenhouse gas emission (GHGE) (kg CO_2 equivalents) and blue water use (m^3) per kg based on median consumption of participants of the Dutch National Food Consumption Survey.

Main Groups	Aggregated Groups	GloboDiet Groups	Median (IQR) GHGE per kg [b]	Median (IQR) Blue Water Use per kg [c]
		Animal-based foods		
Processed meat	Meat, fish and eggs	"07-04" meat products and processed meat and "red", and "07-04" meat products and processed meat and "white" [a]	13.15 (10.44–17.95)	0.13 (0.1–0.16)

Table A2. *Cont.*

Main Groups	Aggregated Groups	GloboDiet Groups	Median (IQR) GHGE per kg [b]	Median (IQR) Blue Water Use per kg [c]
Red unprocessed meat	Meat, fish and eggs	"07-00" meat miscellaneous; "07-01" fresh meat; "07-03" game and "07-05" oval meat	21.91 (12.42–30.03)	0.19 (0.12–0.24)
White unprocessed meat	Meat, fish and eggs	"07-02" poultry	10.87 (10.87–10.87)	0.15 (0.15–0.15)
Dairy	Dairy and cheese	"05" dairy (excl. "05-05" cheese; "05-02" dairy replacers and "05-07-02", "05-08-02" both non-dairy-based products)	2.19 (2.03–2.45)	0.1 (0.09–0.1)
Cheese	Dairy and cheese	"05-05" cheese	12.53 (10.72–13.09)	0.02 (0.02–0.02)
Fish	Meat, fish, and eggs	"08" fish, shellfish, and amphibians	6.95 (5.42–13.36)	0.04 (0.03–0.06)
Eggs	Meat, fish, and eggs	"09" eggs and egg products	4.32 (4.32–4.32)	0.06 (0.03–0.14)
		Plant-based foods		
Potatoes and cereals	Potatoes and cereals	"01" potatoes and other tubers and "06" cereals and cereal products	1.27 (1.11–1.5)	0.03 (0.02–0.05)
Vegetables	Vegetables, fruits, and legumes	"02" vegetables	1.62 (1.3–1.97)	0.07 (0.05–0.09)
Legumes	Vegetables, fruits, and legumes	"03" legumes	1.93 (1.93–1.93)	0.07 (0.07–0.07)
Fruits	Vegetables, fruits, and legumes	"04" fruits, olives (excl. 04.02)	0.85 (0.69–1.3)	0.14 (0.07–0.26)
Nuts and seeds	Nuts and seeds	"04-02" nuts, peanuts, seeds and nut spread	6.32 (4.28–8.68)	0.17 (0.17–1.72)
		Beverages		
Fruit and vegetable juice	Non-alcoholic beverages	"13-01" fruit and vegetable juice	1.42 (1.1–1.5)	0.45 (0.24–0.47)
Soft drinks	Non-alcoholic beverages	"13-02" lemonade, soft drinks	0.6 (0.56–0.65)	0.01 (0.01–0.02)
Coffee and tea	Non-alcoholic beverages	"13-03" coffee, tea, and herbal tea	0.26 (0.21–0.3)	0.02 (0.01–0.03)
Water	Non-alcoholic beverages	"13-04" water	0 (0–0)	0 (0–0)
Alcoholic beverages	Alcoholic beverages	"14" alcoholic beverages	2.02 (0.71–2.21)	0.05 (0.01–0.09)
		Miscellaneous		
Sweets and snacks	Miscellaneous	"11" sugar and confectionery; "12" cakes and sweet biscuits, and "18" savory snacks	2.98 (2.29–3.73)	0.06 (0.04–0.09)
Fats and oils	Fats and oils	"10" fats and oils	4.95 (3.59–6.04)	0.1 (0.08–0.55)
Broth, sauces, and condiments	Miscellaneous	"15" condiments, spices, sauces, and yeast and "16" soups and stocks	1.81 (0.8–3.24)	0.04 (0.02–0.06)
Other	Miscellaneous	"17" miscellaneous; "07-06" meat replacers; "05-02" dairy replacers and "05-07-02", "05-08-02" both non-dairy-based products	0.01 (0.01–1.06)	0 (0–0.01)

[a] GloboDiet group "07-04" processed meat was categorized as red or white meat. Poultry was considered as white meat, remaining meats were categorized as red meat. [b] Calculation: GHGE per food group per 2000 kcal per person * 1000 g/consumption of that food group in gram per 2000 kcal, excluding consumption of zero gram of a food group. [c] Calculation: blue water use per food group per 2000 kcal per person * 1000 g/consumption of that food group in gram per 2000 kcal, excluding consumption of zero gram of a food group. IQR: interquartile range.

References

1. United Nations General Assembly. *Resolution 70/1 Transforming Our World: The 2030 Agenda for Sustainable Development United Nations*; United Nations General Assembly: New York, NY, USA, 2015.
2. United Nations. *Paris Climate Agreement*; United Nations: Paris, France, 2015.
3. Garnett, T. Where are the best opportunities for reducing greenhouse gas emissions in the food system (including the food chain)? *Food Policy* **2011**, *36*, S23–S32. [CrossRef]
4. Tukker, A.; Huppes, G.; Guinée, J.; Heijungs, R.; De Koning, A.; Van Oers, L.; Suh, S.; Geerken, T.; Van Holderbeke, M.; Jansen, B. *Environmental Impact of Products (EIPRO) Analysis of the Life Cycle Environmental Impacts Related to the Final Consumption of the EU-25*; European Commission, Joint Research Centre, Institute for Prospective Technological Studies: Brussels, Belgium, 2006.
5. Foley, J.A.; De Fries, R.; Asner, G.P.; Barford, C.; Bonan, G.; Carpenter, S.R.; Chapin, F.S.; Coe, M.T.; Daily, G.C.; Gibbs, H.K. Global Consequences of Land Use. *Science* **2005**, *309*, 570–574. [CrossRef]
6. Steffen, W.; Richardson, K.; Rockström, J.; Cornell, S.E.; Fetzer, I.; Bennett, E.M.; Biggs, R.; Carpenter, S.R.; De Vries, W.; De Wit, C.A. Planetary boundaries: Guiding human development on a changing planet. *Science* **2015**, *347*, 1259855. [CrossRef] [PubMed]
7. Burlingame, B.; Dernini, S. *Sustainable Diets and Biodiversity*; Food and Agriculture Organization of the United Nations Rome: Rome, Italy, 2010.
8. Willett, W.; Rockström, J.; Loken, B.; Springmann, M.; Lang, T.; Vermeulen, S.; Garnett, T.; Tilman, D.; DeClerck, F.; Wood, A. Food in the Anthropocene: The EAT–Lancet Commission on healthy diets from sustainable food systems. *Lancet* **2019**, *393*, 447–492. [CrossRef]
9. National Institute for Public Health and Environment. Voedselconsumptie 2012–2016 Wat, Waar en Wanneer. 2018. Available online: https://www.wateetnederland.nl/documenten/voedselconsumptie-2012-2016-wat-waar-en-wanneer (accessed on 16 October 2019).
10. Ocké, M.C. Evaluation of methodologies for assessing the overall diet: Dietary quality scores and dietary pattern analysis. *Proc. Nutr. Soc.* **2013**, *72*, 191–199. [CrossRef]
11. Hoffmann, K.; Boeing, H.; Boffetta, P.; Nagel, G.; Orfanos, P.; Ferrari, P.; Bamia, C. Comparison of two statistical approaches to predict all-cause mortality by dietary patterns in German elderly subjects. *Br. J. Nutr.* **2005**, *93*, 709–716. [CrossRef] [PubMed]
12. Schulze, M.B.; Hoffmann, K. Methodological approaches to study dietary patterns in relation to risk of coronary heart disease and stroke. *Br. J. Nutr.* **2006**, *95*, 860–869. [CrossRef] [PubMed]
13. Weikert, C.; Schulze, M.B. Evaluating dietary patterns: The role of reduced rank regression. *Curr. Opin. Clin. Nutr. Metab. Care* **2016**, *19*, 341–346. [CrossRef]
14. Hoffmann, K.; Schulze, M.B.; Schienkiewitz, A.; Nöthlings, U.; Boeing, H. Application of a new statistical method to derive dietary patterns in nutritional epidemiology. *Am. J. Epidemiol.* **2004**, *159*, 935–944. [CrossRef]
15. Biesbroek, S.; Verschuren, W.M.; van der Schouw, Y.T.; Sluijs, I.; Boer, J.M.; Temme, E.H. Identification of data-driven Dutch dietary patterns that benefit the environment and are healthy. *Clim. Chang.* **2018**, *147*, 571–583. [CrossRef]
16. van Rossum, C.; Buurma-Rethans, E.; Dinnissen, C.; Beukers, M.; Brants, H.; Ocké, M. *The Diet of the Dutch: Results of the Dutch National Food Consumption Survey 2012–2016*; National Institute for Public Health and the Environment, RIVM: Bilthoven, The Netherlands, 2020.
17. Slimani, N.; Casagrande, C.; Nicolas, G.; Freisling, H.; Huybrechts, I.; Ocke, M.; Niekerk, E.; Van Rossum, C.; Bellemans, M.; De Maeyer, M. The standardized computerized 24-h dietary recall method EPIC-Soft adapted for pan-European dietary monitoring. *Eur. J. Clin. Nutr.* **2011**, *65*, S5–S15. [CrossRef]
18. The National Insitute for Public Health and the Environment. NEVO Online Version 2016/5.0. Available online: https://nevo-online.rivm.nl/ (accessed on 16 October 2019).
19. Looman, M.; Feskens, E.J.; de Rijk, M.; Meijboom, S.; Biesbroek, S.; Temme, E.H.; de Vries, J.; Geelen, A. Development and evaluation of the Dutch Healthy Diet index 2015. *Public Health Nutr.* **2017**, *20*, 2289–2299. [CrossRef]
20. Kromhout, D.; Spaaij, C.; de Goede, J.; Weggemans, R. The 2015 Dutch food-based dietary guidelines. *Eur. J. Clin. Nutr.* **2016**, *70*, 869. [CrossRef]
21. National Institute for Public Health and Environment. Environmental Impact of Foods; Life Cycle, Product Group; Milieubelasting Voedingsmiddelen; Levenscyclus, Productgroep, n.d. Available online: https://statline.rivm.nl/#/RIVM/nl/dataset/50060NED/table?ts=1587137105547 (accessed on 16 October 2019).
22. Marinussen, M.; Kramer, G.; Pluimers, J.; Blonk, H. *De Milieudruk Van ons eten; Een Analyse op Basis van de Voedselconsumptiepeiling 2007–2010*; Blonk Milieu Advies: Gouda, The Netherlands, 2012; (In Dutch, Summary In English).
23. Vellinga, R.E.; van de Kamp, M.; Toxopeus, I.B.; van Rossum, C.; de Valk, E.; Biesbroek, S.; Hollander, A.; Temme, E.H. Greenhouse Gas Emissions and Blue Water Use of Dutch Diets and Its Association with Health. *Sustainability* **2019**, *11*, 6027. [CrossRef]
24. Vieux, F.; Privet, L.; Soler, L.G.; Irz, X.; Ferrari, M.; Sette, S.; Raulio, S.; Tapanainen, H.; Hoffmann, R.; Surry, Y. More sustainable European diets based on self-selection do not require exclusion of entire categories of food. *J. Clean. Prod.* **2020**, *248*, 119298. [CrossRef]
25. Seconda, L.; Baudry, J.; Alles, B.; Soler, L.-G.; Hercberg, S.; Langevin, B.; Pointereau, P.; Lairon, D.; Kesse-Guyot, E. Identification of sustainable dietary patterns by a multicriteria approach in the NutriNet-Sante cohort. *J. Clean. Prod.* **2018**, *196*, 1256–1265. [CrossRef]

26. Aleksandrowicz, L.; Green, R.; Joy, E.J.; Smith, P.; Haines, A. The impacts of dietary change on greenhouse gas emissions, land use, water use, and health: A systematic review. *PLoS ONE* **2016**, *11*, e0165797. [CrossRef] [PubMed]
27. Hallström, E.; Carlsson-Kanyama, A.; Börjesson, P. Environmental impact of dietary change: A systematic review. *J. Clean. Prod.* **2015**, *91*, 1–11. [CrossRef]
28. Vieux, F.; Perignon, M.; Gazan, R.; Darmon, N. Dietary changes needed to improve diet sustainability: Are they similar across Europe? *Eur. J. Clin. Nutr.* **2018**, *72*, 951–960. [CrossRef] [PubMed]
29. van de Kamp, M.E.; van Dooren, C.; Hollander, A.; Geurts, M.; Brink, E.J.; van Rossum, C.; Biesbroek, S.; de Valk, E.; Toxopeus, I.B.; Temme, E.H. Healthy diets with reduced environmental impact?–The greenhouse gas emissions of various diets adhering to the Dutch food based dietary guidelines. *Food Res. Int.* **2018**, *104*, 14–24. [CrossRef]
30. Biesbroek, S.; Kneepkens, M.C.; Van Den Berg, S.W.; Fransen, H.P.; Beulens, J.W.; Peeters, P.H.; Boer, J.M. Dietary patterns within educational groups and their association with CHD and stroke in the European Prospective Investigation into Cancer and Nutrition-Netherlands cohort. *Br. J. Nutr.* **2018**, *119*, 949–956. [CrossRef]
31. Giskes, K.; Avendaño, M.; Brug, J.; Kunst, A. A systematic review of studies on socioeconomic inequalities in dietary intakes associated with weight gain and overweight/obesity conducted among European adults. *Obes. Rev.* **2010**, *11*, 413–429. [CrossRef]
32. Van Bussel, L.M.; Van Rossum, C.T.; Temme, E.H.; Boon, P.E.; Ocké, M.C. Educational differences in healthy, environmentally sustainable and safe food consumption among adults in the Netherlands. *Public Health Nutr.* **2020**, 1–11. [CrossRef]
33. Ockene, J.K.; Edgerton, E.A.; Teutsch, S.M.; Marion, L.N.; Miller, T.; Genevro, J.L.; Loveland-Cherry, C.J.; Fielding, J.E.; Briss, P.A. Integrating evidence-based clinical and community strategies to improve health. *Am. J. Prev. Med.* **2007**, *32*, 244–252. [CrossRef]
34. World Health Organization. *Global Strategy on Diet, Physical Activity and Health*; World Health Organization: Geneva, Switzerland, 2004.
35. Serra-Majem, L.; Ribas-Barba, L.; Salvador, G.; Serra, J.; Castell, C.; Cabezas, C.; Plasencia, A. Compliance with dietary guidelines in the Catalan population: Basis for a nutrition policy at the regional level (the PAAS strategy). *Public Health Nutr.* **2007**, *10*, 1406–1414. [CrossRef]
36. Kipnis, V.; Subar, A.F.; Midthune, D.; Freedman, L.S.; Ballard-Barbash, R.; Troiano, R.P.; Bingham, S.; Schoeller, D.A.; Schatzkin, A.; Carroll, R.J. Structure of dietary measurement error: Results of the OPEN biomarker study. *Am. J. Epidemiol.* **2003**, *158*, 14–21. [CrossRef] [PubMed]
37. Kanellopoulos, A.; Gerdessen, J.C.; Ivancic, A.; Geleijnse, J.M.; Bloemhof-Ruwaard, J.M.; van't Veer, P. Designing healthier and acceptable diets using data envelopment analysis. *Public Health Nutr.* **2020**, *23*, 2290–2302. [CrossRef] [PubMed]
38. Mertens, E.; Kuijsten, A.; Kanellopoulos, A.; Dofková, M.; Mistura, L.; D'Addezio, L.; Turrini, A.; Dubuisson, C.; Havard, S.; Trolle, E. Improving health and carbon footprints of European diets using a benchmarking approach. *Public Health Nutr.* **2020**, 1–11. [CrossRef] [PubMed]
39. Cardinale, B.J.; Duffy, J.E.; Gonzalez, A.; Hooper, D.U.; Perrings, C.; Venail, P.; Narwani, A.; Mace, G.M.; Tilman, D.; Wardle, D.A. Biodiversity loss and its impact on humanity. *Nature* **2012**, *486*, 59–67. [CrossRef] [PubMed]
40. Castro-Quezada, I.; Ruano-Rodríguez, C.; Ribas-Barba, L.; Serra-Majem, L. Misreporting in nutritional surveys: Methodological implications. *Nutr. Hosp.* **2015**, *31*, 119–127. [CrossRef] [PubMed]
41. Becker, W.; Foley, S.; Shelley, E.; Gibney, M. Energy under-reporting in Swedish and Irish dietary surveys: Implications for food-based dietary guidelines. *Br. J. Nutr.* **1999**, *81*, S127–S131. [CrossRef] [PubMed]
42. Onwezen, M.; Kunz, M.; Dagevos, H.; Verain, M. *Consumers More Inclined to Eat 'Alternative'proteins Compared to 2015*; Wageningen University & Research: Wageningen, The Netherlands, 2020.
43. Reap, J.; Roman, F.; Duncan, S.; Bras, B. A survey of unresolved problems in life cycle assessment. *Int. J. Life Cycle Assess.* **2008**, *13*, 374. [CrossRef]

nutrients

MDPI

Article

Dietary Intake of *trans* Fatty Acids in the Slovenian Population

Nina Zupanič [1,2], Maša Hribar [1], Hristo Hristov [1], Živa Lavriša [1], Anita Kušar [1], Matej Gregorič [3], Urška Blaznik [3], Barbara Koroušić Seljak [2], Petra Golja [4], Rajko Vidrih [4], Katja Žmitek [1,5,*] and Igor Pravst [1,4,5]

1 Nutrition Institute, Tržaška cesta 40, SI 1000 Ljubljana, Slovenia; nina.zupanic@nutris.org (N.Z.); masa.hribar@nutris.org (M.H.); hristo.hristov@nutris.org (H.H.); ziva.lavrisa@nutris.org (Ž.L.); anita.kusar@nutris.org (A.K.); igor.pravst@nutris.org (I.P.)
2 Jožef Stefan Institute, SI 1000 Ljubljana, Slovenia; barbara.korousic@ijs.si
3 National Institute of Public Health, Trubarjeva 2, SI 1000 Ljubljana, Slovenia; matej.gregoric@nijz.si (M.G.); urska.blaznik@nijz.si (U.B.)
4 Biotechnical Faculty, University of Ljubljana, Jamnikarjeva 101, SI 1000 Ljubljana, Slovenia; petra.golja@bf.uni-lj.si (P.G.); rajko.vidrih@bf.uni-lj.si (R.V.)
5 VIST–Higher School of Applied Sciences, Gerbičeva cesta 51A, SI 1000 Ljubljana, Slovenia
* Correspondence: katja.zmitek@vist.si; Tel.: +386-590-68871; Fax: +386-310-07981

Abstract: Consumption of *trans* fatty acids (TFAs) has been unequivocally linked to several adverse health effects, with the increased risk of cardiovascular disease being one of the most well understood. To reduce TFA-related morbidity and mortality, several countries have imposed voluntary or mandatory measures to minimize the content of industrial TFAs (iTFAs) in the food supply. In 2018, Slovenia introduced a ban on iTFAs on top of preceding voluntary calls to industry to reduce its use of partially hydrogenated oils (PHOs) as the main source of iTFAs. To investigate the consumption of TFAs, data available from the nationally representative dietary survey SI.Menu were analyzed. The survey consisted of two 24-h non-consecutive day recalls from 1248 study participants from three age groups (10–17, 18–64, 65–74 years old), combined with socio-demographic, socio-economic, and lifestyle parameters. The analyses demonstrated that, on average, TFAs accounted for 0.38–0.50% of total energy intake (TEI). However, 13% of adolescents, 29.4% of adults, and 41.8% of the elderly population still consumed more than 0.50% TEI with TFAs. The main sources of TFAs in the diet were naturally present TFAs from butter, meat dishes, and meat products, regardless of the age group. Results indicate that following the reformulation activities, the major sources of TFAs in the diets of the Slovenian population now represent foods which are natural sources of TFAs.

Keywords: *trans* fatty acids; partially hydrogenated oils; dietary intake; 24-h recall; EU Menu; Slovenian population

Citation: Zupanič, N.; Hribar, M.; Hristov, H.; Lavriša, Ž.; Kušar, A.; Gregorič, M.; Blaznik, U.; Koroušić Seljak, B.; Golja, P.; Vidrih, R.; et al. Dietary Intake of *trans* Fatty Acids in the Slovenian Population. *Nutrients* **2021**, *13*, 207. https://doi.org/10.3390/nu13010207

Received: 8 December 2020
Accepted: 6 January 2021
Published: 12 January 2021

1. Introduction

Trans fatty acids (TFAs) are fatty acid isomers with one or more double bonds in *trans* instead of *cis* configuration. TFAs can be produced in the process of industrial partial hydrogenation of unsaturated fats (i.e., vegetable oils), which was the previously used method for production of margarine and shortening, or can occur naturally as a result of bacterial biohydrogenation of unsaturated fatty acids in rumen and can therefore be found in meat, milk, and dairy products from ruminants [1].

The intake of industrially produced TFAs (iTFAs) has been found to negatively influence blood cholesterol profile, increase triglycerides [2], stimulate inflammatory responses [3], and increase mortality, particularly from coronary heart disease (CHD) [4]. According to a meta-analysis of four prospective cohort studies, with every 2% of total daily energy gained from TFAs, coronary heart disease incidence increases by 23% [1]. A substantial body of scientific evidence on adverse health effects of iTFAs has led public health organizations to establish a recommendation of an upper tolerable limit for TFA intake at 1% of total energy intake (TEI), although, according to Ref. [5], "the intake should be as low as possible". The intake of TFAs has also been recognized as a key risk parameter in the Global Burden of Disease Study [6], in which optimal daily TFA intake (taking into account TFAs from all sources) was set at 0.5% of TEI or below. Moreover, the WHO's action package called "REPLACE", released in 2018, aimed to completely eliminate iTFAs from the global food supply by 2023 [7].

As one of the first countries to recognize iTFAs as a major public health threat, Denmark set a maximum limit of 2 g TFAs per 100 g of total fat in foods in 2004 [8]. The policy successfully reduced the amount of iTFAs present in the food supply, while, as a result of the action, the cardiovascular mortality rate also declined sharply [9]. By the end of 2018, similar legislation was introduced in 22 other countries, including Switzerland, Austria, Iceland, Hungary, and Norway [10,11]. In Slovenia, a ban on TFAs was introduced in March 2018 with the transitional period of one year [12], although the use of partially hydrogenated oils (PHOs) in Slovenian food supply had already decreased substantially between 2015 and 2017 [13]. During this period, TFAs were a subject of numerous media reports, putting pressure on the food industry to reformulate products containing PHOs [13]. In 2017, the only remaining categories of pre-packaged foods with a notable proportion of items containing PHOs were cakes, muffins, pastries, and biscuits [13]. Expected intake of iTFAs in the general population was therefore low, except for specific population groups, such as regular consumers of cakes, muffins, pastries, and biscuits of certain brands, which were high in iTFAs.

In 2010, reported estimated intakes of TFAs varied from 0.2 to 6.5% TEI across different countries worldwide, with the intakes being higher at younger ages. The estimated mean intake of total TFAs in Slovenian adults was approximately 1% TEI [14]. The objective of our study was to access current mean daily TFA intake levels using data collected within the Slovenian national dietary survey SI.Menu 2017/2018, and to identify major sources of TFAs in people's diets.

2. Material and Methods

2.1. Study Design and Subjects

The data on food intake was obtained during the cross-sectional Slovenian national food consumption survey, SI.Menu 2017/2018, between March 2017 and April 2018, following the European Food Safety Authority (EFSA) Guidance on European Union (EU) Menu Methodology [15]. Methodology and sample characteristics are described in detail elsewhere [16]. In short, 2280 Slovenian residents aged 10–74 years were selected randomly, using the Central Register of Population of Slovenia. All selected participants received an invitation letter with all the information regarding the study and were later visited by the study interviewers who confirmed the eligibility of the participants and collected information from the respondents by interviews. A total of 62.2% of all invited participants joined the survey.

The study design was approved by the National Medical Ethics Committee (KME 53/07/16; approval No. 0120-337/2016 issued on 19.7.2016). Before enrolment in the survey, all participants were asked to sign written informed consent. If a participant was 18 years of age or younger, written consent was signed by the parent(s) or legal guardian(s).

2.2. Food Consumption Data

2.2.1. General Questionnaire and Anthropometric Measurements

During the first interviewer's visit, the participants filled in a general questionnaire, which assessed general socio-demographic and socio-economic determinants such as number of household members, marital status, level of education, monthly net income of the household, as well as habitual frequency and duration of physical activity. For the purpose of this study, the latter was subsequently converted into the International Physical Activity Questionnaire (IPAQ) score [17]. During the first interview, participants' body mass and body height were assessed by the interviewer using portable calibrated scales. Body mass index (BMI; kg/m^2) was determined with the cut-off point for overweightness set at 25 kg/m^2, except for adolescents, where gender/age adjusted cut-off points (>1SD) were applied [18,19].

2.2.2. 24-h Dietary Recalls

Assessment of dietary intake was performed using a two 24-h dietary recall method. The two recalls were carried out on the same day of the week one to three weeks apart and took place at the participant's home. To obtain a sample that would take into consideration the variations in dietary habits between working and weekend days, 71% of the recalls were performed on work days and 29% on weekends. Altogether, 87% of the recalls were repeated within 7 days after the first recall, while the rest were completed within the next two weeks. The recall was structured to follow a daily meal timeline to help the participants systematically recollect foods and beverages consumed during the previous day. To better estimate the portion sizes of reported foods, a nationally adjusted booklet containing 46 pictures of different food products or simple recipes was developed especially for this purpose. Each food product was presented in 6 different portion sizes to help interviewers and participants determine the quantity of the ingested dish. A picture book was validated according to the method of conceptualization in May 2015 at Biotechnical Faculty, University of Ljubljana, Slovenia [16].

Data from the recalls were collected using an application that was based on the Open Platform for Clinical Nutrition (OPEN), developed by the Jozef Stefan Institute (Ljubljana, Slovenia) for the purpose of SI.Menu survey. The OPEN app is an extension of the Slovenian food composition database [20], which includes data on generic and branded foods and provides a list of ingredients for traditional and other recipes frequently consumed in Slovenia. Missing data were completed with data from other European (EuroFIR) and United States Department of Agriculture (USDA) food composition databases [21].

2.3. Assessment of TFA Content

Energy and nutrient content of all reported foods and beverages was calculated based on the data available on the national platform for clinical nutrition, called the Open Platform for Clinical Nutrition (OPEN) [20]. Missing data for TFA content were extracted from the previously established database [22], which was compiled by analytical assessment of foods from Slovenian food. In the OPEN database, the energy value of foods and beverages is calculated based on macronutrients, alcohol, and dietary fiber content, using an approach provided in the EU Regulation 1169/2011 on the provision of food information to consumers [23]. When information on dietary fiber content was available, total energy value was calculated as total available energy using established conversion factors for the calculation of energy (i.e., 17 kJ per g of digestible carbohydrates and protein, 37 kJ per g of fat, and 8 kJ per g of dietary fiber) [23]. In certain foods and beverages, in which a very low content of dietary fiber was expected, total energy content was calculated as 17 kJ per g of total carbohydrates. To enable accurate nutrient profile formation for more complex foods and dishes, a disaggregation method was applied based on the recipes provided by the subjects, when applicable, or traditional recipes collected in OPKP, considering both the yield and retention factors [20]. To differentiate between pre-packaged and non-packaged foods, each food/beverage item was assigned as branded or non-branded. For the purpose of the statistical analyses in this study, each reported food was allotted to one of 17 predefined categories, which were further divided into 96 subcategories. The categorization system was adapted from Dunford et al. [24], with the additional subcategories added only for non-packaged and home-cooked foods.

2.4. Final Sample for Data Analyses

Inclusion and exclusion criteria as well as exclusion of under- and over-reporters in SI.Menu study were previously described [25]. In short, 97 participants were excluded before the final analyses due to incomplete anthropometric data (n = 12), incomplete 24-h recall data (n = 36), or under- and over-reporting (n = 49). The final sample consisted of 1248 subjects: 468 adolescents (mean age 13.4 ± 2.37), 364 adults (mean age 43.6 ± 13.81), and 416 elderly (mean age 68.7 ± 2.7). Under- and over-reporting cut-off points were calculated by Goldberg method [26], considering adaption by Black et al. [27]. Basic metabolic rate (BMR) was calculated based on gender, age, body height, and body mass using the method described by Harris et al. [28] and adapted by Roza and Shizgal [29].

2.5. Statistical Analyses

All analyzed participants completed both 24-h recalls and provided answers to the survey questionnaire. The energy- and TFA-intake estimates for each of the two recalls and per each of the three age categories were normalized using log transformation method. The TFA estimates were further energy adjusted using nutrient residual method [30]. As the 24h recalls measured only short-term consumption patterns, both recalls were combined to estimate the habitual TFA consumption using the multiple source method (MSM) [31]. The algorithm of the MSM method was used to estimate an average TFA intake adjusted for interpersonal variance under the assumption that all survey participants are habitual consumers of TFAs. The data referring to the national representative sample was weighted using the iterative proportional fitting method [32] based on the census data from 2017 to produce representative results according to age, gender, and region of living.

Descriptive statistics for age cohorts and for different socio-demographic-, anthropometric-, and individual-based variables within each age group are shown as frequencies and proportions, or medians and means with standard deviations (SD). Multiple linear and logistic regression models for all three age cohorts were undertaken separately to assess the significant differences between different sub-populations in terms of TFA consumption based on TEI. The unadjusted means of TFA in % TEI were determined by gender, region, BMI, and IPAQ levels for all age groups, education, income for adults and elderly, and employment status for adults only. The prevalence of consumption of more than 0.5% TEI in TFAs was adjusted for socio-demographic, anthropometric, and lifestyle parameters. The logistic regression analysis was used to determine independent predictors for TFA intake greater than 0.5% of TEI with a maximum likelihood as the estimation method for the model parameters. Odds ratios (ORs) with 95% confidence intervals (CIs) were used as a measure of association with exposure to more than 0.5% TEI from TFA. Statistical significance was set at $p < 0.05$. All analyses were performed using STATA version 15.1 (StataCorp LLC, Coledge Station, TX, USA).

3. Results

Considering the study sampling design, we estimated dietary intakes of TFAs in the Slovenian population separately for adolescents, adults, and the elderly (Table 1). Estimates were done for both the amount of consumed TFAs daily (0.68 g, 0.77 g, and 0.89 g, respectively) and for total energy intake (TEI) from TFAs (0.38–0.50%), which is particularly relevant and was therefore used for statistical analyses and modelling. In adolescents, the TFAs represented on average 0.38% (CI: 0.35–0.39) of TEI. The percentage was marginally higher in adults, where 0.42% (CI: 0.40–0.45) of TEI came from TFAs and was the highest in the elderly, whose consumption of TFAs accounted for 0.5% (CI: 0.47–0.53) of TEI (Table 1, Figure 1).

Table 1. Population-weighted *trans* fatty acid (TFA) intake (g/day) and prevalence of total energy intake (TEI) from TFAs > 0.5%/1.0% daily (95% CI), both cumulative and according to gender.

	Adolescents (10–17 years)			Adults (18–64 years)			Elderly (65–74 years)		
	All	Male	Female	All	Male	Female	All	Male	Female
Sample Size									
N (%)	468 (100)	238 (50.85)	230 (49.15)	364 (100)	173 (47.53)	191(52.47)	416 (100)	213 (51.20)	203 (48.80)
Margin of error (%)	4.53	6.36	6.47	5.14	7.45	7.09	4.81	6.71	6.88
TFAs intake									
Mean [g/day]	0.68	0.68	0.68	0.77	0.78	0.75	0.89	0.90	0.87
Median [g/day]	0.67	0.67	0.68	0.73	0.75	0.72	0.84	0.86	0.82
Mean % TEI (95% CI)	0.38 (0.35–0.39)	0.33 (0.31–0.35)	0.41 (0.38–0.45)	0.42 (0.40–0.45)	0.38 (0.35–0.41)	0.46 (0.43–0.50)	0.50 (0.47–0.53)	0.48 (0.43–0.52)	0.51 (0.47–0.56)
TEI from TFAs (%)									
TFA > 0.5% TEI (95% CI)	11.50 (8.34–15.7)	7.5 (4.6–12.2)	15.7 (10.5–23.0)	28.9 (23.1–33.2)	23.0 (16.8–30.5)	32.8 (25.9–40.6)	43.9 (34.1–54.1)	41.6 (24.8–60.6)	45.9 (35.9–56.3)
TFA > 1.0% TEI (95% CI)				2.51 (1.25–4.96)	1.01 (2.47–4.06)	4.02 (1.82–8.66)	3.02 (1.64–5.50)	4.18 (1.89–9.00)	1.96 (0.74–5.14)

Notes: 95% CI: 95% confidence interval; % TEI: percentage of total energy intake; TFA > 1.0% TEI, percentage of participants not adhering to WHO recommendation for intake of TFAs < 1.0% of TEI; TFA > 0.5% TEI, percentage of participants not adhering to Global Burden of Disease Study target value for intake of TFAs < 0.5% TEI.

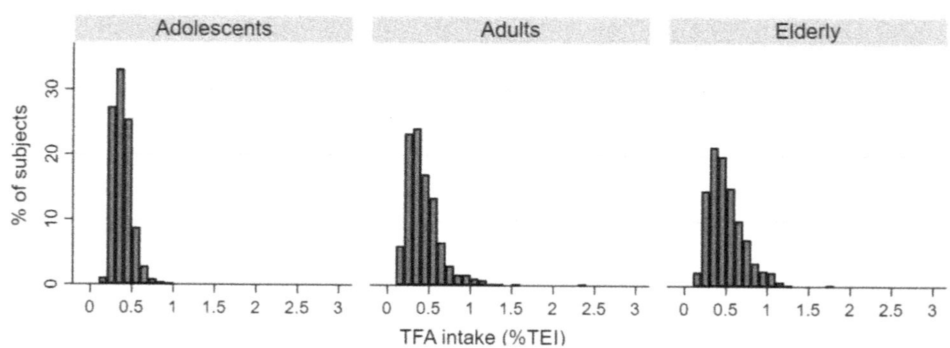

Figure 1. Histograms of daily dietary intake of total *trans* fatty acids (TFAs) expressed in percentage of total energy intake (%TEI) in adolescents, adults, and elderly populations.

On average, women consumed significantly higher % of TEI with TFAs compared to men, regardless of their age. A significant difference was observed in adults with different BMI, with higher TFA consumption recorded in overweight and obese individuals, as compared to those with normal BMI. In the adjusted analysis, there were no other significant differences in TFA consumption in other socio-demographic variables, including region, education, family net income, IPAQ score, or employment status (Table 2).

Additional analyses demonstrated that 13% adolescents, 29.4% adults, and 41.8% of the elderly population consume more than 0.5% TEI with TFAs. Odds for exceeding 0.5% TEI coming from TFAs were significantly higher among adolescent girls and elderly women compared to men. Although not significant, a notable gender-related pattern was observed among adults as well. None of the remaining variables significantly influenced the prevalence exceeding the 0.5% TEI limit (Table 3).

Major sources of TFAs among Slovenian adolescents were butter (14.8%), bread (11.8%), meat dishes (10.9%), and processed meat (10.4%). Biscuits (7.7%), cakes, muffins, and pastry (6.9%), soups (6.6%), and ice cream and edible ices (5.8%), were also notable sources of TFAs among the youth. The majority of TFA intake among adults came from meat dishes (22%) and processed meat (12.8%), but also bread (11.1%), butter (10%), and soups (8.3%). In the elderly, butter was the predominant source of TFAs (24.5%), followed by meat dishes (17.4%), bread (12.4%), processed meat (10.6%), and soups (7.3%) (Figure 2).

Table 2. Adjusted mean (95% CI) levels of *trans* fatty acid (TFA; % TEI) intake by gender, region, body mass index (BMI), international physical activity questionnaire (IPAQ) score, education, income, and employment for different age groups.

Variable		Adolescents (10–17 years)		Adults (18–64 years)		Elderly (65–74 years)	
		n (%)	Adjusted	*n* (%)	Adjusted	*n* (%)	Adjusted
Overall		468 (37.5)		364 (29.2)		416 (33.3)	
Sex	Male	238 (50.9)	0.35 (0.33–0.36)	173 (47.5)	0.39 (0.35–0.42)	213 (51.2)	0.46 (0.43–0.49)
	Female	230 (49.1)	0.41 (0.40–0.43)	191 (52.5)	0.47 (0.44–0.51)	203 (48.8)	0.54 (0.50–0.57)
Place of living	Rural	270 (57.7)	0.37 (0.37–0.39)	202 (55.5)	0.41 (0.38–0.45)	229 (55.1)	0.48 (0.45–0.51)
	Intermediate	76 (16.2)	0.37 (0.34–0.39)	56 (15.4)	0.47 (0.40–0.53)	71 (17.1)	0.51 (0.46–0.56)
	Urban	122 (26.1)	0.40 (0.38–0.42)	106 (29.1)	0.45 (0.40–0.49)	116 (27.9)	0.52 (0.48–0.57)
Education	No university degree	-	-	249 (68.4)	0.44 (0.41–0.47)	342 (82.2)	0.50 (0.48–0.53)
	University degree			115 (31.6)	0.42 (0.38–0.47)	74 (17.8)	0.48 (0.43–0.54)
Family net income	Below average	-	-	118 (38.4)	0.44 (0.40–0.48)	269 (71.5)	0.51 (0.48–0.54)
	Above average			189 (61.6)	0.43 (0.40–0.46)	107 (28.5)	0.47 (0.43–0.51)
BMI	Normal	301 (64.3)	0.37 (0.36–0.38)	148 (40.7)	0.40 (0.36–0.44)	108 (26.0)	0.51 (0.47–0.55)
	Overweight/obese	167 (35.7)	0.39 (0.38–0.41)	216 (59.3)	0.46 (0.42–0.49)	308 (74.0)	0.50 (0.47–0.52)
IPAQ	Low intensity	108 (23.3)	0.37 (0.35–0.39)	127 (35.3)	0.42 (0.38–0.46)	137 (33.4)	0.49 (0.46–0.53)
	Moderate	141 (30.5)	0.39 (0.38–0.41)	108 (30.0)	0.45(0.41–0.50)	133 (32.4)	0.52 (0.48–0.55)
	High intensity	214 (46.2)	0.38 (0.36–0.39)	125 (34.7)	0.44 (0.39–0.48)	140 (34.2)	0.49 (0.45–0.52)
Employment	Employed	-	-	226 (62.1)	0.42 (0.39–0.45)	-	-
	Unemployed			42 (11.5)	0.40 (0.33–0.48)		
	Student			32 (8.8)	0.43 (0.34–0.52)		
	Retired			64 (17.6)	0.50 (0.44–0.56)		

Note: Body mass index (BMI) was considered as normal below 25 kg/m^2, except for adolescents, where gender/age adjusted cut-off points [18,19] were used. Linear regression analysis conducted on samples with excluded missing values (family net income: $n = 57$ (adults) and 40 (elderly); IPAQ: $n = 5$ (adolescents), 4 (adults), 6 (elderly)); identified predictors accounting for difference in the %TEI from TFA: $p < 0.001$ sex (adolescents), $p < 0.001$ sex (adults), $p = 0.0402$ BMI (adults), $p < 0.001$ sex (elderly).

Table 3. Percentage of the population exceeding 0.5% of total energy intake (TEI) from *trans* fatty acid (TFA) intake by sex, place of living, education, family net income, body mass index (BMI), international physical activity questionnaire (IPAQ) score, and employment.

Variable		Adolescents (10–17 Years Old)			Adults (18–64 Years Old)			Elderly (65–74 Years Old)		
		n	>0.5% TEI *n* (%)	Odds Ratio *	*n*	>0.5% TEI *n* (%)	Odds Ratio *	*n*	>0.5% TEI *n* (%)	Odds Ratio *
Overall		468	61 (13.03)		364	107 (29.40)		416	174 (41.83)	
Sex	Male	238	20 (8.40)	1	173	42 (24.28)	1	213	64 (30.05)	1
	Female	230	41 (17.83)	2.45 (1.35–4.46)	191	65 (34.03)	1.65 (0.98–2.78)	203	110 (54.19)	2.45 (1.58–3.81)
Place of living	Rural	270	33 (12.22)	1	202	53 (26.24)	1	229	84 (36.68)	1
	Intermediate	76	6 (7.89)	0.65 (0.25–1.64)	56	19 (33.93)	1.48 (0.73–2.98)	71	31 (43.66)	1.06 (0.59–1.92)
	Urban	122	22 (18.03)	1.71 (0.92–3.15)	106	35 (33.02)	1.30 (0.74–2.30)	116	59 (50.86)	1.68 (1.01–2.80)
Education	No university degree		-	-	249	75 (30.12)	1	342	146 (42.69)	1
	University degree				115	32 (27.83)	0.96 (0.53–1.75)	74	28 (37.84)	0.66 (0.36–1.24)
Family net income	Below average (≤1300 €)		-	-	118	39 (33.05)	1	269	111 (41.26)	1
	Above average (>1300 €)				189	56 (29.63)	1.03 (0.58–1.85)	107	42 (39.25)	0.96 (0.58–1.60)
BMI	Normal	301	33(10.96)	1	148	38 (25.68)	1	108	55 (50.93)	1
	Overweight (including obese)	167	28 (16.77)	1.64 (0.92–2.91)	216	69 (31.94)	1.66 (0.95–2.90)	308	119 (38.64)	0.75 (0.46–1.24)
IPAQ	Low intensity	108	13 (12.04)	1	127	31 (24.41)	1	137	53 (38.69)	1
	Moderate	141	25 (17.73)	1.28 (0.61–2.70)	108	34 (31.48)	1.53 (0.82–2.87)	133	60 (45.11)	1.12 (0.66–1.91)
	High intensity	214	20 (9.35)	0.71 (0.34–1.53)	125	41 (32.80)	1.37 (0.73–2.55)	140	60 (42.86)	1.13 (0.66–1.91)
Employment	Employed		-	-	226	61 (26.99)	1		-	-
	Unemployed				42	13 (30.95)	0.97 (0.40–2.33)			
	Student				32	8 (25.00)	1.51 (0.55–4.13)			
	Retired				64	25 (39.06)	1.98 (0.97–4.02)			

Note: Body mass index (BMI) was considered as normal below 25 kg/m^2, except for adolescents, where gender/age adjusted cut-off points [18,19] were used. Logistic regression analysis conducted on samples with excluded missing values (family net income: $n = 57$ (adults) and 40 (elderly); IPAQ: $n = 5$ (adolescents), 4 (adults), 6 (elderly)). * Odds ratio for exceeding 0.5% TEI from TFA intake; identified predictors accounting for >0.5% TEI from TFA: $p = 0.0026$ sex (adolescents), $p = 0.0543$ sex (adults), $p < 0.0001$ sex (elderly).

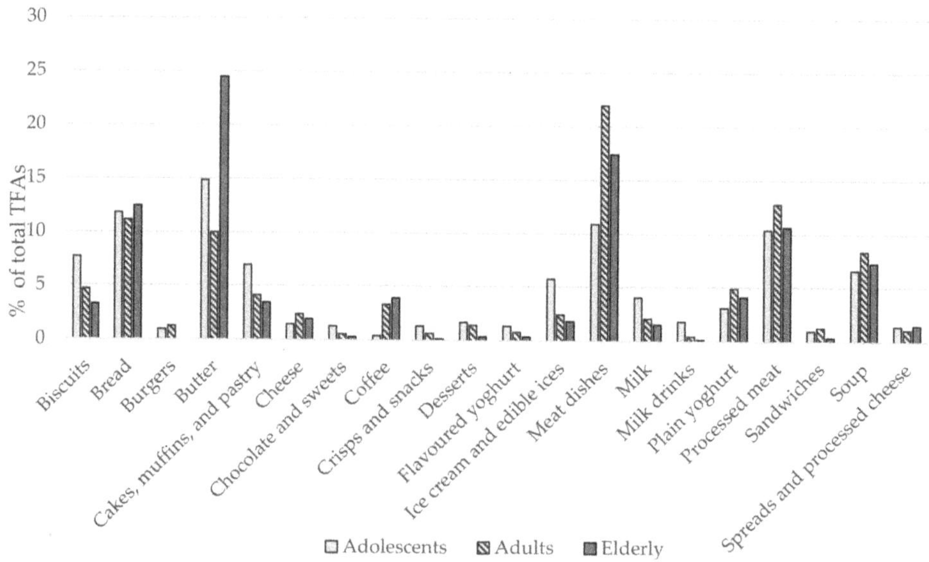

Figure 2. Relative contribution of food categories to *trans* fatty acid (TFA) intake among different age groups (% of total TFAs intake). Note: "Coffee" category also includes coffee drinks that can contain milk or cream.

4. Discussion

With an average value between 0.38 and 0.50% TEI, total intake of TFAs in Slovenia in 2017/2018 was much lower than the estimate of Micha et al. for 2010 [14] and also lower than WHO recommendations [5] of up to 1% TEI. Worldwide analyses of dietary risks among 195 countries demonstrated that in 2017, TFA intake in Central and Western European countries averaged at around 0.4% TEI [6], which is comparable with the TFAs intakes estimated in our study. Much higher intakes were, on the other hand, reported from high-income North America and certain Latin American countries [6].

Several studies have reported a decreasing consumption of TFAs during the last decades due to active public campaigns as well as voluntary or mandatory measures following the cumulative evidence on the adverse health effects of iTFA consumption [33–35]. However, in 2016, Stender and colleagues [36] published a market basket investigation, in which they reported a 3-fold increase in the availability of biscuits containing PHOs in Slovenia between 2012 and 2014. Our subsequent studies demonstrated that by 2017, the proportion of pre-packaged foods containing PHOs had already dropped significantly [13]. The highest proportion of TFA-containing items was found among biscuits, for which additional analyses demonstrated that they would exceed EU regulatory TFA limits in 69% of cases among PHO-containing products [22]. However, the present study revealed that biscuits account only for approximately 7% of all TFAs consumed, which can be either of industrial or natural sources.

Even though we were unable to differentiate between industrially produced and naturally present TFAs, due to limited data about the food composition, relative contribution of different food categories demonstrated that a great proportion of TFAs in Slovenia were consumed in the form of vaccenic acid, a natural TFA found in dairy and beef products. While some evidence suggests that both industrial and ruminant TFAs show similar effects on composition of plasma lipoproteins [2,37], the topic remains controversial [38–42]. Their chemical structure is indeed different and some evidence suggests that compared to ruminant vaccenic acid, iTFAs have much higher potency for promoting inflammation, endoplasmic reticulum stress, and cholesterol synthesis [43]. Nevertheless, the intake of TFAs of animal origin are generally well below the recommended 1% TEI, which greatly limits their possible impact on cardiovascular disease risk. The major sources of naturally present TFAs in Slovenia were butter, meat dishes, and processed meat, which jointly contributed on average 36% of all TFAs consumed among adolescents and up to 53% in the elderly. Especially due to high butter consumption, the intake of TFAs is the highest among elderly Slovenians, which is in contrast to the findings of Micha and colleagues [14], whose systematic analysis showed TFA intake tended to be higher among younger participants. The observed discrepancy most likely results from a shift from artificial towards predominantly natural dietary sources of TFAs, due to the diminished use of PHOs. Adolescents tend to consume more biscuits and other foods which used to be high in iTFAs, while adults and the elderly consume more meat and butter, which now represent the predominant sources of TFAs.

Successful elimination of iTFAs from the food supply is a significant advancement in cardiovascular disease prevention, although concerns have been voiced whether TFAs in processed foods might not necessarily be replaced by healthier alternatives, such as mono- and polyunsaturated vegetable oils [44,45]. Replacement of iTFAs with another solid fat could concomitantly increase the intake of saturated fats, which would hamper the efforts to lower their intake. However, studies from the U.S. and Canada showed that a decrease in TFA content in the food supply was not accompanied by an increase in saturated fat content [46,47]. Whether similar trends took place in Europe and other parts of the world has not been investigated yet. Furthermore, an overview of identified food sources of TFAs in the Slovenian population highlighted that further reduction of TFA intake would be only possible with reduced intake of foods, which are natural sources of TFAs, particularly meat, butter and dairy products. Regarding this is should be noted, that some evidence also suggests some possible beneficial effects of specific types of TFAs, which are naturally present in these foods, such as *trans* vaccenic acid [38–42].

The strengths of this study are in the nationally representative sample for Slovenia (aged 10–74 years), the use of a robust methodological approach, and the fact that the actual data about TFA levels for many foods were available, as they had been collected within the "Trans fats in foods" project [48]. An additional strength of the present study is that we did not only estimate TFAs intakes but also investigated lifestyle and socio-demographic parameters, which might be associated with TFA intakes. However, there are also some important limitations that should be noted. Namely, data collection with two 24 h recalls can result in misreporting and has limited accuracy [49], but other methodological approaches also have considerable limitations. To minimize errors, interviewers were trained to have participants recollect all ingested foods and drinks and corresponding portion sizes as accurately as possible. The remaining under- and over-reporting errors were corrected for during data analyses. Alternatively, using blood biomarkers of TFA plasma levels could increase the precision of TFA intake estimates in the population, but such an approach would not reveal the major dietary sources of TFAs. A limitation of the present study is the quantification of only total TFA intake, due to difficult distinctions between industrial and naturally present TFAs in complex food products. However, the reported adverse health effects tend to be similar for both TFA sources [50] and therefore the used approach should be equally relevant. Another limitation comes from assessing TFA content in foods. Relying on data from food composition databases can introduce errors, as the data can be incomplete, especially when it comes to specific food constituents, such as TFAs. To decrease the magnitude of error, products known to contain the largest amounts of iTFAs had had their TFA content determined analytically in one of our preceding studies [22]. We should also note that the used methodological approach does not enable us to evaluate variability in food consumption and identification of specific risk scenarios, such as TFA intakes in brand-loyal consumers of PHO-containing biscuits and cakes, for which the probabilistic exposure assessment would be an appropriate method.

5. Conclusions

In last decade, the intake of TFAs in Slovenia has dropped under 0.5% of TEI, with butter and meat products becoming a predominant source of TFAs consumed. With additional mandatory restrictions on the use of PHOs in foods in place, more vulnerable specific consumer groups are now also protected. Further reduction of TFA intakes would be only possible with considerable changes in dietary patterns towards less full-fat dairy and high-fat meats, however, a complete elimination of TFAs is not feasible within the scope of a balanced omnivorous diet.

Author Contributions: N.Z. performed the data analyses and wrote the manuscript; B.K.S. was responsible for information technology, and H.H. for the preparation of the database and for data analyses. M.H. conducted food-matching to estimate TFA levels. A.K., K.Ž., Ž.L. and R.V. collaborated in data collection and assessment for determination of TFA content. M.G. and U.B. were responsible for SI.Menu study design and food consumption data. P.G. was responsible for planning anthropometric measurements and physical activity in the SI.Menu study. I.P. was responsible for assuring the set-up and funding of the study, prepared the study design, collaborated in the data analyses, and reviewed the manuscript. K.Ž. made a revision of the final draft and prepared the submission. All authors reviewed the manuscript and agreed to the published version of the manuscript.

Funding: This study was conducted as a part of the research project L3-7538 "The content of trans fats in foods and population intakes", which was funded by the Slovenian Research Agency and Ministry of Health of Republic of Slovenia, and research program P3-0395 "Nutrition and Public Health", funded by the Slovenian Research Agency. The original SI.Menu 2017/18 national food consumption survey was funded by the European Food Safety Authority (EFSA contract No. OC/EFSA/DATA/2014/02-LOT2-CT03), Ministry of Health of the Republic of Slovenia, and Slovenian Research Agency.

Institutional Review Board Statement: The study was conducted according to the guidelines of the Declaration of Helsinki, and approved by the National Medical Ethics Committee, Ljubljana, Slovenia (KME 53/07/16; approval No. 0120-337/2016 issued on 19.7.2016).

Informed Consent Statement: Informed consent was obtained from all subjects involved in the study.

Data Availability Statement: The data presented in this study are available on request from the corresponding author.

Acknowledgments: We acknowledge support of all SI.Menu project partners and their researchers, particularly Nataša Delfar, Metka Zaletel, Darja Lavtar, Ada Hočevar Grom, and Ivan Eržen (National Institute of Public Health, Ljubljana, Slovenia); the Department of Food Science at Biotechnical Faculty and Group of Anthropology at Biotechnical Faculty of University of Ljubljana (Slovenia); Stojan Kostanjevec (University of Ljubljana, Faculty of Education, Ljubljana, Slovenia; Majda Pajnkihar (University of Maribor, Faculty of Health Sciences, Maribor, Slovenia); Tamara Poklar Vatovec (University of Primorska, Faculty of Health Sciences, Koper, Slovenia); Eva Valenčič (Jožef Stefan Institute, Ljubljana, Slovenia); and Evgen Benedik (University Medical Centre Ljubljana, University Children' s Hospital, Ljubljana, Slovenia). We also acknowledge support of Veronika Belec and her team and interviewers at GfK Research Institute (Slovenia) for conducting the survey field work.

Conflicts of Interest: The authors declare no conflict of interests. The funders had no role in the design of the study; in the collection, analyses, or interpretation of data; in the writing of the manuscript; or in the decision to publish the results. I.P. has led and participated in various other research projects in the area of nutrition, public health, and food technology, which were (co)funded by the Slovenian Research Agency, Ministry of Health of the Republic of Slovenia, the Ministry of Agriculture, Forestry and Food of the Republic of Slovenia, and in case of specific applied research projects, also by food businesses. I.P., M.G., and U.B. are members of a workgroup for food reformulation at Ministry of Health of Republic of Slovenia.

References

1. Mozaffarian, D.; Katan, M.B.; Ascherio, A.; Stampfer, M.J.; Willett, W.C. Trans fatty acids and cardiovascular disease. *N. Engl. J. Med.* **2006**, *354*, 1601–1613. [CrossRef] [PubMed]
2. Brouwer, I.A.; Wanders, A.J.; Katan, M.B. Effect of animal and industrial trans fatty acids on HDL and LDL cholesterol levels in humans—A quantitative review. *PLoS ONE* **2010**, *5*, e9434. [CrossRef]
3. Mozaffarian, D. Trans fatty acids—Effects on systemic inflammation and endothelial function. *Atheroscler. Suppl.* **2006**, *7*, 29–32. [CrossRef] [PubMed]
4. De Souza, R.J.; Mente, A.; Maroleanu, A.; Cozma, A.I.; Ha, V.; Kishibe, T.; Uleryk, E.; Budylowski, P.; Schünemann, H.; Beyene, J. Intake of saturated and trans unsaturated fatty acids and risk of all cause mortality, cardiovascular disease, and type 2 diabetes: Systematic review and meta-analysis of observational studies. *BMJ* **2015**, *351*, h3978. [CrossRef] [PubMed]
5. World Health Organization. *European Food and Nutrition Action Plan 2015–2020*; World Health Organization: Copenhagen, Denmark, 2015; Available online: https://www.euro.who.int/__data/assets/pdf_file/0008/253727/64wd14e_FoodNutAP_140426.pdf?ua=1 (accessed on 28 December 2020).
6. Afshin, A.; Sur, P.J.; Fay, K.A.; Cornaby, L.; Ferrara, G.; Salama, J.S.; Mullany, E.C.; Abate, K.H.; Abbafati, C.; Abebe, Z. Health effects of dietary risks in 195 countries, 1990–2017: A systematic analysis for the Global Burden of Disease Study 2017. *Lancet* **2019**, *393*, 1958–1972. [CrossRef]
7. Ghebreyesus, T.A.; Frieden, T.R. REPLACE: A roadmap to make the world trans fat free by 2023. *Lancet* **2018**, *391*, 1978–1980. [CrossRef]
8. L'Abbé, M.R.; Stender, S.; Skeaff, C.M.; Ghafoorunissa, R.; Tavella, M. Approaches to removing trans fats from the food supply in industrialized and developing countries. *Eur. J. Clin. Nutr.* **2009**, *63*, S50–S67. [CrossRef]
9. Restrepo, B.J.; Rieger, M. Denmark's policy on artificial trans fat and cardiovascular disease. *Am. J. Prev. Med.* **2016**, *50*, 69–76. [CrossRef]
10. World Health Organization. *Eliminating Trans Fats in Europe: A Policy Brief*; WHO Regional Office for Europe: Copenhagen, Denmark, 2015.
11. Zuchowska-Grzywacz, M.; Kowalska, M. Trans fatty acids in food—Current legal regulations as protections for consumers and food manufacturers. *Acta Aliment.* **2019**, *48*, 105–114. [CrossRef]
12. Republike Slovenije. Pravilnik o Največji Dovoljeni Vsebnosti Transmaščobnih Kislin v Živilih. Available online: https://www.uradni-list.si/glasilo-uradni-list-rs/vsebina/2018-01-0801/ (accessed on 28 December 2020).
13. Zupanič, N.; Hribar, M.; Pivk Kupirovič, U.; Kušar, A.; Žmitek, K.; Pravst, I. Limiting trans fats in foods: Use of partially hydrogenated vegetable oils in prepacked foods in Slovenia. *Nutrients* **2018**, *10*, 355. [CrossRef]
14. Micha, R.; Khatibzadeh, S.; Shi, P.; Fahimi, S.; Lim, S.; Andrews, K.G.; Engell, R.E.; Powles, J.; Ezzati, M.; Mozaffarian, D. Global, regional, and national consumption levels of dietary fats and oils in 1990 and 2010: A systematic analysis including 266 country-specific nutrition surveys. *BMJ* **2014**, *348*, g2272. [CrossRef] [PubMed]
15. EFSA. Guidance on the EU Menu methodology. *EFSA J.* **2014**, *12*, 3944. [CrossRef]
16. Gregorič, M.; Blaznik, U.; Delfar, N.; Zaletel, M.; Lavtar, D.; Seljak, B.K.; Golja, P.; Kotnik, K.Z.; Pravst, I.; Mis, N.F. Slovenian national food consumption survey in adolescents, adults and elderly. *EFSA Supporting Publ.* **2019**, *16*, 1729E.
17. Lee, P.H.; Macfarlane, D.J.; Lam, T.H.; Stewart, S.M. Validity of the international physical activity questionnaire short form (IPAQ-SF): A systematic review. *Int. J. Behav. Nutr. Phys. Act.* **2011**, *8*, 115. [CrossRef] [PubMed]
18. De Onis, M.; Onyango, A.; Borghi, E.; Siyam, A.; Blössner, M.; Lutter, C. Worldwide implementation of the WHO child growth standards. *Public Health Nutr.* **2012**, *15*, 1603–1610. [CrossRef] [PubMed]
19. De Onis, M.; Onyango, A.W.; Borghi, E.; Siyam, A.; Nishida, C.; Siekmann, J. Development of a WHO growth reference for school-aged children and adolescents. *Bull. World Health Organ.* **2007**, *85*, 660–667. [CrossRef]
20. Korošec, M.; Golob, T.; Bertoncelj, J.; Stibilj, V.; Seljak, B.K. The Slovenian food composition database. *Food Chem.* **2013**, *140*, 495–499. [CrossRef]
21. Bodner-Montville, J.; Ahuja, J.K.; Ingwersen, L.A.; Haggerty, E.S.; Enns, C.W.; Perloff, B.P. USDA food and nutrient database for dietary studies: Released on the web. *J. Food Comp. Anal.* **2006**, *19*, S100–S107. [CrossRef]
22. Kušar, A.; Hribar, M.; Lavriša, Ž.; Zupanič, N.; Kupirovič, U.P.; Hristov, H.; Abramovič, H.; Vidrih, R.; Zlatić, E.; Kokalj, D. Assessment of trans-fatty acid content in a sample of foods from the Slovenian food supply using a sales-weighting approach. *Public Health Nutr.* **2020**, *24*, 12–21. [CrossRef]
23. EC. Regulation (EU) No 1169/2011 of the European Parliament and of the Council of 25 October 2011 on the Provision of Food Information to Consumers. Available online: https://eur-lex.europa.eu/legal-content/EN/AUTO/?uri=CELEX:02011R1169-20180101&qid=1547797296601 (accessed on 28 December 2020).
24. Dunford, E.; Neal, B.; Macgregor, G.; Czernichow, S.; Ni Mhurchu, C.; Food Monitoring Group. International collaborative project to compare and track the nutritional composition of fast foods. *BMC Public Health* **2012**, *12*, 559.
25. Zupanič, N.; Hristov, H.; Gregorič, M.; Blaznik, U.; Delfar, N.; Koroušić Seljak, B.; Ding, E.L.; Fidler Mis, N.; Pravst, I. Total and Free Sugars Consumption in a Slovenian Population Representative Sample. *Nutrients* **2020**, *12*, 1729. [CrossRef] [PubMed]
26. Goldberg, G.; Black, A.; Jebb, S.; Cole, T.; Murgatroyd, P.; Coward, W.; Prentice, A. Critical evaluation of energy intake data using fundamental principles of energy physiology: 1. Derivation of cut-off limits to identify under-recording. *Eur. J. Clin. Nutr.* **1991**, *45*, 569–581. [PubMed]

27. Black, A.E. Critical evaluation of energy intake using the Goldberg cut-off for energy intake: Basal metabolic rate. A practical guide to its calculation, use and limitations. *Int. J. Obes.* **2000**, *24*, 1119–1130. [CrossRef] [PubMed]
28. Harris, J.A.; Benedict, F.G. A biometric study of human basal metabolism. *Proc. Natl. Acad. Sci. USA* **1918**, *4*, 370. [CrossRef]
29. Roza, A.M.; Shizgal, H.M. The Harris Benedict equation reevaluated: Resting energy requirements and the body cell mass. *Am. J. Clin. Nutr.* **1984**, *40*, 168–182. [CrossRef]
30. Willett, W.C.; Howe, G.R.; Kushi, L.H. Adjustment for total energy intake in epidemiologic studies. *Am. J. Clin. Nutr.* **1997**, *65*, 1220S–1228S. [CrossRef]
31. Harttig, U.; Haubrock, J.; Knüppel, S.; Boeing, H. The MSM program: Web-based statistics package for estimating usual dietary intake using the Multiple Source Method. *Eur. J. Clin. Nutr.* **2011**, *65*, S87–S91. [CrossRef]
32. Kolenikov, S. Calibrating survey data using iterative proportional fitting (raking). *Stata J.* **2014**, *14*, 22–59. [CrossRef]
33. Wanders, A.J.; Zock, P.L.; Brouwer, I.A. Trans fat intake and its dietary sources in general populations worldwide: A systematic review. *Nutrients* **2017**, *9*, 840. [CrossRef]
34. Temme, E.H.; Millenaar, I.L.; Van Donkersgoed, G.; Westenbrink, S. Impact of fatty acid food reformulations on intake of Dutch young adults. *Acta Cardiol.* **2011**, *66*, 721–728. [CrossRef]
35. Hutchinson, J.; Rippin, H.L.; Jewell, J.; Breda, J.J.; Cade, J.E. Comparison of high and low trans-fatty acid consumers: Analyses of UK National Diet and Nutrition Surveys before and after product reformulation. *Public Health Nutr.* **2018**, *21*, 465–479. [CrossRef] [PubMed]
36. Stender, S.; Astrup, A.; Dyerberg, J. Artificial trans fat in popular foods in 2012 and in 2014: A market basket investigation in six European countries. *BMJ Open* **2016**, *6*. [CrossRef] [PubMed]
37. Gebauer, S.K.; Destaillats, F.; Dionisi, F.; Krauss, R.M.; Baer, D.J. Vaccenic acid and trans fatty acid isomers from partially hydrogenated oil both adversely affect LDL cholesterol: A double-blind, randomized controlled trial. *Am. J. Clin. Nutr.* **2015**, *102*, 1339–1346. [CrossRef] [PubMed]
38. Kuhnt, K.; Baehr, M.; Rohrer, C.; Jahreis, G. Trans fatty acid isomers and the trans-9/trans-11 index in fat containing foods. *Eur. J. Lipid Sci. Technol.* **2011**, *113*, 1281–1292. [CrossRef]
39. Aldai Elkoro-Iribe, N.; Dugan, M.; Rolland, D.; Kramer, K. Survey of the fatty acid composition of Canadian beef: Backfat and longissimus lumborum muscle. *Can. J. Anim. Sci.* **2009**, *89*, 315–329. [CrossRef]
40. Wang, X.; England, A.; Sinclair, C.; Merkosky, F.; Chan, C.B. Trans-11 vaccenic acid improves glucose homeostasis in a model of type 2 diabetes by promoting insulin secretion via GPR40. *J. Funct. Foods* **2019**, *60*, 103410. [CrossRef]
41. Lim, J.-N.; Oh, J.-J.; Wang, T.; Lee, J.-S.; Kim, S.-H.; Kim, Y.-J.; Lee, H.-G. Trans-11 18: 1 vaccenic acid (TVA) has a direct anti-carcinogenic effect on MCF-7 human mammary adenocarcinoma cells. *Nutrients* **2014**, *6*, 627–636. [CrossRef]
42. Bassett, C.M.; Edel, A.L.; Patenaude, A.F.; McCullough, R.S.; Blackwood, D.P.; Chouinard, P.Y.; Paquin, P.; Lamarche, B.; Pierce, G.N. Dietary vaccenic acid has antiatherogenic effects in $LDLr^{-/-}$ mice. *J. Nutr.* **2010**, *140*, 18–24. [CrossRef]
43. Oteng, A.-B.; Kersten, S. Mechanisms of Action of trans Fatty Acids. *Adv. Nutr.* **2019**, *11*, 697–708. [CrossRef]
44. Stender, S.; Astrup, A.; Dyerberg, J. What went in when trans went out? *N. Engl. J. Med.* **2009**, *361*, 314–316. [CrossRef]
45. Eckel, R.H.; Borra, S.; Lichtenstein, A.H.; Yin-Piazza, S.Y. Understanding the complexity of trans fatty acid reduction in the American diet: American Heart Association Trans Fat Conference 2006: Report of the Trans Fat Conference Planning Group. *Circulation* **2007**, *115*, 2231–2246. [CrossRef] [PubMed]
46. Mozaffarian, D.; Jacobson, M.F.; Greenstein, J.S. Food reformulations to reduce trans fatty acids. *N. Engl. J. Med.* **2010**, *362*, 2037–2039. [CrossRef] [PubMed]
47. Ratnayake, W.; L'Abbe, M.; Mozaffarian, D. Nationwide product reformulations to reduce trans fatty acids in Canada: When trans fat goes out, what goes in? *Eur. J. Clin. Nutr.* **2009**, *63*, 808–811. [CrossRef] [PubMed]
48. NUTRIS. The Content of Trans Fats in Foods and Population Intakes—Public Health Implications. Available online: https://www.nutris.org/en/projects/trans-fats-in-foods (accessed on 28 December 2020).
49. Poslusna, K.; Ruprich, J.; de Vries, J.H.; Jakubikova, M.; van't Veer, P. Misreporting of energy and micronutrient intake estimated by food records and 24 hour recalls, control and adjustment methods in practice. *Br. J. Nutr.* **2009**, *101*, S73–S85. [CrossRef]
50. Remig, V.; Franklin, B.; Margolis, S.; Kostas, G.; Nece, T.; Street, J.C. *Trans* fats in America: A review of their use, consumption, health implications, and regulation. *J. Am. Diet. Assoc.* **2010**, *110*, 585–592. [CrossRef]

MDPI
St. Alban-Anlage 66
4052 Basel
Switzerland
www.mdpi.com

Nutrients Editorial Office
E-mail: nutrients@mdpi.com
www.mdpi.com/journal/nutrients

www.ingramcontent.com/pod-product-compliance
Lightning Source LLC
Chambersburg PA
CBHW040743250926
51850CB00007BB/1393
9783725803033